The Quest for the Quantum Computer

JULIAN BROWN

A TOUCHSTONE BOOK
PUBLISHED BY SIMON & SCHUSTER
New York London Toronto Sydney Singapore

Touchstone
Rockefeller Center
1230 Avenue of the Americas
New York, NY 10020

First Touchstone Edition 2001

Designed by Leslie Phillips
Manufactured in the United States of America

1 3 5 7 9 10 8 6 4 2

The Library of Congress has cataloged the Simon & Schuster edition as follows:
Brown, J. R. (Julian Russell), 1957–
Minds, machines, and the multiverse: the quest for
the quantum computer / Julian Brown
p. cm
Includes bibliographical references and index.
1. Quantum computers. I. Title
QA76.889 B76 2000
004.1—dc21 99-056638
ISBN 0-684-81481-1
0-684-87004-5 (Pbk)

Originally published as: *Minds, Machines, and the Multiverse*

Acknowledgments

"Anything you can do in classical physics, we can do better in quantum physics." I heard that remark in Boulder, Colorado, a few years ago when Dan Kleppner, a distinguished MIT quantum physicist, was giving a lecture to a group of scientists on the subject of quantum chaos. His comment struck me at the time as particularly apposite because I saw it as being in many ways a leitmotif for this book.

Quantum physics appears to be a supreme overarching theory of the universe. Nuclear physics, atomic spectroscopy, chemistry, electronics, materials science, the physics of stars and black holes, and even the structure of the universe, all depend crucially on the laws of quantum theory. Yet in our everyday lives most of us are normally only aware of the classical physics that Newton brought to the fore. That quantum physics is built upon strange unworldly behavior at the atomic level has been appreciated by scientists for much of the twentieth century. There have been hints that quantum physics can engender unusual states at the macroscopic level too: Witness the phenomena of superconductivity and superfluidity. But for a profound theory, such things would appear to be curiously limited in scope.

Now, a century after the first faltering steps in the development of quantum theory and a half century after the invention of the digital computer, we are witnessing a scientific revolution that could put the theory into greater public prominence, while transforming our understanding of computing science and physics. For the realization is that quantum mechanics offers completely new ways to process *information.* Indeed, quantum physics is not simply better than classical physics in this regard, it appears to be unimaginably more powerful.

The purpose of this book is to tell the story of how this scientific revolution has come about. My intention has not been simply to explain the ideas but to articulate the motivations, actions, and grand visions of some of the leading scientists involved. Inevitably, I will have overlooked some important contributions and perhaps misinterpreted others, and for that I can only apologize to the people concerned.

For a book aimed at the general reader, I have not completely avoided offering some mathematically demanding material. In his book *A Brief History of Time,* Stephen Hawking revealed that his editor had warned him that for each equation he added to the manuscript he could expect his sales figures to halve. I have, foolishly perhaps, ignored that advice by including some semimathematical treatments and explanations here and there, though I have made the concession of consigning most of them to the back of the book, in the appendices, where they can perhaps be safely ignored. The reason I was keen to include at least some mathematical descriptions was simply that in my own study of quantum computation the only time I *really* felt that I understood what was happening in a quantum program was when I examined some typical quantum circuits and followed through the equations. The good news is that the circuits and equations themselves (or at least the ones I have included here) are not all that complicated. The most difficult aspect of them is the notation employed, which to the uninitiated will, no doubt, look rather strange. For the mathematically challenged, my advice is simply to skip any sections that prove too difficult. Gaps here or there in one's knowledge shouldn't spoil the big picture.

Quantum physics certainly represents an alien world to most people. Couple that with the hard-nosed logic of computation and you might think you are in for a difficult journey through a rarefied and abstract terrain. Fortunately, that is very far from the case. Quantum computation is, in itself, astonishingly compelling, but the subject also has a very human face, populated as it is by many deep-thinking and charismatic characters, some of whom I have had the privilege of meeting and talking to at length during the course of writing this book.

Among the people I would particularly like to thank for their stimulating and hugely enlightening conversations are David Deutsch and Artur Ekert, both of whom read through the text and offered many invaluable suggestions; Ed and Joyce Fredkin, who were kind enough to put me up at their house in Boston when I attended the PhysComp '96 symposium; and Charles Bennett, Peter Shor, David Di Vincenzo, Seth Lloyd, Tom Toffoli, Norman Margolus, Gilles Brassard, John Preskill, William Wootters, John Smolin, Jeff Kimble, Neil Gershenfeld, David Cory, Andrew Steane, David Wineland, Chris Monroe, Richard Hughes, Paul Townsend, Arjen Lenstra,

William Athas, Tom Knight, and Rolf Landauer, who sadly died earlier this year.

I would also like to express my gratitude to Paul Davies, who has been a constant source of advice, wit, and uncommonly clear-sighted wisdom. Finally, I would like to thank my agents, John Brockman and Katinka Matson, and my editors at Simon & Schuster, who have patiently steered this long-running project through: Bob Asahina, who agreed to my idea for the book; Bob Mecoy, who oversaw things for a while; Bill Rosen, who provided many invaluable comments on the final manuscript, and his assistant Sharon Gibbons, for her diligence in bringing everything to a conclusion. Notwithstanding all of these contributions, I would stress that any mistakes, omissions, and oversights are entirely my own.

Julian Brown
San Francisco 1999

Contents

Foreword

Science during the last few centuries has had unparalleled success both at predicting the behavior of physical processes and at harnessing them to meet human needs. Yet focusing solely on these objectives is a fairly efficient way of bringing scientific discovery to a halt. Since the only effective engine of fundamental discovery is the desire to understand the world better, the primary objective of scientific research must be to obtain that understanding.

The widespread rejection of this objective is, I believe, responsible for the relative dearth of fundamental discoveries during the closing decades of the twentieth century. It is not that we are running out of things to discover! On the contrary, it seems to me that there has never been a time of greater opportunity for fundamental discovery. The trouble is that our culture has come to value science only for its accurate predictions and its technological spinoffs. It denigrates explanation as a mere matter of "interpretation" or taste, and its education system ruthlessly selects for those who are most proficient at applying existing ideas faithfully. It is no wonder, then, that young scientists end up feeling they are not authorized to question the standard assumptions of their fields. And what one does not seek, one is unlikely to find.

Happily, this trend has not yet reached its logical conclusion. New, fundamental branches of science continue to emerge, and quantum computation is a prominent example. The story of this field to date is a particularly stark refutation of the fin de siècle cynicism of those commentators who foresee nothing fundamentally new under the sun. As you will read in this book, the universal model of computation that was laid down by Alan Turing and others in 1936 and considered for decades to be so self-evidently final as to qualify as a branch of mathematics has now turned out to be neither universal nor final, and certainly not a branch of mathematics. It has been superseded by a much deeper and more accurate physical theory that not only allows distinctively new modes of computation but also has implications for many other branches of physics and beyond.

Most of the tools—the basic principles of computation and quantum physics—needed to open up this field have existed since the 1930s, so people sometimes ask why it took so long to take off. I am not sure why. But the key issue is not how long it took but *what* it took. The early pieces in the jigsaw puzzle were rather diverse—or seemed so at the time. Each of the early contributors was coming from a different philosophical direction and addressing a different set of scientific problems. They were each hoping for different, sometimes conflicting results. Nevertheless, all those who were successful had one vital motivation in common: They were determined to use their best theories not just as compact descriptions of how objects behave but as the explanatory basis of their personal models of what the world is like.

I invite the reader to look out for this motivation as it repeatedly surfaces, at every moment of fundamental discovery, in Julian Brown's absorbing account.

David Deutsch
Centre for Quantum Computation,
University of Oxford
September 1999

The Quest for the Quantum Computer

1

Late-Night Quantum Thoughts

> Any technology that is sufficiently advanced is indistinguishable from magic.[1]
>
> ARTHUR C. CLARKE

Life in Other Universes

Gaunt, hair unkempt, and skin a ghostly shade, David Deutsch cuts a strange figure even by the standards of the eccentrics and oddballs that so often inhabit the upper reaches of science. Yet appearances can be deceptive. Talk to Deutsch and you won't find the withdrawn, head-in-the-clouds academic you might expect but someone who is articulate, personable, and exceedingly bright. You'll also quickly discover this is an individual whose ideas about life and the universe are bigger and more radical than those of anyone you are ever likely to meet.

As I drove from London to visit him at his home in Oxford, a fatal accident on the road alerted me to the bizarre implications of some of Deutsch's ideas, in particular his resolute belief[2] in the existence of an infinite set of parallel universes—the multiverse, as he calls it. Within the multiverse there are endless alternate realities in which, supposedly, a huge army of mutant copies of everyone on this and other planets plays out an almost limitless variety of dramas. According to this world view, each time we make the smallest decision the entire assemblage divides along different evolutionary paths like some vast ant colony splitting off in divergent directions. In the process, large numbers of our copies suddenly embark on different lives in different universes.

Presumably, I thought, in some of those fabled other-universes the truck in front hadn't jackknifed, the accident hadn't happened, the crash victims were still alive, and I was now on my way instead of stuck in a jam. But did that mean the victims would therefore feel as though they had survived regardless of the accident because they would only be aware of the universes in which they had continued to live?

A foreshadow of this all-possible-worlds idea was intriguingly captured by Jorge Luis Borges in a short story now often cited by quantum physicists. In "The Garden of Forking Paths," written back in the 1950s, an illustrious Chinese governor is said to have written a strange kind of novel constructed as a kind of labyrinth. "In all fictional works, each time a man is confronted with several alternatives, he chooses one and eliminates the others; in the fiction of Ts'ui Pen, he chooses—simultaneously—all of them," Borges wrote. By writing a novel that pursued all possible story lines *simultaneously,* Borges's fictitious author, Ts'ui Pen, could have been writing a script for the multiverse.[3]

But why, I wondered, should we care about other universes? There's surely enough to worry about in this one without concerning ourselves over the fate of the poor lost souls living in others. With such thoughts running through my mind, I was ready to be briefed by this world's leading advocate for the existence of those other worlds. When I finally arrived and entered Deutsch's house, it took me a moment to persuade myself that I hadn't stepped right into another universe. His place was an unbelievable mess. Papers, books, boxes, computer equipment, mugs, magazines, videos, and numerous other items were strewn everywhere. Even a trip to the bathroom entailed climbing over a mound of cardboard boxes. How does this man function, I wondered?

Okay, so this wasn't exactly a parallel universe, more a parallel mode of living. Yet out of the chaos there were definite, albeit eclectic, signs of order. A whiteboard in the corner of his living room was covered with equations, a reference to Oliver Stone's *Natural Born Killers* here, and notes about time travel there. On a piano in another corner of the room lay open a score of Beethoven's *Waldstein* Sonata. On the floor, littered among the detritus, were strange and elaborate drawings of spidery-looking aliens, designs for an animated film that Deutsch was making with some friends.

But it wasn't long before our conversation leapt far beyond the confines of Deutsch's domicile. "Somewhere in the multiverse there are Earths that were not hit by a comet 60 million years ago and on which the dinosaurs evolved into intelligent beings," Deutsch informed me. "Now those dinosaurs might look completely different from us, and their houses might look different from ours, and their cars probably look different from ours—but if you look at their airplanes and spaceships, they probably look much more similar to ours, and if you look at their microchips, they're going to look *very* similar to ours. The more knowledge that is embodied in a physical object, the more alike it will look in different universes because the more it will have been subjected to the same tests of efficiency, validity, truth, and so on."

"Do you really think there are such universes, in which dinosaurs have all those things?" I asked incredulously. "Undoubtedly," Deutsch replied without hesitation. "That's what the laws of physics tell us."

Deutsch is a physicist, winner of the 1998 Paul Dirac prize for theoretical physics and a researcher at the Centre for Quantum Computation at Oxford University's Clarendon Laboratory. Owing to his unusual lifestyle, though, he rarely sets foot in the university's buildings, preferring instead to inhabit his own private world. If you call Deutsch on the telephone before 2 P.M., you are unlikely to get an answer. He prefers to sleep during the day and work uninterrupted through the night.

During those long nocturnal hours Deutsch has deliberated on some extraordinary ideas. Among them is one of the biggest ideas in physics and mathematics since the development of quantum theory, a breakthrough that could also lead to the most significant invention since the digital computer. It is the concept of a quantum computer, a machine that has begun a revolution in computing—and far beyond—that looks set to be as momentous as the upheaval in Newtonian physics brought about in the early twentieth century. As *Discover* magazine once put it,[4] a quantum computer "would in some sense be the ultimate computer, less a machine than a force of nature."

Why so? Because a quantum computer would make any existing computer—even the fastest Cray supercomputer—seem exceedingly puny. It would solve problems that will never be cracked by any conceivable non-quantum successors of current computers. It would make possible virtual-reality simulations of things we know can never be simulated with current technology, even in principle. But perhaps most important, it would throw open doors to a totally new kind of laboratory in which we could explore those alternate universes of which Deutsch is so fond. Imagine living in a small house for many years and then discovering in the basement a trapdoor that opened onto a colossal subterranean world of rooms that appeared to stretch on into infinity. For physicists and computer scientists that, in some sense, is what the arrival of the first quantum computer would be like.

○ ○ ○ ○ ○

Exciting though the idea of an extraordinary new kind of computing machine might be, the quest for the quantum computer touches upon issues that are much grander than simply a new technology. The study of the ideas involved has already led to remarkable and strange new insights into the nature of our universe.

From the early 1970s to the present day, Deutsch has seen his ideas grow from an apparently small ripple in the world of academic physics to what is now an intellectual maelstrom. By the nineties, the subject of quantum computing had become one of the most exciting and rapidly moving research disciplines in physics and computing. Chaos theory and superstring theory—the scientific gold rushes of the 1980s—suddenly seemed far less enticing in the light of the new Klondike: quantum computation. Theoretical and experimental physicists, mathematicians, and computer scientists realized that this subject offered one of the remaining big opportunities within physics to mine the secrets of nature.

Following the announcement of the first experimental demonstrations of simple quantum programs, the media began to catch on. The inevitable headline QUANTUM COMPUTING TAKES A QUANTUM LEAP, in the *International Herald Tribune,* was occasioned when researchers at MIT, IBM, Oxford University, and the University of California at Berkeley reported in 1998 that they had succeeded in building the first working computers based on quantum mechanics. The story was all the more compelling because these scientists had fashioned their first quantum processors from the unlikeliest of materials. The quantum calculations weren't performed, as you might imagine, by manipulating tiny specks of matter, nor were they the product of elaborate banks of electronic hardware or even huge contraptions the size of particle accelerators. There were no laser beams, no smoke and no mirrors.

No, the quantum logic was all executed within the modest confines of a small flask of *liquid.* No ordinary liquid, either, but a specially prepared version of the common anesthetic and solvent chloroform. Who would have predicted such a thing a few years before?

Of course, these early demonstrations of experimental quantum computing were very simple, and to be fair, other demonstrations have used the more familiar apparatus of the modern physics lab, such as lasers and mirrors. But none of these hinted at the truly extraordinary powers a fully fledged machine would possess. If you imagine the difference between an abacus and the world's fastest supercomputer, you would still not have the barest inkling of how much more powerful a quantum computer could be compared with the computers we have today. More remarkable yet, this amazing discovery originated not from some breakthrough in electronics or software design but in large measure from pure theory—from Deutsch's highly individual take on the laws of physics.

The Quantum AI Experiment

The origins of this quantum information revolution are to be found in the pioneering work of Rolf Landauer, a physicist who became an IBM Fellow in 1969 and who, until his death in 1999, worked at IBM's Thomas J. Watson Research Center in Yorktown Heights, near New York City. "Back in the 1960s, when I was still in school," Deutsch said, "Landauer was telling everyone that computation is physics and that you can't understand the limits of computation without saying what the physical implementation [i.e. type of hardware] is. He was a lone voice in the wilderness. No one really understood what he was talking about—and certainly not *why.*"

The prevailing view at the time was that computation was ultimately an abstract process that had more to do with the world of mathematical ideals than with the physics of machines. But Landauer's view began to take hold when he, and subsequently his IBM colleague Charles Bennett, discovered a crucial link between physics and computation, which we'll explore in the next chapter.

"The next thing that happened was in 1977–78," Deutsch continued. "I proposed what today would be called a quantum computer, though I didn't think of it as that. I thought of it as a conventional computer operating by quantum means that had some additional quantum hardware that allowed it to do extra things. But I didn't think of the additional hardware as being part of the computer. The purpose of this was not to do computations, it was to test the many-universes theory. It had been thought that the theory was untestable. I realized that if you had a quantum computer, and the quantum computer could count as an observer, then it could observe things that would distinguish between the many-worlds interpretation of quantum mechanics and other so-called interpretations."

Many worlds, many universes—or nowadays, "many minds" or "many histories"—take your pick. All are variants of a particular interpretation of quantum theory, the theory of the subatomic realm. Ever since the development of quantum theory at the beginning of the twentieth century, scientists have known that on the atomic scale matter behaves very differently from what we see on everyday scales. Particles such as electrons and atoms do not behave like the billiard balls of Newton's classical physics but seem to exhibit characteristics more in keeping with fuzzy wavelike entities. These and other properties led to a great deal of puzzlement and angst over how quantum theory should be understood as a description of reality—or realities. So much so that today there are a multitude of different interpretations of what goes on at the atomic level.

In the early 1980s, Deutsch's proposed experiment (described more fully in Chapter 3) sounded like the stuff of science fiction. To test the existence of multiple universes, he envisaged the construction of a thinking, conscious artificial intelligence whose memory worked "at the quantum level." Such a machine, he claimed, could be asked to conduct a crucial experiment inside its own brain and report back to us whether Deutsch was indeed right to believe in the existence of parallel universes. The idea of assembling a conscious machine took some swallowing, but what exactly did Deutsch have in mind when he talked of "quantum memory"?

Well, nearly 20 years later we have the answer because quantum computer memory is on the verge of becoming an experimental reality. But even if Deutsch's ideas looked far-fetched at first sight, they were compellingly presented and, more important, appeared to be scientifically testable. As the philosopher of science Karl Popper pointed out, only those theories that are open to the possibility of experimental refutation or falsifiability are worthy of being considered good science. The trouble was that

Deutsch's test would have to be carried out on a new kind of machine that no one had yet built or knew how to build.

But he was following a well-established tradition in physics: dreaming up thought experiments to question fundamental assumptions about the world. Such experiments, even if they cannot be carried out in practice, have often acted as a powerful spur for creative thinking in physics. Albert Einstein, for example, devised several that were instrumental in the development of relativity and quantum theory. Sure enough, Deutsch's thought experiment, although nearly impossible to carry out then as now, convinced him that the many-universes interpretation had to be correct.

The "experiment" didn't make much of an impression on other scientists, though. This was largely because the work remained unpublished for many years. Deutsch originally described the idea in 1978 in a paper he sent to the journal *Physics Review.* The editors had a policy of not publishing papers on the interpretation of quantum theory, but Deutsch felt his idea was sufficiently exceptional to warrant the breaking of that rule. The review's editors thought otherwise. Unfortunately, Deutsch did not submit his paper to any other journal and thereby lost his claim to being the first to publish something close to the concept of a quantum computer. Why didn't he publish elsewhere?

Deutsch explained that he was not especially interested in claiming priority for his ideas. "I usually lose interest in a paper when it's about three-quarters finished," he said. "Finishing papers is quite a chore. I've got a whole stack of unfinished papers. If I do finish a paper, I send it off and it's good-bye, I'm already working on other things. It was only when I happened to meet David Finkelstein at the Quantum Gravity II conference [in 1984] that I mentioned this paper to him—we were all having a riotous argument, as often happens over dinner, about the many-universes interpretation. Somebody said that it couldn't be tested, and I said, 'Yes, it can, I have a proof of that!' and Finkelstein said, 'Where's it been published?' I said, 'It hasn't. Here's a preprint,' and he said, 'Well, can I publish it then?' "

Finkelstein was one of the editors of the *International Journal of Theoretical Physics,* which duly published Deutsch's paper in 1985, some seven years after it had been written. Moreover, Deutsch published another paper[5] in the same year that, though it didn't prove an overnight sensation, is now widely regarded as a landmark in physics. Among other things, it highlighted a property of quantum computers that no one else had noticed: quantum parallelism. It is this crucial feature that now causes so much excitement in the study of quantum computation.

Exploring Hilbert Space

Interpretations aside, it's long been known that at the atomic level waves can behave like particles, and particles have waves associated with them. A single entity such as an electron, for example, can travel along many different routes simultaneously as if it were really a spread-out phenomenon like a wave. The essential idea of quantum parallelism advanced by Deutsch was this: If an electron can explore many different routes simultaneously, then a computer should be able to calculate along many different pathways simultaneously too.

Despite the enormous significance of that idea, Deutsch admits his 1985 quantum parallelism paper "didn't make a big splash" at first. Though he had shown quantum computers could do certain tasks more efficiently than classical computers, the tasks in question were hopelessly limited and rather contrived. One, for example, involved solving a rather simple mathematical problem in which you could get two answers in one shot rather than having to do the calculation twice. This meant you could do parallel calculations in a way that had never been realized before, but the irony was that calculation worked only 50 percent of the time. So superficially there appeared to be no practical gain, on average. Not surprisingly perhaps, the subject maintained a low profile.

Nevertheless, a few other scientists took note. Charles Bennett at IBM and, a little later, Artur Ekert at Oxford University, immediately saw the theoretical importance and started working flat out on quantum computation and information theory, producing some of the first practical applications. According to Deutsch, it was this theoretical fuse that later ignited the current explosion of interest—but only after a dramatic intervention from another quarter.

In 1994 Peter Shor, a computer scientist working at AT&T's Bell Labs in New Jersey, discovered how a quantum computer could solve a very important mathematical problem, one that had long been known to be beyond the reach of ordinary computers. He showed how a quantum computer could calculate the factors, or divisors, of very large numbers extremely rapidly. Solving this particular problem had implications that went far beyond mathematics.

Ever since the 1970s, the difficulty of finding the factors of large numbers had formed the basis of an extremely important and now widely used method of protecting information with secret codes. The code works a little like the secret numbers you find on a so-called onetime pad, only instead of having to carry around a special pad that might be stolen, all you need to send a message is a large number, *n,* that can safely be given to anybody. The key to unlocking the code depends on knowing which numbers when multiplied together produce *n,* and normally this would be known only by the intended recipient. By showing how a quantum computer could factor very large numbers quickly, Shor threatened to blow a devastating hole through the world's most sophisticated forms of secrecy.

The Shor program (or algorithm) not only offered powerful evidence for the claim that a quantum computer could far exceed the capabilities of a conventional computer, it revealed how that extra resourcefulness could be applied to an interesting mathematical problem, and as an added bonus, it offered a powerfully important real-world application. It was this third issue that did most to energize efforts to build a quantum computer: because of the huge repercussions such a machine would have for military communications, government secrecy and surveillance, data protection, e-commerce, and the privacy of ordinary citizens. If anyone could build a full-scale quantum computer, it's possible that he or she would be able to access everything from your bank account to the Pentagon's most secret files. It's no surprise, then, that significant funds backing this line of research have come from such organizations as the U.S. Department of Defense, the National Security Agency, NATO, and the European Union.

If anybody is going to build one of these machines, the intelligence agencies are making certain they will know about it first. It's not without reason, then, that Shor's factorization program was quickly recognized as the quantum computer's "killer app."

For Deutsch, though, the power of Shor's algorithm signified even more clearly the reality of parallel universes. Consider the problem of factoring the number 15. It's 3 times 5. Now try factoring a five-digit number, like 24,287. Not so easy. Even with a calculator it would probably take a while to find the factors, which are 149 and 163. Now notice two things. First, the opposite process, multiplying the two numbers, is straightforward. Second, it's only when the factors are quite large, and indivisible by smaller numbers (149 and 163 are *prime numbers*), that the problem is hard. If, on the other hand, I asked you to factor 24,288, you could easily get the answer by trial division because the factors—2, 3, 11, and 23—are small.

Now imagine how difficult it would be to factor a 250-digit number that was the product of two large primes. It turns out that even with today's most advanced supercomputers, it's very unlikely that we would *ever* be able to solve such a problem—with the fastest known classical algorithm, it would take longer than the age of the universe. Yet a quantum computer using Shor's algorithm, if ever such a machine could be built, could crack the problem in seconds or minutes because it would be able to compute simultaneously along as many as 10^{500} or more different pathways. As Deutsch argues in his book *The Fabric of Reality,* such an unimaginably large number presents believers in a single universe with a gargantuan problem:

> There are only about 10^{80} atoms in the entire visible universe, an utterly minuscule number compared with 10^{500}. So if the visible universe were the extent of physical reality, physical reality would not even remotely contain the resources required to factorize such a large number. Who did factorize it, then? How, and where, was the computation performed?[6]

For Deutsch the only answer that makes any sense is that different parts of the calculation are performed in different universes, all 10^{500} of them. Furthermore, the existence of Shor's algorithm is powerful testimony even though it's never actually been run on any hardware. "That argument—'Where was it done?'—is already valid today even before we have ever built a factorization engine," Deutsch said. "We can look in *theory* at the design of the machine, never mind whether we can actually build it. To me it's no more convincing for somebody to come and tell me, 'Well, look we've factored this number,' than to look at the equations that say the machine *would* factor the number if you could only build it."

To understand nature in its entirety, Deutsch believes, we must accept the existence of an almost limitless number of universes. We normally experience reality only within one of those universes. But in certain kinds of quantum events, parallel universes can differentiate themselves from one another and, under special circumstances, can later interfere with one another to produce effects that according to him are impossible to explain if you cling to the idea of a single universe.

Despite Deutsch's claims for the manifest truth of the many-universes interpretation, the majority of other physicists remain unpersuaded. They would claim that it is possible to accept the idea of quantum parallelism without buying into the concept of many universes. The more neutral language physicists prefer is to say that quantum states occupy a vast, abstract, multidimensional space known as Hilbert space, named after the German mathematician David Hilbert.

This space is not space in the conventional sense but a space of "states." The classical-state space of an object such as a ball is a kind of map that describes all of its possible states and motions, including its trajectory through the air if it is thrown and its rotation about its own axis if it is given some spin. Although classical-state space is unbounded spatially, it is severely curtailed by the fact that time flows in only a single direction—along a single vector. As the physicist Roger Penrose has put it, "Hilbert space becomes that same universe, with time and every other possible vector flowing in all possible directions." That extra freedom pays big dividends and is the reason a quantum computer would be able to compute certain things very quickly: because it has access to an infinitely larger space—Hilbert space—than a classical machine (see Figure 1.1). This raises the puzzling question of why classical computers and the rest of the classical world in which we apparently live are confined to such a minuscule aspect of the full quantum reality.

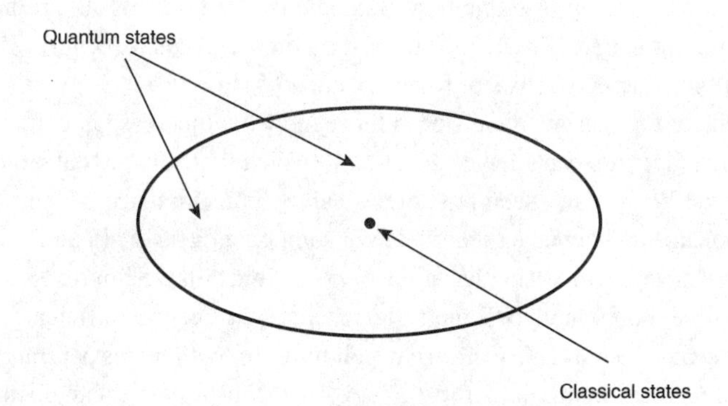

Figure 1.1. The hugeness of Hilbert space

Hilbert space, an abstract mathematical space spanned by all possible quantum states, is unimaginably larger than the sum total of all possible classical states.

Let's put these philosophical worries to one side for the time being. Although Deutsch developed his ideas about quantum computers through his explorations of physics, we're going to take an excursion down into the depths of the quantum world via a less precarious route. On the way, we'll get our first glimpse of what a quantum computer might be like. Don't worry too much about understanding everything first time around, because we'll be exploring this strange and unfamiliar terrain in greater detail in subsequent chapters. For now, just sit back and enjoy the ride.

The End of Moore's Law?

Over the past four decades, ever since integrated circuits were invented, the number of switches or transistor devices crammed onto individual silicon chips has increased at a fiendish rate. In fact, it has more or less doubled every eighteen months, which explains why computers become obsolete at such an alarming speed. This doubling trend was first spotted back in 1964 by Gordon Moore, cofounder and chairman emeritus of Intel, and is widely known in the industry as Moore's law.

Moore's observation was based on just a few years' experience in producing the first integrated circuits, before microprocessors had even been invented. Yet his prediction has held up well (see Figure 1.2). In the fall of 1997, to give an idea how far the industry had come, Moore informed an Intel developers' forum in San Francisco that Intel's entire production output for the year would amount to nearly 10^{17} transistors. This figure, he pointed out, was comparable to an estimate by the famous Harvard biologist and ant expert, E. O. Wilson, of the number of ants living on Earth. "We're making one transistor for every ant," Moore said.

One of the reasons there's been this exponential increase in computing power is that with each device generation (which typically lasts about three and a half years), it's been possible to halve the feature size of individual components and hence to quadruple the component density. In addition, as circuits have become smaller, a lot of other things have improved at the same time: chips have become faster, increased in energy efficiency, and become more reliable. Costs too have plummeted, with prices per transistor dropping a millionfold in forty years. In the history of manufacturing there's been nothing else like it.

But it would be wrong to regard Moore's law as a sacrosanct law of nature written on tablets of silicon. There's no particular reason it has to re-

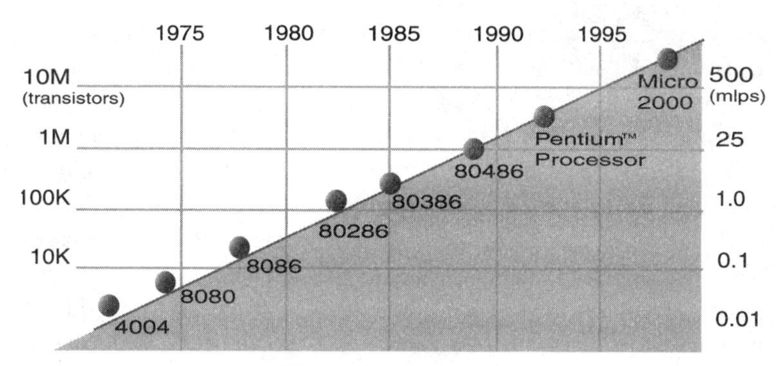

Figure 1.2. Moore's law

Moore's law states that the number of transistors on silicon chips doubles approximately every 18 months. Such doubling has produced an exponential growth in computing power since the 1960s and is projected to continue until about the year 2010. Copyright 1999 Intel Corporation.

main true indefinitely, and in fact, growth has varied somewhat over the years: Moore originally stated his law as a doubling *every year,* but he changed it in the 1970s to a doubling every two years. (Now, when people refer to the law, they usually settle for something in between, i.e., a doubling every eighteen months.) Also, whenever we meet an exponential curve in real life, it invariably hits a wall at some point. It's like the old problem of how much money you would need to place on a chessboard if you put one penny on the first square, two on the second, four on the third, and so on, doubling each time until you reached the last square of the board. Initially, you might think you wouldn't need very much. But when you actually calculate the amount, it stacks up to 10^{19} pennies, or $100 quadrillion ($100,000,000,000,000,000), more than the combined wealth of all the civilizations in the history of the world.

When numbers are doubled at regular intervals, the results very quickly blow up beyond all reasonable bounds. So in the case of Moore's law it seems only likely that it will break down at some point. But prophesying the imminent demise of Moore's law has proved to be a loser's game because virtually since it was proposed people have said that there were obstacles just around the corner that would cause it to go off track. So far, all such predictions have turned out to be wrong.

In the late 1980s, for example, some industry commentators talked about the 1 micron barrier. They were referring to the potential difficulty of designing circuit elements smaller than a micron (1 micron is one-thousandth of a millimeter). The reason 1 micron was thought to be a barrier was that manufacturers of silicon chips were rapidly reaching the limits of resolution imposed by the wavelength of light. Silicon chip circuits are produced using a photographic technique in which a laser shines through a mask onto a silicon wafer covered with a photosensitive resist. For the same reason that you can't paint a picture through a pinhole with a brush larger than the pinhole, the smallest size of the components that can be produced in this way is limited by the wavelength of the laser light. Given that the wavelength of visible light is about 0.5 micron, a practical limit of 0.5–1 micron looked unavoidable.

However, chip manufacturers have been resourceful in squeezing that limit down, partly through improved focusing techniques and partly by using shorter-wavelength lasers. In 1999 manufacturers began production of 0.18 micron chips with the aid of lasers working in the ultraviolet region of the spectrum and contemplated further reductions in size down to 0.13 and 0.08 microns. Now people talk about the "point one" barrier.[7]

In parallel with this reduction in size, the semiconductor business has encountered another trend, sometimes known as Moore's second law. This states that as the sophistication of chips goes up, the cost of the manufacturing plant doubles roughly every four years. That's why it now typically costs $2 billion to build a manufacturing plant. If you consider the two laws together, it doesn't take long to realize that something is going to have to give unless companies like Intel can grow exponentially, which seems unlikely. If costs continue rising exponentially, it seems inevitable that economics will cut into Moore's first law.

What do industry experts think will happen? Microsoft founder Bill Gates clearly didn't envisage economics curtailing the future rapid development of semiconductor chips when he predicted in 1995 in his book *The Road Ahead* that Moore's law would hold for another twenty years.[8] On the other hand, in 1997 Stan Williams, who is in charge of fundamental physics research at Hewlett Packard, predicted trouble at about the year 2010 on both economic and physical grounds. Speaking at a conference on quantum computation held at the Royal Society in London, Williams said that by then individual transistors on chips will be turned on or off by as few as eight electrons, compared with about 500 today (see Figure 1.3).

Sometime beyond 2010, Williams said, chip manufacturers will reach a point at which each transistor device will have shrunk to atomic pro-

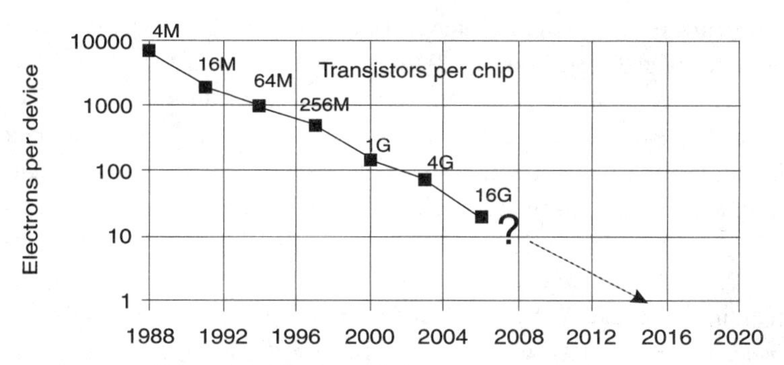

Figure 1.3. Number of electrons per transistor

Switching logic states in today's silicon chips involves the movement of hundreds of electrons, but the number is rapidly diminishing with time. The labels 4M to 16G refer to memory chips ranging from 4 million to 16 billion bits.

portions and will be switched on or off by a single electron. Clearly, it will not be possible to subdivide the components any further at this point, so this stage will represent a physical limit. Given this physical limit and the colossal cost of building chip manufacturing plants, Williams predicted that around the year 2010 progress will slow as we enter the quantum era.

However, there may still be ways in which the rule of Moore's law could be maintained. In 1997 Ed Fredkin, a former MIT physics professor and a leading thinker on the fundamental aspects of computing, gave a talk[9] about Moore's law in which he argued that it would continue for 100 years because chip manufacturers will be able to extend their devices into three dimensions. At the moment, they make use only of the surface of silicon chips. If they could stack circuits on top of one another, then they would have a whole new dimension to move into. By this reasoning, Fredkin predicted that sometime in the twenty-first century we would see a CPU with an Avogadro's number of processing elements (approximately 10^{24}). Avogadro's number refers to the number of molecules in a "mole" of a substance, which in Avogadro's day meant some 22 liters of gas but today is defined by the number you get in a small chunk of carbon—12 grams, to be precise. It is a useful indication of the huge number of molecules you get in a macroscopic amount of matter. Fredkin, therefore, anticipated the time when nearly all the atoms or molecules in lump of matter can be put to worthwhile computational use.

Such a device would be unimaginably powerful: As a memory device alone it would be able to store 100 trillion copies of the *Encyclopaedia Britannica* or a million copies of the entire contents of the Library of Congress. As a processor, apart from opening up new possibilities in artificial intelligence, such a device might be used, Fredkin suggested, for ultrahigh-resolution simulations of the weather, the Earth, the oceans, and even societies.

During Fredkin's lecture he cited an article that had appeared with perfect timing on the front page of that morning's *International Herald Tribune*. INTEL TRASHES AN AXIOM OF THE COMPUTER AGE, the headline exclaimed. The story was about the possible cessation of Moore's law—but not because Intel had seen the end of the road. No, the company had found a way to *beat* Moore's law.

The story featured Intel's development of a new kind of flash memory chip that could store not one but two bits of information on each transistor, potentially doubling the storage capacity of Intel's chips at a stroke. Meanwhile, IBM announced that it had found a way of using copper instead of aluminum as the "wire" connecting transistors on silicon chips. Because of copper's superior properties as an electrical conductor, its use in chip manufacturing would help ameliorate the problem of delivering enough power to the transistors as they diminish in size.

In 1999 there was further excitement when researchers at Hewlett Packard and UCLA announced that they had discovered a way of configuring tiny circuits assembled from a single molecular layer of a chemical known as rotaxane. The rotaxane molecules can operate as switches and offer the possibility of a further significant leap in miniaturization.

Whether these developments will really enable chip manufacturers to overtake Moore's law remains to be seen, but they will certainly help keep the technology motoring forward for the next few years. Even so, by around 2010 chip manufacturers look set to enter the twilight zone of the atomic realm, whereupon they will find themselves increasingly contending with the laws of quantum physics.

From Bill Gates to Quantum Gates

What differences will arise as semiconductor manufacturers encroach upon the quantum realm? Researchers already have some clues from the results of experiments on so-called nanocircuits. Special fabrication techniques have made possible the construction of semiconductor devices in which electrons are confined to atomically thin layers, narrow wires, and tiny blobs known as quantum dots. One important finding is that it's proved possible to make transistors that switch on or off according to the presence or absence of a single electron. There's keen industrial interest in the idea of the single-electron transistor (SET) because it could open the way to ultra high-density memory chips.

Although the SET works at the quantum level by responding to an individual electron, its properties are still basically classical. The flow of electrons in an electric current passing through a conventional semiconductor can also be regarded as a classical phenomenon, rather like the diffusion of molecules in a gas. As the electrons travel through a crystal of silicon, for example, they are frequently scattered by the atoms within the lattice, causing them to behave like particles. However, as circuits are made smaller, electrons travel shorter distances and are subject to less disturbance. In these circumstances the wavelike aspects of electrons can begin to exert their influence.

These wavelike properties are directly exploited to switch on and off another nanoscale device, known as the quantum interference transistor, or QUIT. This too is the subject of intense research interest. At the moment, though, these devices are a long way from being practical—not only are they tricky to make but they also have to be cooled to liquid helium temperatures. At room temperatures the devices would be overwhelmed by "hot" electrons and would fail. It's possible, though, that the temperature problem could be overcome by making these devices smaller still. Despite their "nano" label, which refers to atomic scales of 1 nanometer—a billionth of a meter—these devices typically measure hundreds of atoms across.

Other problems will, doubtless, arise in building circuits on the atomic scale. One of the biggest will be supplying enough energy to all the switches without burning up the circuit. We'll explore that problem in the next chapter. But let's assume the obstacles can be overcome. Should we limit ourselves to imitating the switching behavior used in conventional computer chips? Are quantum transistors capable of doing anything other than simply switching on and off?

It turns out that quantum devices can indeed perform new tricks, which form the basis of quantum computation. The first novelty is known as *superposition.* This is connected with the idea that just as individual subatomic particles cannot necessarily be pinned down to one place, so individual logic states may not necessarily be tied to one value. To understand this, consider the digital nature of conventional computers.

Digital computers use binary logic switches, known as gates, to process information. Information is coded in strings of binary bits, each of which takes the value of 0 or 1. In a PC, a 0 is normally represented by a low voltage (usually 0 volts), and a 1 by a signal of typically a few volts. So information is carried by the pattern of high and low voltages pulsating across the circuit board. At any one time, all inputs and outputs will be in one or the other of these two states. No other values are allowed. As an example of a logic gate, consider one of the simplest, the NOT gate. This has one input and one output. If the input is 0, the output is 1, and if the input is 1, the output is 0. The gate simply reverses the value of the input, as its name implies.

Now consider an atom that has a single electron in its outermost orbit. This electron can be moved—"excited"—into a higher orbit by shining light of a particular frequency on it. The electron makes a quantum leap into a higher energy state. If this excited state is reasonably stable, we can use it and the atom's ground state to represent the numbers 1 and 0, respectively. If an excited atom is given a similar pulse, the electron drops back down into its ground state again, releasing its extra energy in the form of a photon of light. So the pulse of light flips the state whatever its value, performing, in effect, the action of a standard NOT gate.

But what happens if we shine light on an atom in its ground state for only half the time needed to excite it? Quantum mechanics dictates that in an atom the electron can occupy only one of a set of discrete energy levels that are spaced, albeit unevenly, like rungs on a ladder. So if there are no other energy levels between the ground state and the excited state, where can the electron go? As it turns out, the electron finds itself in both orbits *simultaneously.* The electron is said to be in a superposition of the ground and the excited states.

The term superposition comes from the study of wave phenomena. When waves of water, for example, come from different directions, their combined effect can be calculated by adding the waves together—in other words, superposing one wave on the other. The superposition of electron states comes about because according to quantum mechanics all particles such as electrons have wavelike aspects. There are, in theory, an infinite number of different superpositions we can make, because when the light is shone for different lengths of time, the electron takes on a range of different superposition states.

Used in this way, the atom can store a single unit of quantum information, known as a quantum bit, or *qubit.* A qubit therefore differs from a conventional digital bit in that it can store values "intermediate" between 0 and 1. Superficially, a qubit might appear to be similar to a classical *analog* bit of information carried by an electrical signal that takes any value between the voltages representing 0 and 1 (see Figure 1.4). But there is a fundamental difference between a qubit and an analog bit: Whenever a measurement is made of a qubit, the answer can *only* be either 0 or 1—not some intermediate answer, as we would expect for an analog signal. This difference, as we will see, has profound consequences.

Classical digital bits are grouped together in computers to represent larger numbers. A two-bit *register* can represent any number between 0 and 3 because, together, the two bits can take the values 00, 01, 10, and 11, which are the base 2 representations for the numbers we know better as 0, 1, 2, and 3. Two qubits could similarly represent each of these values, just like ordinary bits. However, two qubits can also be put into a superposition of any or all of these states. Thus a quantum register of two qubits can represent the numbers 0, 1, 2, and 3 *simultaneously.* If we consider a register of eight qubits, we could obtain a superposition representing $2^8 = 256$ numbers; a register of 1,000 qubits could represent $2^{1,000}$ (approximately 10^{300} numbers) simultaneously. In contrast, a classical register of 1,000 bits

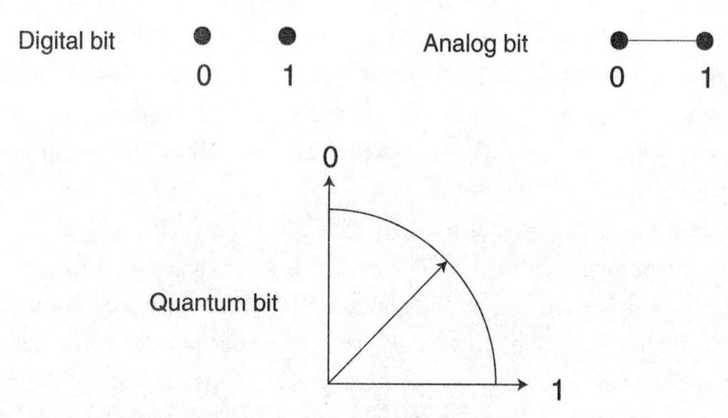

Figure 1.4. Bits and qubits

Digital information comes in bits represented by 0s and 1s. An analog bit can take any value between 0 and 1. A qubit can exist as a superposition of 0 and 1, which can be represented by a vector pointing in a direction intermediate between those representing 0 and 1.

could represent any of the integers between 0 and approximately 10^{300}, but only one at a time.

It is tempting to attribute the potentially enormous power of a quantum computer to this capacity to hold astronomically large numbers of values simultaneously in a relatively small number of qubits. Actually, this is only part of the story. Superposition is closely related to the quantum phenomenon of *entanglement.*

Entanglement is the ability of quantum systems to exhibit correlations between states within a superposition. If we have two qubits each in a superposition of 0 and 1, the qubits are said to be entangled if the measurement of one qubit is always correlated with the result of the measurement of the other qubit. According to some physicists it is specifically these entangled superpositions that open up extraordinary new possibilities in information processing. Entanglement, in this view, is seen as the secret to quantum computation rather than superposition per se.

To illustrate, imagine you were a mathematics student and were given a list of eight difficult mathematical problems to solve. You could do the task serially, by solving each problem in turn, but it would take you a long time. Alternatively, if you and seven fellow students agreed to share the problems you could solve one each. Such parallel processing would greatly speed up the task. Now suppose you asked each of your friends to write the answer to his or her problem on a small piece of paper, which would then be placed in a hat. After everyone had finished, the hat would contain all the answers. You might then pick a piece of paper at random out of the hat and examine the answer. You then realize the answers are meaningless because nobody has written the particular question he or she was solving. What use, after all, is an answer when you don't know the question that prompted it?

Similar reasoning applies to the quantum computer. If you had a superposition of numbers from 0 to 7 stored in a three-qubit register and performed a complicated series of operations on these numbers to do some mathematical calculation, you would be in a similar predicament. Reading off an answer would merely tell you one possible answer but not the number that generated it. Quantum entanglement, though, enables us to link quantum registers so that whenever an answer appears in one register, we can always look in the other to find out what number generated it. Without quantum entanglement, the quantum computer would be like the hat full of answers without the questions. With entanglement, it becomes more like a hat full of answers each carefully labeled with a note of the question.

The implication of quantum entanglement is that we can perform mathematical operations on a potentially enormous superposition of numbers in parallel without requiring any extra circuitry for each part of the superposition. From a many-universes perspective, quantum computation affords the possibility of massive parallelism by taking advantage of parallel universes instead of parallel processors.

But there's something else we need in order to exploit the power of quantum computation, and it's known as *interference.* This phenomenon helps overcome a severe restriction imposed by quantum mechanics: We're not actually allowed to look at each and every answer individually. It is as if the quantum rules say that whenever we examine one piece of paper in the hat, we inevitably destroy the rest.

Interference arises from the fact that the wavelike aspects of quantum particles can overlap one another to cause unusual and distinctive patterns of behavior. In quantum computation, we use interference to read off a new result that depends mathematically on all those intermediate results without revealing what any of them are.

David Deutsch actually disagrees with those who ascribe the power of quantum computation to entanglement and instead sees interference as the crucial phenomenon. The reason is that in a many-universes perspective, entanglement is a natural by-product of having a system of parallel realities. The surprise is interference because it allows these different realities to *overlap and collaborate.* It is, indeed, interference that makes the notion of multiple realities conjured up by such stories as Borges's "Garden of the Forking Paths" and, more recently, the movie *Sliding Doors* very different from the real thing.

We will see more precisely in Chapters 3 and 4 how entanglement and interference can be used together to solve problems far beyond the capabilities of any present-day computers.

The Hunter-Gatherers Take a Quantum Leap

To David Deutsch, the possibility of developing a new kind of superpowerful computer is certainly interesting, but it's very far from his overriding concern. What excites him is that quantum computing offers a totally new world view. "Building quantum computers is the least of it," he told me. "I'm not in this business in order to make better computers. Although I would still be fascinated by this work, I really wouldn't be interested in working on it if it weren't for the implications for physics. I want to understand physics fundamentally."

As we will see, there are formidable obstacles to the possibility of constructing a quantum computer. But there's nothing in the laws of physics that says it won't be possible, as far as we know. And as Deutsch stresses, you don't even have to build a quantum computer to be able to glimpse its darkest secrets. "What is really important about quantum computers is that they show us that there's a deep and unsuspected connection between physics and computation. Computation is connected to all sorts of human things like thought, knowledge, life, and so on, whereas physics is the most fundamental description of nature. So here we have an unsuspected, very deep connection between human-type things and fundamental-type things. I think philosophy is going to take a long time to assimilate this."

Deutsch made a substantial start in doing just that in his book *The Fabric of Reality.* In it he attempted to weave together four strands of scientific and philosophical thinking into an integrated whole. Quantum physics, evolution, computation, and knowledge are, he claims, all somehow inextricably linked. Much of his argument is predicated on the existence of multiple universes, but even people who reject that idea have been impressed by Deutsch's insights. It is an undeniably grandiose scheme, to which we will return.

There's another sense in which this story of the quest for the quantum computer is more than just a glimpse of a new technology. It is that, as Deutsch puts it in *The Fabric of Reality,* quantum computation is "nothing less than a distinctively new way of harnessing nature." At key points in human history, civilization took a leap forward because people discovered a new way of exploiting nature. Toolmaking, farming, the industrial revolution, and the information revolution were all triggered by the discoveries of new ways of manipulating nature. All of these advances transformed the way humans live. Quantum computation, Deutsch argues, could turn out to be as significant in its effects on human civilization.

On hearing Deutsch's grand vision for the future of quantum computation, I'm reminded of the scene in Stanley Kubrick's *2001: A Space Odyssey* when those strange-looking apes discover for the first time the power of wielding a bone as a weapon. What a powerfully symbolic moment that was, with the future of humanity hanging in the balance between aggression and creativity. These hominids had taken their first step toward understanding and exploiting nature. Suddenly the picture cuts from a bone spinning in the air to a futuristic spinning space station orbiting the Earth. Modern humans had arrived and were now beginning their odyssey into space, all thanks to that first spark of imagination in those apes.

Should we attach portentous significance to the arrival of the first quantum computer—should it ever happen? "People in the nineteenth century imagining the late twentieth century imagined a world of immensely sophisticated steam engines," Deutsch said. "They couldn't conceive of an information age and the Internet and all those kinds of things. Now, a computer is just a machine, but on the other hand it is more than just a machine because the essential part of a computer is not that it manipulates forces and energies—although it does do that. The important thing is that it manipulates information. The quantum computer is a fundamental new step in that it doesn't just manipulate information, it allows different universes to *cooperate.* This is as fundamental a difference in principle as any of the previous steps. In fact, numerically it potentially gives much greater leverage and a much greater increase in power for human beings to do things than any of the previous steps. It must change the way we think about ourselves even if we never use it."

Arthur C. Clarke, who co-wrote the screenplay for *2001* and who also wrote *The Sentinel,* the short story from which the screenplay evolved, once commented that Kubrick had wanted a science fiction tale of "mythic grandeur." If Deutsch's late-night thoughts are right, *Homo sapiens* is in for another leap forward. Whether it will add up to something of mythic grandeur, though, is a question I hope you will find illuminated in this book.

2

God, the Universe, and the Reversible Computer

Computers are useless. They can only give us answers.

PABLO PICASSO

Any universal computer could simulate the universe.

ED FREDKIN

The Computer That Just Coasts

When flying from London to New York a few years ago, I took a notebook computer to do some writing during the journey. Unfortunately, after what couldn't have been much more than an hour, the machine suddenly issued a series of warning beeps and promptly expired. The batteries were dead. Now, on a long flight this struck me as not a little inconvenient, especially as American Airlines' idea of movie entertainment was the tedious kind of fodder you'd only ever watch on a plane—because there's no escape.

So there I was with hours to kill and nothing but a blank screen on my computer to gaze at. Why, I thought, could they not make the batteries on notebook computers last longer? Despite decades of research, rechargeable batteries remain hopelessly limited in their energy capacity. Lithium technology may have extended the capacity of notebook machines somewhat, but they still cannot run for much longer than a few hours without power from the wall. It is for the same reason that electric powered cars remain a rarity rather than a mass-market phenomenon. Even the models that are available rely mostly on lead-acid batteries, a technology that has been with us for more than a century.

Undoubtedly, battery technology will slowly improve, but another possible solution exists for extending the running time of portable computers: Reduce the energy consumption. By cutting the amount of juice consumed by their machines, computer manufacturers could automatically increase the running time using existing batteries. What are the prospects for cutting energy consumption in this way?

In contrast to battery technology, computer technology has evolved at an incredible pace since the 1940s. Modern computers are certainly hugely more energy efficient than the early vacuum-tube behemoths, which typically were large enough to fill a whole room. Such machines posed considerable problems, not least because they generated so much heat. Some of the female computer operators who worked at Bletchley in the U.K. on the famous Enigma code-breaking project during the Second World War even felt obliged to strip down to their underwear because of the excessive heat.

With today's silicon technology, of course, it's a different story: How often have you seen IT workers taking their clothes off? But although silicon technology has transformed the scale of the heat output, energy consumption remains a surprisingly important consideration in the design of chips. Microprocessors typically run as hot as a cooking surface, which is why many need small fans to keep them from overheating. Nevertheless, with the appearance of new, lower-power microchips, it would seem that it should prove possible to reduce the power consumption of portable computers and thereby significantly increase their running time.

Except there's a catch. As we saw from Moore's law in the last chapter, chip manufacturers keep increasing the number of switching devices on their microprocessors and winding up the running speed. Such increases drive up the energy consumption and have in recent years more or less wiped out savings made by additional miniaturization. Furthermore, because there's only so much heat you can remove from a given volume of space, the problem of energy consumption could prove overwhelming as manufacturers strive to pack exponentially larger amounts of computing power into the same chip packages.

This problem of waste heat in computing therefore raises an important theoretical question: What is the minimum amount of energy required to perform a computation? If there were an absolute minimum energy required to work out, say, a calculation such as 2 + 2 = 4, it would clearly impose strict limits on what was possible no matter what technological improvements were made.

As it happened, some of the people I was about to visit in New York and Boston had discovered the answer to this question in the 1970s. To this day, their answer seems quite extraordinary. They found that computation could, in principle, be done without expending *any* energy and that it should be possible to build a computer that, when you switch it on, starts running and from there *just coasts.* The discovery of this remarkable fact completely changed the way physicists looked at computation.

Shannon's Information Theory

Strange though it might seem, answering the problem of energy consumption in computers is what finally pointed the way to ideas that now underpin quantum computation. To understand how, it's useful to take a step back in history for a moment.

Early in the twentieth century, as telegraphy and telephony began to take off, scientists began to ponder what limitations physics might impose on communication systems. In 1927 Ralph Hartley, working at Bell Labs, took an important step by proposing a measure for the amount of information in a message. He concluded that an appropriate choice was to take the *logarithm* of the total number of possible messages because the logarithm tells you the total number of digits or characters required to convey a message. (For a brief discussion of what logarithms are and how they work, see Appendix A.)

Although the intrusion of logarithms might have been a surprise in those days, today it actually seems quite natural. When we talk about computer memories in terms of bytes, kilobytes, and megabytes, we are, in fact, implicitly using a logarithmic measure of information. One byte can store 256 different numbers or messages, but few people think of computer memory that way. Rather we think of one byte as eight *bits* of information—a string of eight 1s and 0s. (Eight is the logarithm of 256 using base 2 because $2^8 = 256$.)

Nevertheless, the explicit use of logarithms helped Claude Shannon, an

American mathematician and electrical engineer, to make a big connection in the 1940s when he showed how the theory of heat—thermodynamics—was applicable to information transmission. Shannon, like Hartley, also worked at Bell Labs, which was clearly *the* place to be for pondering the growing problems caused by the crowding of communication channels. In a famous and much-cited article entitled "A mathematical theory of communication," he presented a unifying theory of the transmission and processing of information.[1]

Shannon's theory took the logarithm idea further by taking into account the likelihood or *probability* of any particular message. Consider a rain forecast for tomorrow morning in the Sahara Desert. Let's assume the forecast is a simple, clear-cut prediction: rain or no-rain. Simplistically, one could say that either forecast contained the same amount of information. However, if you think about it, a prediction of no-rain is not particularly informative, because a lack of rain in the Sahara is what you would expect most of the time. In contrast, a prediction of rain contains considerable information (assuming it to be reliable) because it seldom happens.

To take such effects into account, Shannon's theory related information content to its probability according to the formula[2]:

$$\textit{information} = -\log \textit{probability}$$

What this says is that the information content of a message is proportional to the logarithm of its probability. This relationship observes the logarithmic idea proposed by Hartley, while the minus sign ensures that the answers come out the right way round: If a message has a low probability and hence is very unlikely, then the information content is high, and vice versa. (Logarithms of numbers less than 1 are always negative. For a message with a probability of one half, or 50 percent, for example, log ($^1/_2$) is -1, so the formula gives an information value of 1. For a message with a probability of one quarter, or 25 percent, we have log ($^1/_4$) $= -2$, so the formula gives an information value of 2. Since all probabilities are, by definition, less than certain, the formula above always gives a *positive* number for the information content.)

So where does thermodynamics come into the picture? The answer is in the math. The above relationship turns out to resemble a famous formula in physics found on the tombstone of the man who inspired it[3], Ludwig Boltzmann. Tragically, Boltzmann took his life in 1906, convinced that his ideas had fallen by the wayside of physics. Today, they actually form part of its bedrock. His ideas, which revolved around the formula, $S = k \log W$, opened up a totally new way to understand the thermodynamic concept of *entropy*.

Earlier, in the nineteenth century, the German physicist Rudolph Clausius had studied the theory of heat engines and established entropy as a kind of measure of unavailable energy—energy that is unavailable for doing useful work. His definition of entropy was expressed as a simple ratio between heat and temperature.

With this ratio, Clausius showed that the total entropy of a closed system can never decrease. This conclusion is a version of the famous second law of thermodynamics, which states that the entropy of the universe is always increasing—synonymous with the idea that the universe is inexorably running down toward its own "heat death."

Boltzmann's definition of entropy was quite different. The formula $S = k \log W$, where S is the entropy, W is the probability of the given state, and k is a fixed constant of nature (called, in homage to the man, Boltzmann's constant) related entropy to the amount of atomic disorder in a system. The importance of Boltzmann's work was that it provided an explanation for the second law of thermodynamics, which had previously been understood as a macroscopic phenomenon, in terms of the *microscopic* behavior of atoms. In fact, it helped confirm the very existence of atoms, which until the early twentieth century had remained hypothetical.

Using purely statistical arguments, it is possible to show that systems have a natural tendency to become more and more disordered. This tendency can be illustrated by taking a deck of playing cards, sorted into suits and sequence, and shuffling it. You are, of course, exceedingly unlikely to finish with the cards in the same regular order. Likewise, if you shake a jar partially filled with a layer of red marbles on top of a layer of blue marbles, you will expect the red and blue marbles to end up randomly scattered, with all trace of the separate layers lost.

The reason disorder almost always increases is that disordered states hugely outnumber highly ordered states, so when a system is shaken and stirred it almost inevitably settles upon one of the more disordered states. Boltzmann showed that the same arguments applied to heat energy. At the atomic level, heat is expressed in the relentless motion of atoms and molecules. If we could look at the molecules in a liquid or gas, we would see that they are seething with energy, bouncing off the walls of the container, and colliding with one another. An important feature here is the temperature: the higher it is, the faster the molecules vibrate and jostle.

A heat engine extracts useful energy from heat by taking two heat sources at different temperatures and allowing them to mix in a controlled way. A steam engine, for example, uses both steam and the cooling effect of the atmosphere. Now, if you mix a hot gas with a cold gas, what you get, as happens when mixing the colored layers of marbles, is an increase in the disorder, or entropy. Boltzmann's formula quantifies this process precisely in terms of the logarithm of the probability of the system state before and after any change.

What does all this have to do with information theory? Shannon's formula, *information* = −log *probability*, and Boltzmann's $S = k \log W$ have the same form because they both involve logarithms of probabilities. Shannon saw the connection but apparently decided to appropriate the use of the term *entropy* in information theory only on the advice of the brilliant Hungarian-born mathematician John von Neumann, who reputedly said, "It is already in use under that name . . . and besides, it will give you great edge in debates because nobody really knows what entropy is anyway."[4]

Because of the minus sign in the Shannon formula, information in this new viewpoint was seen as a negative form of entropy. (In the 1950s a new term was coined for negative entropy, *negentropy,* although this is rarely used nowadays.) Boltzmann himself had recognized entropy as a form of missing information. This is because the higher the entropy of a system, the more disordered it is and the less information we will have about its internal state.

This link between information and thermodynamics was intriguing and suggested a fundamental connection between knowledge and physics. But although Shannon's theory of information helped establish the link, it didn't yet resolve the question of energy consumption in computation. That depended on another line of research—one that, oddly enough, led people down the wrong path for many years.

The Puzzle of Maxwell's Demon

Among the outstanding achievements of the nineteenth-century British physicist James Clerk Maxwell were the laws of electromagnetism and the kinetic theory of gases. In 1867 he also devised a thought experiment, later nicknamed "Maxwell's demon," that turned out to be very influential, with ramifications that extended into quantum mechanics and information theory, neither of which was known in his time.

Maxwell imagined a vessel divided into two regions by a wall in which there is a tiny hole. The vessel is filled with air, and sitting by the hole is a small creature—the demon—who is able to watch individual molecules of air pass back and forth through the hole. The demon can also open and close a shutter in front of the hole without expending any energy. Although a practical shutter would certainly require some energy for its operation, for the purposes of this thought experiment, physicists begin by assuming that this energy expenditure can be made arbitrarily small.

The idea of the thought experiment is that the demon decides to open and close the shutter at such moments that ensure only fast-moving gas molecules are allowed to pass from chamber A into chamber B, while only slow-moving ones are permitted in the opposite direction (see Figure 2.1). The demon would certainly have to be rather clever to do this and very quick-witted, but again we assume that he requires no (or very little) energy. The result would be that the air in chamber A would gradually cool down, while the air in B would heat up, in defiance of the second law of thermodynamics. With Maxwell's demon we could extract heat energy

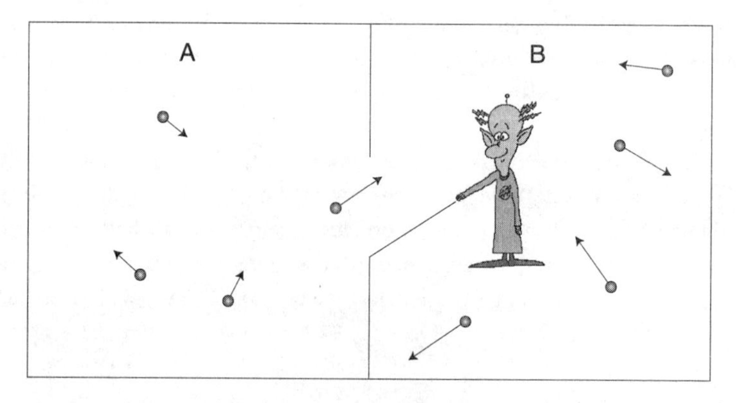

Figure 2.1. Maxwell's demon

By opening and closing a shutter between two chambers full of gas, a tiny demon could arrange for fast-moving molecules to gather on the right-hand side and slow-moving molecules to gather on the left-hand side. One chamber would then heat up with respect to the other, apparently in defiance of the second law of thermodynamics.

from the air or sea and generate unlimited power for all our needs. It would, in short, provide what is known as a perpetual motion machine of the second type. (A perpetual motion machine of the first type would generate energy from nothing, contravening the first law of thermodynamics.)

Maxwell suggested his thought experiment not to undermine the second law but rather to highlight the fact that it was a statistical principle and therefore true only probabilistically. There is, after all, a small probability that for a brief moment fast-moving (hot) molecules of air could gather at one end of a vessel while slow-moving (cold) molecules clustered at the other end simply by chance. The probability of significant fluctuations of this kind is, in practice, vanishingly small—to all intents and purposes, zero. Nevertheless, the idea of the demon served as a reminder that the second law of thermodynamics had only a statistical rather than an absolute certainty.

In their book *Maxwell's Demon,*[5] an invaluable collection of the many and varied papers written on this subject, Harvey Leff and Andrew Rex point out that though Maxwell did not intend his hypothetical character to be a serious challenge to the second law, many researchers subsequently saw it as a puzzle that had to be solved.

One of the most famous attempts to do this was by the physicist Leo Szilard in 1929. He simplified the thought experiment by imagining a cylinder in which there is only *one* molecule of gas rattling around and by replacing the shutter with a piston, which can be made to do useful work. Consider that the piston starts out maximally extended, while the molecule flies around in the space in front of it (see Figure 2.2). If the demon waits until the molecule is on the right-hand side (a), he can move the piston halfway into the cylinder (b) without expending any energy. He now attaches a weight to the piston via a pulley (c). As the molecule rebounds off the piston, the piston moves, lifting the weight (d). The molecule loses some speed in the process but absorbs some heat from the walls of the cylinder to restore its temperature and hence speed. The system thereby returns to its starting condition. The overall effect of this experiment is that the demon has converted a small amount of external heat into useful work.

Once again we seem to have a clear violation of the second law of thermodynamics. Szilard attempted to resolve this paradox by looking more closely at how the demon decides when to move the piston. In particular, he focused on the need for the demon to make a measurement of the molecule and argued that the entropy decrease in shutting the molecule off from

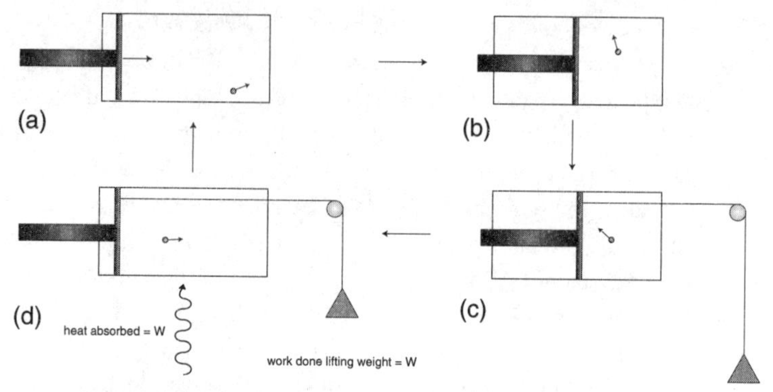

Figure 2.2. Szilard's simplified model for Maxwell's demon

By considering only one molecule of gas, Leo Szilard showed how Maxwell's demon could be reduced to a binary decision process. If the demon knows when the molecule is in the right half of the cylinder, he can push the piston halfway in without expending any energy. Afterward, energy can be extracted by causing the piston to lift a weight.

one side of the cylinder (and hence halving its volume) was compensated by an increase in entropy associated with measuring the molecule's position. According to this argument, the second law would be saved by the fact that the demon had to expend energy and entropy by making measurements.

Szilard's analysis, although it turned out to be not wholly correct, is now regarded as having been a breakthrough in information theory. In their book, Leff and Rex concluded:

> The ingenuity of Szilard's engine is striking. His tractable model allows thermodynamic analysis and interpretation, but at the same time entails a binary decision process. Thus, long before the existence of modern information ideas and the computer age, Szilard had the foresight to focus attention on the "information" associated with a binary process. In doing so he discovered what is now called the binary digit—or "bit"—of information. . . . While he did not fully solve the puzzle, the tremendous import of Szilard's 1929 paper is clear: he identified the three central issues related to Maxwell's demon as we understand them today—measurement, information, and memory—and he established the underpinnings of information theory and its connections with physics.[6]

The correct resolution of Maxwell's demon did not arrive until many years later. We will see why shortly.

Much Ado About *kT*

Once people began to build digital computers in the 1940s, it was only natural to apply some of the ideas about information and entropy to the new paradigm of computing. In 1949 Von Neumann gave a lecture in which he identified a minimum amount of energy required "per elementary act of information—that is, per elementary decision of a two-way alternative and the elementary transmittal of one unit of information."[7] His minimum was close to a value physicists like to represent with the letters *kT*. Here *k* is Boltzmann's constant, and *T* is the temperature of the system in question.

Found in virtually any physics textbook, *kT* quantifies a kind of atomic unit of heat because it represents the amount of heat energy typically carried by all atoms in solids, liquids, and gases. At room temperature, *kT* is about 3×10^{-21} joules (a joule is about one quarter of a calorie, the amount of heat required to raise the temperature of a gram of water by one degree Celsius).

On a human scale *kT* is obviously very small, though translated into atomic units it's actually very significant: Molecules of oxygen at room temperature typically move at speeds of around a thousand miles per hour. Yet *kT* is still approximately 1 million times smaller than the amount of energy used by logic gates today. Actually, Von Neumann expressed his minimum energy for an elementary act of computation as *kT* log 2, where log 2 is the logarithm to base *e* of 2, which is around 0.69. We'll see how the factor of log 2 arises, but numerically it scarcely changes things.

Von Neumann's energy minimum flowed in part from Szilard's work, which had suggested precisely the same quantity, *kT* log 2, as the minimum energy required to make a measurement. Nevertheless, Von Neumann's claim was that this amount was involved in *manipulating* one bit of information. This idea seemed plausible at the time because if you wanted to store one bit of information by charging a capacitor, for example, it would seem that you would need to use enough energy to ensure that the signal was not swamped by noise. Noise energy is generated by thermal effects, which makes *kT* the obvious choice for a lower limit on the energy required.

In 1950 Leon Brillouin of IBM made these ideas more concrete by applying quantum theory and Shannon's theory of information to the analysis of Maxwell's demon. He argued that the demon would be unable to see the position of a molecule against the ambient heat radiation emitted from the walls of the chamber unless it was equipped with a torchlight. The demon would have to expend at least one photon that was more energetic than the background photons to see the molecule. Thus the measurement would, according to Brillouin's calculations, entail dissipating an energy of the order of kT.

So during the 1950s the received wisdom was that each act of computation required a basic minimum of energy corresponding to kT, the thermal energy of individual molecules and atoms. And there the matter might have rested. But as it turned out this consensus was wrong.

Landauer's Principle

The clouds began to break in 1961, when IBM's Rolf Landauer made a crucial discovery, which he described in an article in IBM's *Journal of Research and Development.*[8] In the article he tackled head-on the question of the minimum energy involved in computation. While recognizing Brillouin's claim that the measurement process might require an energy dissipation of around kT, he pointed out that it was difficult to ascribe the kind of measurements Brillouin had considered to the computing process. Instead, Landauer advanced the notion of *logical irreversibility,* in which *physical* irreversibility would arise only from irreversible mathematical operations.

With the notions of reversibility and irreversibility, we are once again taken to the nineteenth century, because it was then that scientists discovered that heat engines would function with maximum efficiency only when they were operated reversibly. To operate reversibly, an engine needs to be cranked, in theory, infinitely slowly and smoothly so that the temperature of its internal parts does not get out of equilibrium with the environment or its heat inputs. Under these circumstances no useful energy is wasted as heat.

Landauer's argument was that, by analogy, computation might need to involve heat dissipation only when you did something irreversible with the information. Furthermore, he showed that the only processes that were irreversible in a computation were those in which information was discarded. The role of *erasure* was a fundamentally new idea.

At first sight this idea, now known as Landauer's principle, seems counterintuitive. If we carry out a series of arithmetical calculations on a blackboard, for example, the strenuous part of the process would seem to be all the adding and multiplying. Rubbing out one calculation to perform another would seem to be the easy part! But according to Landauer it was only this process of erasure that demanded energy. Everything else could, in principle, be done for free. Of course, in practice, addition and multiplication do require energy whether they are performed by means of silicon chips or our brains, but not when we are theorizing about the ultimate limits of computing. Assume a frictionless blackboard . . .

Part of Landauer's argument was to consider a ball bearing placed in a model landscape consisting of two wells, labeled 0 and 1, as shown in Figure 2.3. The ball will naturally come to rest in either well and can represent a single bit of information. To manipulate the bit, you could move the ball from one well to the other. To do this you would need to apply a force to roll the ball uphill and over the peak separating the wells. This would require some energy initially, but as the ball passed the peak it would start to roll down the other side, giving up that energy. As there is no energy difference between wells 0 and 1, in theory, it should be possible to flip the ball between the states using no net energy. So a simple bit manipulation such as a NOT function need not require any energy at all.

What happens, though, if you want to reset the memory? If the ball is already at 0, you need do nothing, and if it is at 1, you need to push it over the hill. However, as Landauer argued, this is not how computer circuits work. A computer typically pushes data around regardless of the exact data that are being handled. For a working circuit, what we need is a single type of action that will perform the reset function in either case. Imagine that

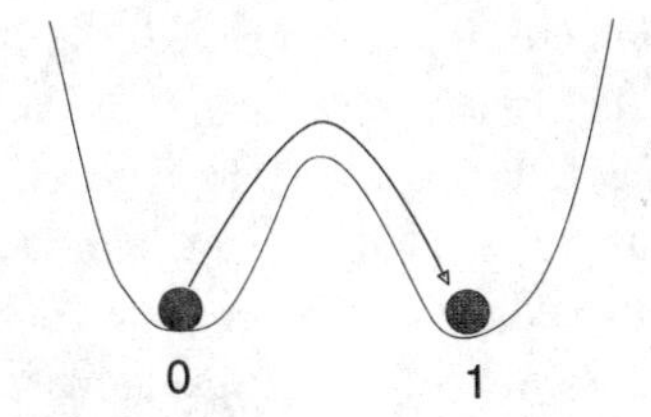

Figure 2.3. Information storage in a double well

A ball in a double well can represent a single bit of information. Flipping the state of the bit need not require any net energy, provided the initial state of the ball is known.

there was such an action. Its effect would be to produce the same result—that is, 0, given the two different starting positions 0 and 1. If such a process is not going to incur any energy cost over the ball's complete journey, then according to the second law of thermodynamics the process must be reversible. But if you reversed the ball's trajectory, you would need an action capable of producing two different outcomes from the same initial state. Such a situation is impossible because the only way you could achieve two outcomes is by doing two different things. Therefore, resetting the system to 0 must entail some energy cost.

One way of achieving the reset operation, for example, would be to fill the wells with a viscous fluid such as oil. The effect of the oil would be to slow the ball and damp any tendency to bounce back and forth within either of the wells. To reset the ball to 0, we could then apply the same type of action, irrespective of the ball's position; a leftward force large enough to get the ball over the middle hump would do the trick. If the ball were already in the left-hand well, it would climb the left-hand slope, and once the force was removed it would fall back again. Provided the damping effect of the oil was sufficient, the ball would come to rest in well 0 rather than bouncing back over the hump to well 1. So we now have one action that could reset the ball to 0—but at the expense of dissipating energy.

As for how much energy must be expended, Landauer showed that it was given by the expression $kT \log 2$, a quantity Ed Fredkin has dubbed the Landauer constant. (For an explanation, see Appendix B.) So the correct answer to the minimum energy cost of computation was not that processing information inevitably costs energy for each manipulation of a single bit, but that we need to spend energy each time a bit of information is *thrown away.*

This discovery by Landauer was a major step forward in understanding the physics of information processing, but he didn't quite get the whole answer. In his 1961 paper he argued that though it might be possible to construct a computer using fully reversible logic, such a machine would require additional memory circuits at each stage to keep intermediate results of calculations that could not be thrown away. He concluded that such a machine would be too unwieldy for practical purposes and that it would also be incapable of running a program for an unlimited time because it would need an unlimited amount of memory to store its intermediate results. The machine would, in effect, choke on its own garbage.

The Reversible Computer

Landauer correctly concluded that a computer had to use some energy each time memory was reset. And he was correct about a reversible machine suffering from a garbage problem. What he didn't see was that there might be a way to overcome this problem by getting the computer to clear up its own garbage. That insight came in 1973 from his IBM colleague Charles Bennett.

Bennett, who is now an IBM fellow and is generally recognized as one of the leading authorities on the physics of computation, came up with an ingenious solution to the garbage problem. He had experimented with some simple computer programs to see whether it was possible to write them without having to use any logically irreversible operations. Initially, he had expected that it wouldn't be possible to do anything computationally interesting. However, he managed to produce some reversible programs, including one that used repeated subtraction to test whether one integer was divisible by another. He then noticed a remarkable feature that was common to these programs:

> The computation consisted of two halves, the second of which almost exactly undid the work of the first. The first half would generate the desired answer (e.g., divisible or not) as well as, typically, some other information (e.g., remainder and quotient). The second half would dispose of the extraneous information by reversing the process that generated it, but would keep the desired answer. This led me to realize that any computation could be rendered into this reversible format by accumulating a history of all information that would normally be thrown away, then disposing of this history by the reverse of the process that generated it.[9]

So to make any program reversible, what we do is this: We must first rewrite the program so that no intermediate results are discarded or overwritten. We then let the computer run until it produces a result that appears at its output. We print that output and then run the computer program *backwards.* Running it backwards simply undoes everything the program did in the forward run, returning the machine to its original state. The result is that the input, output, and all intermediate states are reset to their starting values.

Of course, if we hadn't printed the output halfway through, the net result would be that the computer would have achieved absolutely nothing. However, by printing the output, we ensure that we get the answer we want while we finish with the computer completely reset, waiting to perform another calculation. All of the intermediate results have been erased, so there is no garbage to worry about. What's more, we haven't expended any energy.

Of course, if you tried running the program on any conventional computer, you *would* expend energy, because all conventional computers work irreversibly. Bennett had shown how to make any program logically reversible. How feasible would it be to implement a logically reversible program in a *physically* reversible way?

In his 1973 paper Bennett offered an example of a naturally occurring system that could, in principle, "compute" with arbitrarily small amounts of energy: the biochemical responsible for the replication, transcription, and translation of the genetic code. The system used by living cells to read their genes depends on an enzyme known as RNA polymerase. It is responsible for synthesizing RNA copies of the gene sequences stored in DNA. These RNA copies then serve to direct the synthesis of the proteins encoded by those genes.

DNA normally consists of two strands zipped together in the shape of a double helix. To read the information stored in its genetic code, the helix unwinds, forcing the strands apart. An exposed region of single-stranded DNA can then serve as a template for the construction of RNA copies. A copy is assembled by pairing chemical molecules known as nucleotide bases, which form part of the RNA molecule, with complementary bases of the DNA. The process is orchestrated by the RNA polymerase enzyme, which takes small molecules known as nucleoside triphosphates—ATP, GTP, CTP, and UTP—from the surrounding solution and attaches them to the growing RNA strand (see Figure 2.4). The DNA template ensures that the correct molecules are inserted at each location along the strand. If the next DNA base is adenine, only UTP will slot into place because energetically these bases form the most stable pair. Likewise for thymine and ATP, cytosine and GTP, and guanine and CTP. As each molecule is added, the enzyme shifts forward one notch along the DNA.

The amazing thing about this complicated process is that it all happens reversibly. This is true, in fact, of all reactions facilitated by enzymes or catalysts. If you want the reaction to go in one particular direction, however, you need to give the system some assistance. In living cells the

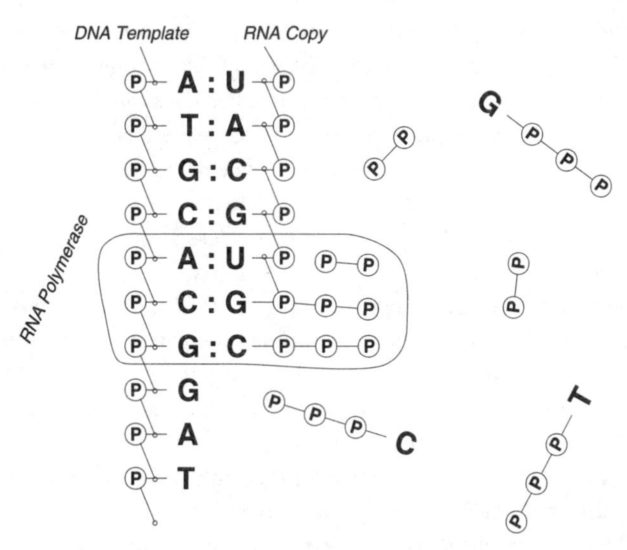

Figure 2.4. DNA transcription by RNA polymerase

The biochemical system responsible for copying and translating DNA sequences depends on reversible chemical reactions orchestrated by RNA polymerase. These reactions are normally driven forward by supplying fresh ATP, GTP, CTP, and UTP molecules to replace the ones consumed in assembling RNA strands. If the reactions are driven sufficiently slowly, the energy consumed can be made arbitrarily small.

process is driven forward by supplying fresh molecules of ATP, GTP, CTP, and UTP, while other metabolic reactions mop up the spent pyrophosphate (PP in Figure 2.4) molecules that are released when ATP, GTP, CTP, and UTP attach themselves to the lengthening RNA strand.

In bacteria, RNA polymerase copies around thirty nucleotide bases per second, dissipating about $20kT$ per nucleotide—some thirty times the Landauer limit. But if the system isn't forced so hard, the energy consumption can be greatly reduced. In 1982 Bennett published a wide-ranging paper that explored the whole subject of the thermodynamics of computation.[10] In it he argued that if ATP, GTP, CTP, and UTP were each present at just 10 percent above the equilibrium concentration (the concentration at which the forward reaction's rate is the same as the reverse reaction's), the energy dissipation would be as little as $0.1kT$ per nucleotide, well below the old kT "barrier." The disadvantage of running the process under these conditions is that RNA synthesis would be much slower, and for each eleven steps forward, the system would slip ten steps back. Nevertheless, such a chemical system offered tangible evidence for the possibility of computing using arbitrarily small amounts of energy. Bennett called the system an example of a *Brownian computer* because it relies on the same thermal process that generates Brownian motion, the ceaseless chaotic molecular motion seen when small particles of matter suspended in a fluid are examined under a microscope. Bennett's choice of DNA and RNA as a computational system was particularly prescient in the light of more recent discoveries showing how DNA can be used to compute solutions to mathematical problems. We'll examine the capabilities of DNA-based computers in Chapter 9.

In his 1982 paper, Bennett also cleared up the remaining confusion over the problem of Maxwell's demon. He argued against previous claims by Brillouin and others that each of the demon's measurements unavoidably dissipates energy. On the contrary, Bennett suggested that the demon could perform its measurements *reversibly* and therefore with no energy expenditure. Where the demon did need to dissipate energy, he said, was in *resetting its memory* after each measurement.

If we consider Szilard's version of the demonic experiment, in which there is just one molecule, Bennett's argument went as follows: At first the molecule wanders freely throughout the apparatus. The demon inserts a partition in the middle, trapping the molecule on one side or the other. The demon now performs a reversible measurement to discover whether the molecule is on the left or the right. The demon then uses this information to extract energy from the molecule by inserting a piston against the partition on the empty side and allowing the molecule to push against the piston. The piston can do useful work until it reaches its end and the molecule fills the whole apparatus again. The apparatus is now restored to its original state, except that the demon still has a record of the measurement in its brain. If the whole process were to be repeated, the demon would have to erase its record before performing the next measurement. It is this erasure that entails logical irreversibility and a minimum energy expenditure that balances the energy extracted from the piston. This, then, was the true resolution of Maxwell's demon: The energy cost is not in making measurements but in helping the demon forget.

Although Bennett was the first to publish the complete solution to the idea of a reversible computer, it later became apparent that a Belgian mathematician, Yves Lecerf, had advanced the notion of a reversible computing machine as early as 1963.[11] His paper went almost unnoticed for many years, probably because it was in French; Bennett didn't learn of it until around 1989. Lecerf only used reversibility for a purely mathematical purpose, without drawing any connections to the thermodynamics of computation. Of greater significance to the latter was the work of two other researchers who independently deduced methods for building reversible computers within a year or so after Bennett. These were Ed Fredkin, formerly of MIT, and Tom Toffoli of Boston University.

Reversibility and the Laws of Physics

The idea that one can compute using no energy at all may sound too good to be true, but remarkably enough, it seems that it is possible, in theory at any rate. This conclusion not only has important practical implications but also offers a new vista of the relationship between computation and physics. To see this, it's important to know that all fundamental physical processes are reversible. That may seem counterintuitive: If you burn toast, spill red wine over the carpet, or crash your car, the results would seem to be fairly irreversible. However, at the atomic level all physical processes are fully time symmetric. This is clear when you look at the equations of quantum mechanics and classical mechanics. If you reverse the sign of time in the equations, they look essentially the same.

In his book *About Time,* Paul Davies describes the time symmetry of the fundamental laws as "an almost sacred principle of physics."[12] On a macroscopic scale the principle is far less obvious because of the statistical effect of the second law of thermodynamics. If a system starts off in a highly ordered state, it is likely to evolve into a more disordered state simply as a matter of chance. The fact that in everyday life we see so many things decaying and falling apart appears to be the result of our universe having started off in an exceedingly ordered state. The reason for this special initial state remains one of the great unsolved puzzles of physics.

However, macroscopic systems do not always have to degrade irreversibly. In the nineteenth century the French engineer Sadi Carnot showed how a heat engine can behave reversibly. The Carnot engine consists of some (idealized) gas enclosed in a cylinder with a piston. The cylinder is alternately heated and cooled. When it is heated, the pressure of the gas rises. The natural tendency of the gas is to expand by pushing the piston outward, which can be translated into useful work (by means of a crankshaft, for example, as in a conventional car engine). When the gas is cooled, the pressure drops and the gas contracts, pulling the piston back part of the way. However, restoring the piston to its most compressed state requires some additional effort, so this part of the cycle absorbs work from the outside world. The overall effect, though, as the engine cycles between its hot and cold states, is for it to convert heat into a positive amount of work.

To ensure reversibility, the motion of the piston must be sufficiently slow so that the gas is always in equilibrium with its surroundings. Reversibility guarantees that the engine works at its maximum theoretical efficiency, so that no useful energy is wasted. This means that the engine can, in principle, be operated in reverse, and it then converts work into heat at the same rate as it converts heat into work when operating in the forward direction.

So having seen that the laws of physics are microscopically reversible and that they can be made to operate reversibly even on everyday scales, assuming perfect equilibrium with the environment, the idea that computation could also be reversible might seem only natural if we regarded computation as a physical rather than abstract process. It was precisely this kind of thinking that motivated Ed Fredkin to invent his reversible scheme for computers. But he had other reasons for believing that computers ought to be reversible.

Is the Universe a Computer?

Ed Fredkin is, by all accounts, an unusual scientist. "An undisciplined genius," Rolf Landauer told me. Fredkin flunked college, and in 1954 he joined the Air Force and trained as a fighter pilot. After earning his wings a year later, he was grounded because of asthma and was sent to work on a defense project in which he got his first taste for programming computers. He was immediately hooked by their capabilities, and shortly after leaving the Air Force he became a computer consultant for Bolt Beranek & Newman.

During his time at the company he was taken by what looked like an intriguing trend in physics: More and more aspects of nature were turning out to be atomic, or discrete. Within the previous hundred years, physics and chemistry were revolutionized by the atomic theory, which posited that matter was made from discrete building blocks rather than continuous stuff, and quantum theory, which showed that the forces of nature and many properties of matter were discrete too. Toward the end of the nineteenth century, for example, it was shown that electricity was not a fluid but consisted of particles called electrons. Not long after, light was shown to consist not of continuous waves but of individual particles called photons. In the 1920s it was shown that even something as apparently continuous as the motion of a flywheel had to be ultimately quantized.

Fredkin wondered where this trend might lead. His answer was radical: *Everything,* he thought, must be quantized. Space, time, and all properties of matter, in his scheme, must come in single, indivisible units so that nothing is continuous. Part of the attraction of this idea was that it put a finite cap on the amount of information in the universe. If some aspects of nature were continuous, this would imply that they contained an infinite amount of information. There are, after all, an infinite number of numbers between 0 and 1. If spatial positions can be defined with unlimited accuracy, then even something as simple as the location of a single billiard ball on a table, for example, would contain an infinite amount of information.[13]

Fredkin saw that a universe in which everything was quantized would be analogous to a digital computer because in both, everything would be discrete and finite. In effect, the difference between Fredkin's view of the universe and the conventional view was like the difference between digital and analog computers. The idea that the universe is a gigantic digital computer processing information was a view for which Fredkin became famous, if not notorious. At PhysComp '96, a physics and computing workshop, in Boston, Fredkin reflected on his view of "digital mechanics" and recalled that the physicist Philip Morrison remarked that the only reason Fredkin thought the universe was a computer was because he was a computer scientist, in the same way that if he had been a cheese merchant, he would have claimed that it was made of cheese.

Yet back in the 1950s there appeared to be a major flaw in Fredkin's conception: computational irreversibility. If the universe was a gigantic digital computer, the laws of physics had to be part of the computer's program. It was known that the laws of physics at a microscopic level were reversible, yet all known computers were irreversible. This disparity did not discourage Fredkin. Indeed, it spurred him on to question the very assumption that computers had to be irreversible. He faced another obstacle, though, in pursuing his ideas: He was not a trained scientist, and getting new ideas in science accepted is usually extremely difficult if you're not established in the academic circuit.

Instead, Fredkin made use of his early expertise in computing to found a company—Information International Incorporated, Triple-I for short. It started out as a computer consultancy and led Fredkin into producing programmable film readers, machines that used computers to automate the analysis of data stored on photographic film. It grew from a one-man operation to a multimillion-dollar industry by the late sixties. In the process Fredkin became wealthy enough to buy a Caribbean island in 1968.

By then Fredkin's ideas had begun to have an impact. Marvin Minsky, an MIT professor and a leading figure in artificial intelligence research, was involved in setting up a program funded by the Defense Department that was the precursor to MIT's Laboratory for Computer Science. Fredkin had connections with the project through his company, and in 1966 he began discussing with Minsky the possibility of becoming a visiting professor at MIT.

Minsky had to overcome resistance from his MIT colleagues, who resented the idea of a college dropout being appointed as professor. Nevertheless, Fredkin was appointed and had such an electrifying effect on his colleagues and students that within a year he was promoted to full professor—when he was still only thirty-four. Not long after, he was appointed director of the Laboratory for Computer Science.

It was also partly because of Minsky that Fredkin met Richard Feynman, the famous Nobel Prize–winning physicist with whom he became good friends. Robert Wright's book *Three Scientists and Their Gods,* which features Fredkin as the first of its three subjects, recounts the tale of how Fredkin and Minsky had traveled west on business and found themselves in Pasadena, home of the California Institute of Technology, with time to kill. After deciding to call up some "great people" on the off chance of meeting them, Fredkin suggested Linus Pauling, but he wasn't home. Minsky suggested Feynman. "So we just called him out of the blue," Fredkin recalled. "He had never met either of us or heard of us, and he invited us over to his house and we had an amazing evening. I mean, we got there at, like, eight or something and stayed till three in the morning and discussed an amazing number of things."[14]

In 1974 Fredkin took a year off from MIT to spend time at Caltech working with Feynman. "The deal was that I would teach Feynman about computers, and he would teach me about quantum mechanics," Fredkin told me when I saw him at the PhysComp '96 meeting. "Feynman later complained that I got the better deal, but actually I think we both did pretty well."

The Fredkin Gate

During his time at Caltech, Fredkin solved the big problem with his world view. Unaware of Charles Bennett's result on reversible computing, he found a way of constructing logic gates that could work reversibly. His approach differed from Bennett's in that he sought a detailed description of how a reversible computer might be built, logic gate by logic gate. Bennett's argument had been more of a proof of principle. The key to Fredkin's vision was a new kind of device that became known as the Fredkin gate.

Like Bennett, Fredkin realized that if logic circuits were to be reversible, they had to avoid throwing away information. Yet such a standard logic gate as an AND gate, which has two inputs and one output, clearly lost information. An AND gate can have either 0 or 1 as each of its inputs; the output is always 0, unless both inputs are 1. If you tried to reverse such a gate, you could only know for sure what the inputs were if the output was 1. If the output was 0, there would be three possible states for the inputs. Hence information is lost by an AND gate. This behavior is summarized in the gate's truth table (see Figure 2.5).

The Fredkin gate avoids information loss by having the same number of outputs as inputs. In fact, it has *three* inputs and *three* outputs because it turns out that gates with only two inputs and two outputs are either not reversible or don't do enough to be interesting. This latter point is rather crucial because it relates to a deep aspect of computers discovered by the British mathematician Alan Turing in the 1930s.

Turing showed that computers are *universal* in the sense that any one computer can do the same thing as any other. This might not seem obvious, because we tend to use large supercomputers for complex tasks like weather forecasting and aerodynamic modeling, while word processing and spreadsheet calculations are left for desktop machines. However, the only real difference between supercomputers and desktop PCs is their

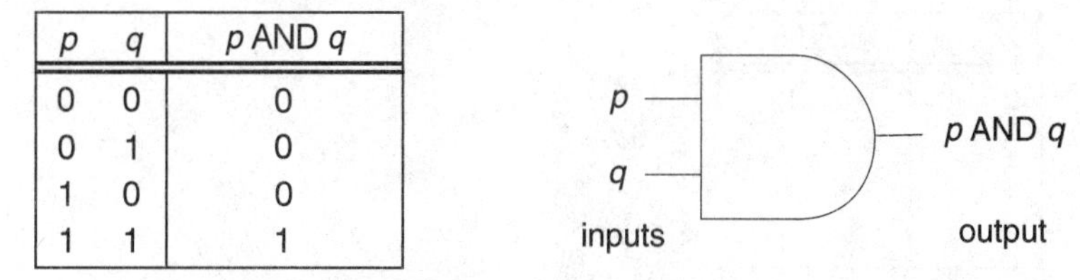

p	*q*	*p* AND *q*
0	0	0
0	1	0
1	0	0
1	1	1

Figure 2.5. Truth table and circuit symbol for the AND gate.

speed and memory capacity. If you gave a PC enough memory and enough time, it could do anything that a supercomputer could. Computers are fundamentally the same in a way that most machines are not. A coffee grinder, for example, cannot be reprogrammed to bake bread. A CD player will not play vinyl discs. A jumbo jet cannot fly like a fighter plane. But any computer can mimic the behavior of any other computer, as evidenced by the increasing number of emulators you can find these days for running Unix software on NT platforms, and Windows on Apple Macs, and for resurrecting classic old computer games that once ran on now-defunct platforms, such as the Atari and Sinclair Spectrum.

This concept of *universality* is what makes the computer so powerful and so versatile. To achieve universality, though, the computer needs a certain minimum level of internal machinery. Surprisingly, though, when you look at the level of individual logic gates, it turns out that you don't need a huge variety of different types. It is possible to show that using conventional irreversible logic elements, any binary logic circuit can be implemented by a collection of AND, OR, and NOT gates. What's more, each of these gates can be implemented by one, two, or three NAND gates. A NAND gate is equivalent to an AND gate followed by a NOT gate.

The NAND gate is said to be universal because it alone can be used to construct any possible binary logic circuit—including the complete circuit for a computer.[15] AND, OR, and NOT gates are not individually universal because you cannot construct a computer from any one of those gates alone. The NAND gate is therefore rather special and, indeed, some early computers were constructed from nothing but NAND gates. But like the AND function, the NAND gate loses information and is therefore irreversible. If we only know the output and it's 1, there is no way of reversing the gate to reproduce the inputs with any certainty because there are three possible sets of values. The Fredkin gate overcomes this limitation.

At first sight the Fredkin gate's truth table looks rather complicated. But

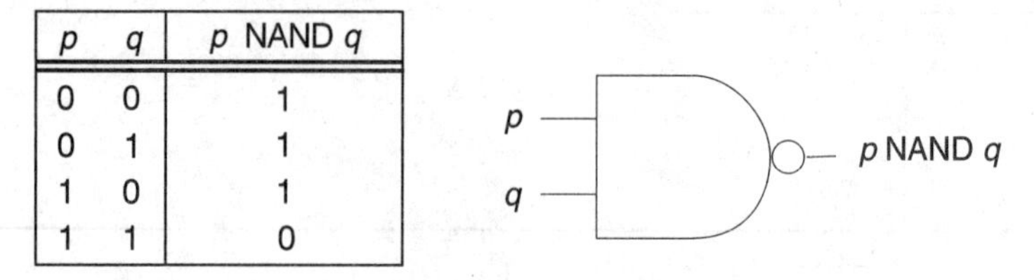

p	*q*	*p* NAND *q*
0	0	1
0	1	1
1	0	1
1	1	0

Figure 2.6. Truth table and circuit symbol for the NAND gate.

c	p	q	c	p'	q'
0	0	0	0	0	0
0	0	1	0	0	1
0	1	0	0	1	0
0	1	1	0	1	1
1	0	0	1	0	0
1	0	1	1	1	0
1	1	0	1	0	1
1	1	1	1	1	1

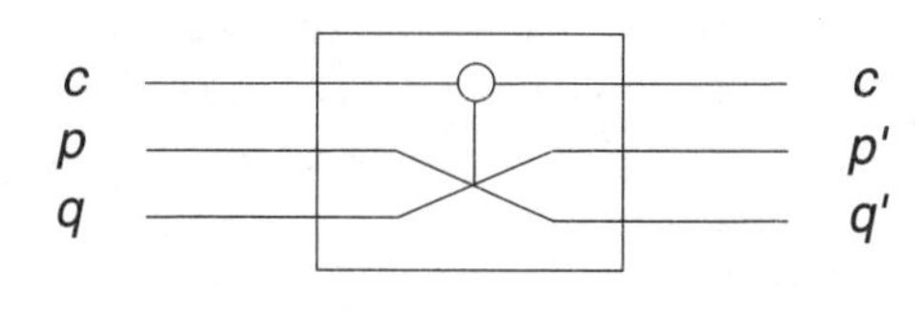

Figure 2.7. Truth table and circuit symbol for the Fredkin gate.

the gate actually does a very simple thing. All three inputs—*c, p,* and *q*—go straight through unchanged if *c* is 0. Otherwise—if *c* is 1—*p* and *q* are swapped. That's it. You'll notice that the value of *c* is unchanged whatever the circumstances. It is described as the *control input* because it controls the exchange of the other inputs without itself being changed (this controlled swapping is suggested by the circuit symbol shown in Figure 2.7).

So what's so special about the Fredkin gate? First, it's fully invertible: You can actually undo its effect simply by passing its outputs through another Fredkin gate. Second, it's universal because it is possible to construct any binary logic circuit using just this gate. Third, the gate always has as many 0s and 1s in its outputs as its inputs. Fredkin believed this last feature would be important in devising physical implementations.

The gate formed the basis of what Fredkin called *conservative logic* because it could be used to construct any kind of logic circuit, which, in principle, could function without expenditure of energy. Proving its universality, though, was harder than for the NAND gate. There are, in fact, thirty-six possible conservative three-input gates, but most of them are equivalent to one another if you relabel input and output signals at will. (Because the gates are reversible, there is no real difference between inputs and outputs.) But when Fredkin eliminated these variations on the Fredkin gate, he was left with an odd man out known as the SMP gate—short for *symmetric majority parity.* At first there didn't seem to be any way to construct an SMP gate out of Fredkin gates. "It was a great puzzle," recalled Fredkin. "My thought was they must be equivalent. But we could not find an implementation of one of the gates in terms of the other. Equivalence should mean that I can take this gate and make a circuit out of it and that circuit should do exactly what the other gate does. But we tried and tried and couldn't do it."

This was somewhat disconcerting, because if the Fredkin gate was to be universal, there had to be a way of implementing any circuit, including one that could emulate the SMP gate. When Fredkin returned from Caltech to work at MIT again, he set up a research group in what he called information mechanics. Rather than tie up all the loose ends himself, he offered this problem as a task for his students to solve. "I had some fantastic students," said Fredkin. "I told them, I'm going to write down a set of open questions that no one in the world knows the answers to. These are going to be your homework."

Sure enough, several of his students produced answers. One student in particular, Guy Steel, produced the optimal solution by using a computer to search exhaustively every possible combination of gates. It took a minimum of six Fredkin gates to make a circuit that would replicate the SMP gate, and the same number of SMP gates to make one Fredkin gate. "The circuit was too hard to guess or even to work out by hand. There were too many combinations. Guy Steel's solution was impressive," Fredkin recalled. The computer search therefore confirmed that Fredkin's gate was indeed a universal building block for a reversible computer.

The Billiard Ball Computer

Fredkin's breakthrough in finding a system of conservative logic that could be used to construct a reversible computer did not satisfy some of his colleagues at MIT. The problem was that though Fredkin had produced an idealization of a reversible logic gate, he had not provided a physical model.

"When I came back from Caltech it was assumed that I would plunge into some senior management position of some kind because I had been the director of the Laboratory for Computing Science before I left," Fredkin told me ruefully. "I didn't want to do that. I wanted to concentrate on this physics stuff, which was viewed very skeptically by the department since I was the only one in the world who was interested in it."

So disbelieving were some of his MIT colleagues that one of them challenged his whole scheme. A professor in the electrical engineering department produced a "proof" that purported to show that there could never be a realistic physical model of a conservative logic gate because it would violate a law of physics.

"This person," Fredkin continued, "had gone through a funny kind of argument that said that if there could be a conservative logic gate, it would imply that you could then make a perfect diode, which would let you rectify noise and make energy from noise, violating the second law of thermodynamics. That's like making high-grade energy out of nothing but pure heat. The point was, it was a way to argue that what I was doing really didn't make a lot of sense. So I was highly motivated to find a physical model."

Thus was born the *billiard ball computer,* a hypothetical machine that provided an elegant refutation of the reductio ad absurdum proposed by Fredkin's critic. One of the appealing aspects of this model was that it didn't require any fancy electronics but instead depended on the simplest of mechanical technologies: billiard balls and reflecting barriers.

In the model, billiard balls roll across a frictionless table and pass through a hall of "mirrors," a set of barriers strategically placed to bounce the balls back and forth as if in a demented game of bagatelle. The inputs and program are encoded into the pattern of balls injected into the machine, and the result appears in the pattern of balls that emerge from output slots.

How would it actually compute? Fredkin first imagined a pair of billiard balls. He saw that if their paths crossed, one of two things could happen at the intersection. If the balls crossed at different times, they would simply carry on moving unperturbed by each other. If they arrived at the same time, they would collide and bounce off each other, following new paths, as illustrated in Figure 2.8.

With these simple ingredients Fredkin saw that he had the essence of a reversible logic gate. Suppose we represent the signal 1 at any point by the

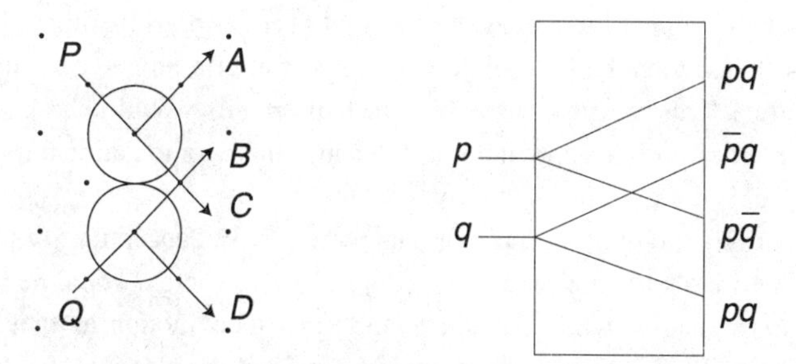

Figure 2.8. Billiard ball model of the interaction gate

The interaction gate consists simply of two colliding billiard balls *(left).* In the symbolic representation of the interaction gate *(right), p* and *q* represent input signals, while *p* and *q* represent their inverses (i.e., NOT *p* and NOT *q*). The product *pq* for two of the outputs represents *p* AND *q,* so that a ball emerges at these outputs only if both *p* and *q* are 1 (i.e., balls enter both inputs).

presence of a billiard ball, and the signal 0 by its absence. If a single ball comes along P, it will continue uninterrupted along path C. If one comes along Q, it emerges on path B. If they arrive together, the balls collide and bounce away from each other following paths A and D. Fredkin noted that this behavior was equivalent to a logic gate with two inputs and four outputs.

The *interaction gate,* as Fredkin called it, has the ability to perform all of the logical operations required for digital logic. For example, it is only when two balls arrive simultaneously that we see balls emerge along A and D. The result is an AND operation. The gate will also work backwards in the sense that if the balls approach from the opposite direction, they will emerge along the input paths. Full reversibility (with respect to time as well as space), however, also demands that no energy is wasted by the gate. Provided the components are perfectly elastic and frictionless, the gate will be fully reversible.

So the interaction gate had some important features Fredkin needed for a physical model of conservative logic. But on its own such a gate would not be much use because once the balls have interacted they would fly away from each other, never to meet again. So Fredkin added mirrors or reflectors to the balls' paths. With these the balls could be redirected to other gates at the same velocity with no delay. The collisions of the balls depended on precise timing.

Fredkin imagined a grid of points superimposed on his circuits (as shown on the left in Figure 2.8). Mirrors would be placed only at 45-degree and 90-degree angles, and gates would be aligned with grid points. These features ensured that balls would always appear at grid points at set intervals of time. This meant that the billiard ball circuits would behave as if they were clocked like computer chips because their computational trajectories would be locked in synchrony.

Of course, such billiard ball circuits were purely conceptual. In practice, billiard balls are not perfectly elastic, so some energy would be consumed in collisions, while friction would slow their motion as they ran across the table. But Fredkin drew inspiration from the classical theory of gases, which depended on gas molecules being regarded as perfectly elastic spheres that travel without friction through empty space.

Pretty soon, after playing around with various layouts for the balls and mirrors, Fredkin realized that he had sufficient functionality to construct the conservative logic gates he had conceived during his period at Caltech.

With his students, he set out to elucidate some detailed implementations to prove his case.

"My philosophy was, whenever you think of a brand-new idea, ten ideas follow it immediately," Fredkin recalled. "I thought, 'Hold it! Don't think that through. I have some presents I can give people.' So I went to my students and told them about it. The other thing I did was, the day I thought of this, I called Feynman and I described on the telephone how he could draw these pictures. He got it all. About a week later I got a wonderful letter back from him; he had been playing with ideas the whole week, and he said he'd found this great circuit."

The circuit was for a device that acted like a switch. The switch gate was interesting because with it you could get one ball to act as a control signal, switching the path of a second ball, apparently without affecting the path of the first (see Figure 2.9).

The way the switch gate works is rather clever. If a single ball enters along path A representing the control input, *c,* it passes through uninterrupted and emerges at E. Nothing emerges at C or D. If, however, two balls arrive simultaneously along A and B, they collide and rebound from each other and are then reflected by barriers. After traveling a short distance

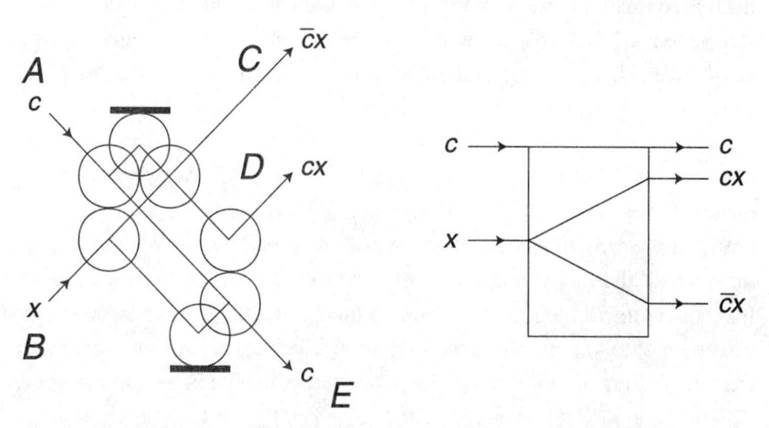

Figure 2.9. Billiard ball realization of the switch gate

In the switch gate *(left)* the control ball, *c,* appears to switch the path of another ball, *x.* In fact, when both balls enter the gate at the same time, they swap roles so that the target ball emerges along the expected path (E) of the control ball. The logical functions performed by this gate are shown in the circuit representation *(right).* The switch gate is a useful building block for the construction of more complex logic circuits.

from the barriers, the balls collide once more and emerge along paths D and E. Notice that in this case the ball that emerges from E, which represents the control signal, is, in fact, the ball that started out on path B. In other words, the control ball has swapped places with the target ball. So in either case, provided a ball enters along A, a ball will emerge along path E. On the other hand, if a ball enters along path B, the target input, then the presence or otherwise of a ball on path A determines whether a ball emerges along path C or D. So the control input, *c*, in effect, redirects *x* to one of two different outputs.

As often happens in science, by the time Feynman had figured out his circuit, so had one of Fredkin's students, Andrew Ressler. "The switch gate was a very neat thing for doing logic," remembered Fredkin. "We had a meeting, me and my students. I said, 'Look. One thing is important for us. We've got to keep Feynman interested.' We agreed. 'Okay. Let's give this to Feynman even though Ressler also thought of how to use it. We'll call it the Feynman gate.' "

As a way of keeping the world's most famous physicist on board, it was a cunning move. Fredkin knew from experience that Feynman sometimes liked to indulge his vanity—a fact that prompted the following anecdote from Fredkin: "When I was at Caltech, Feynman wanted to buy a [kind of sleeper] van. I had a friend who was a pilot who had a company that did van conversions. So I introduced him to my friend, and they came to an agreement. My friend said that he could paint any kind of stuff on it. Feynman said, 'Nah, I'm not interested.' So later I saw Feynman and I said to him, 'You know he could paint anything on it.' He said, 'Well, what do you mean?' I said, 'He could paint Feynman diagrams on it.' [16] You might know Feynman's license plate read PHOTON. Anyway, he took this idea up and he made up all these Feynman diagrams, and the van was covered with them. It was a white van with these black Feynman diagrams all over. His point was that not too many people would know what they meant. But those that did would come up to him and say, 'Why are all these Feynman diagrams all over your van?' He would say, 'Because I'm Richard Feynman!' "

As an example of how the Feynman-Ressler gate, as Fredkin now prefers to call it, can be used, Figure 2.10 shows how four switchgates can be connected to construct a Fredkin gate.

As it happens, few people refer to the Feynman-Ressler gate anymore. Sadly, it doesn't even get a mention by name in Feynman's posthumous

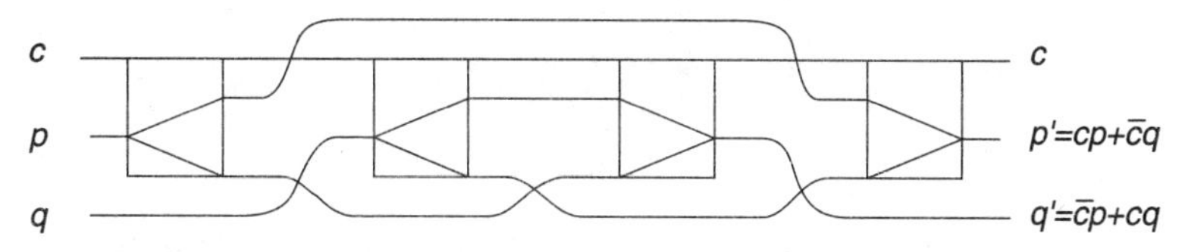

Figure 2.10. Realization of the Fredkin gate using switch gates.

collection of lectures on computation, though there is a billiard ball representation of the Fredkin gate, which ingeniously keeps track of all of the timings (see Figure 2.11).

Nevertheless, a related gate captured everyone's interest (including Feynman's) with the advent of quantum computing. This is the controlled-NOT gate, which we will examine in Chapter 4.

Having established that billiard balls and mirrors could be fashioned into Fredkin gates, Fredkin and his students began to work out how these components could be interconnected to construct more complex circuits still, ranging from serial adders to a complete billiard ball computer.

As we saw earlier, the important proviso for a reversible machine was to avoid erasing information. The trouble with conservative logic elements like the Fredkin gate, though, was that they generated three outputs when typically you needed only one for the next stage of the calculation. What was to be done with the other outputs?

You might think it would be acceptable to forget about the unwanted signals, rather like leaving some pins on a computer chip unconnected. However, losing information is just as bad as actively erasing it because if one of these gates is left in an unknown state, it has to be reset before it can be used again. Actually, in the billiard ball model, the issue is even more basic: losing information means losing billiard balls. If you built a whole computer without worrying about all of the intermediate outputs, you would pretty soon find that the machine had literally lost its balls. The only way it could be sustained would be by injecting an unlimited supply of fresh ones.

Fredkin realized that he had to create circuits that didn't lose track of their balls. As it was, his original presumption that a conservative logic gate (such as the Fredkin gate) ought to conserve 0s and 1s from the inputs to the outputs was useful because in the billiard ball model this constraint was automatically built in. There was no way a gate could create or destroy

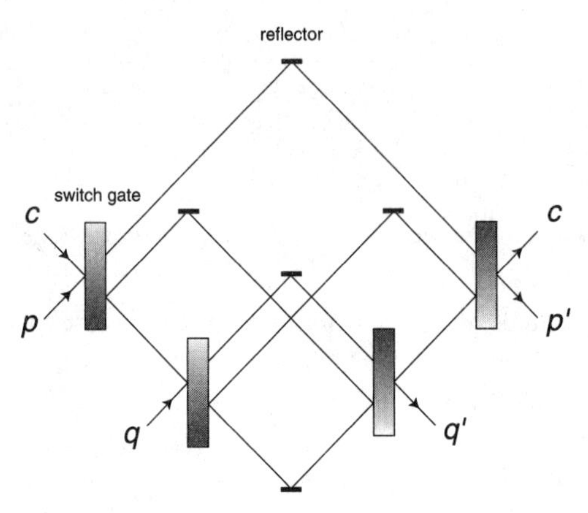

Figure 2.11. The Fredkin gate realized by billiard ball switch gates.

billiard balls.[17] However, this requirement didn't prevent the buildup of unwanted intermediate signals. Fredkin's answer was to create "garbageless" conservative logic.

The idea resembled the solution Charles Bennett produced in his treatment of reversible logic. Imagine a small conservative logic network. It produces the answer to a calculation on one output but presents several additional outputs carrying unwanted garbage signals. Fredkin saw that for every such network it would be possible to construct an inverse circuit that reversed the calculation, reproducing the original inputs.

By connecting a circuit to its inverse, it would be possible to perform a calculation and then turn everything back into the original inputs again. In the billiard ball model, the result would be that all of the balls supplied as inputs would be returned at the end of the calculation and could then be passed on for use in another circuit. Regenerating inputs could then serve as a method of recycling billiard balls. To benefit from the calculation, though, Fredkin needed a way to extract the results calculated in the middle of the process before the calculation was reversed. He did this by adding "spy" circuits: Each consisted of one Fredkin gate and two fixed inputs (see Figure 2.12).

Any signal directed toward the control input of the Fredkin gate is duplicated by this circuit so one copy can be passed on to the inverter circuit, leaving the other and its complement for use in the next step of a larger computation.

Using these techniques, it was only a matter of time before Fredkin and

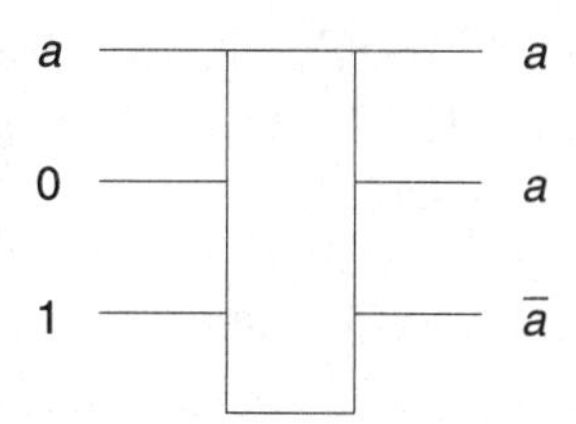

Figure 2.12. Spy circuit using a Fredkin gate

Spy circuits are needed to duplicate the final result of a computation before the computation is reversed to clear up any garbage. The above circuit uses a single Fredkin gate to produce a duplicate of the input *a* and a single copy of its complement, *a*.

his students had worked out a complete design for a billiard ball computer. At this stage it remained, of course, a pure thought experiment. There was no question of anyone's building a real billiard ball computer, because it would be impossible to keep the balls moving exactly along their trajectories even if frictional and inelastic losses could be overcome. As a concept, though, it proved that Fredkin had been right: Reversible conservative logic was a theoretical possibility. Another attractive feature of his approach to garbageless computation was that it explicitly removed garbage a step at a time rather than letting it mount up, though to be fair Bennett's reversible approach could also be made equally attentive to its garbage duties.

Of course, the billiard ball computer did not prove that the universe was a computer. But it certainly added substance to Fredkin's argument that physics and computation were alike. And although the model may originally have been a thought exercise, its most convincing demonstration came when it was brought to life on a machine known as a cellular automaton.

The God Game

John von Neumann may have been in error over the question of the minimum energy involved in computation, but he was unquestionably one of the great scientific thinkers of the twentieth century. He made important contributions to quantum theory, invented a branch of mathematics known as game theory, designed one of the first computers to use a flexible stored program, and served as a consultant on the Manhattan Project to build the first atomic bomb. He also invented the idea of the cellular automaton.

In the 1940s Von Neumann became fascinated by the connection between computers and biology. He began thinking about how one could produce artificial forms of life and in 1948 gave a series of lectures at Princeton on the subject of self-reproducing machines. Over the next few years he developed his ideas further, but he died before he was able to complete the full theory. However, at the suggestion of his colleague Stanislaw Ulam, he had produced the basic idea for the cellular automaton, a name that was later suggested by Arthur Burks, who edited Von Neumann's papers after his death.

Von Neumann's cellular automaton was a "creature" that lived on a two-dimensional checkerboard. Each square, or "cell," of the checkerboard was programmed to follow certain mathematical rules whereby cells could live, die, or reproduce. The overall behavior of the automaton was therefore governed by the behavior of the cells. Von Neumann's particular automaton was a monster of complexity because each cell could have as many as 29 different states.

Ed Fredkin and others studied simpler cellular automata (CAs) intensively, but the idea didn't come to public attention until the 1960s, when the Cambridge mathematician John Conway developed a version, which he called the Game of Life. Thanks to widespread popularization in magazines such as *Scientific American,* the "game" proved a huge success in capturing people's interest. Here was something virtually anyone could understand.

It works as follows: Think of a checkerboard with counters in only some of the squares. Now follow these rules. If an empty square has three occupied neighbors, it too acquires a counter. The cell comes to "life," nurtured by its neighbors. If a square has two occupied neighbors, then it remains unchanged. Finally, if an occupied square has any other number of occupied neighbors (0, 1, 4, 5, 6, 7, or 8), then its counter is lost. In other words, the cell dies either because of a lack of neighborly support or because of overcrowding.

You then start out with some 2-D pattern of checkers on the board, apply the rules to every square, and see how the pattern changes. You repeat the process again and again and watch how the shapes evolve. Although it is perfectly possible to play the game using pencil and paper, it is, of course, much easier to sit back and watch it on a computer, with the shapes displayed on the screen.

Over the years people discovered a whole zoo of initial patterns that could evolve in spectacular ways. They also experimented with different rules. The resemblance to the emergence of complexity in the biological world was uncanny. Computer scientists and mathematicians realized that cellular automata were a powerful tool for studying how complexity can naturally arise from very simple laws. One of the scientists who took note was Tom Toffoli.

In the early 1970s Toffoli was studying CAs for his doctoral thesis at the University of Michigan. He became interested in the way certain patterns would enter into cycles in which the same set of shapes would recur indefinitely. Sometimes these cycles were very long, and it was not always clear when a particular sequence had entered the cycle. Toffoli decided he needed a way to backtrack through sequences to find out how they had arisen.

Toffoli therefore began to wonder whether there was a way of making CAs reversible. Conway had proved that they were capable of universal computation, but there was some Russian research that suggested reversible logic would not be able to perform universal computation. If this were true, CAs could not be made to work reversibly. But Toffoli found an answer that showed the Russian claim must be incorrect. His argument depended on a method of translating *any* CA into a reversible CA—but one that also performed the job of the original automaton.

The essence of his "constructive proof" was to imagine a two-dimensional cellular automaton that stores its history on a stack of checkerboards piled up in the third dimension. Each time the cellular automaton evolves through one time step, it places a copy of its previous state on the stack. Reversing the CA is then simply a matter of looking back through the stack.

Because cellular automata were capable of universal computation, this proof also demonstrated that it was, in principle, possible to construct a reversible computer—a serendipitous discovery for which Toffoli therefore shares credit with Bennett and Fredkin. Toffoli's proof, though, was less appealing than the methods suggested by Bennett and Fredkin because it did not offer any way of dealing with the garbage created by a calculation. All it did was delay the moment when information had to be erased.

Nevertheless, Fredkin had long had a deep fascination with cellular automata, regarding them as the most natural model for his view of the universe as a computer. Cellular automata, after all, share three very important properties with the physical world. Computations are local, and they occur in parallel; and just as the laws of physics appear to be the same wherever you are in the universe, so too are the rules for cellular automata. With reversibility came another precise correspondence.

Fredkin therefore immediately recognized the importance of Toffoli's work and invited him to MIT as the first recruit for his newly created information mechanics research group. This proved a fruitful move not only because of the work that was achieved but also because it was with Toffoli's assistance that Fredkin eventually published his own work on conservative logic and the billiard ball computer, in the early 1980s.[18]

Another researcher who joined Fredkin's group was Norman Margolus. Encouraged by Fredkin, who acted as his thesis adviser, Margolus was the first to devise a reversible CA that incorporated techniques for cleaning up unwanted garbage. His cellular automaton emulated Fredkin's billiard ball model and therefore became its first working demonstration. The rules governing the behavior of the individual cells needed to be more complicated than the simple rules used in the Game of Life. In fact, getting a CA to work reversibly (without resorting to Toffoli's stack) is trickier than it may first appear. The problem is that a single cell cannot tell you about its past, because its fate is also influenced by its neighbors. Therefore, for a CA to be reversible, the whole neighborhood must somehow retain a memory of each individual cell's previous state.

Margolus came up with two different types of rules. The first approach partitioned the checkerboard into groups of four cells so that the state of each block would reversibly determine its own state at the next time step. Each block then acted as a kind of four-input and four-output gate. The blocks were still able to interact with *their* neighbors because on the next time step the lattice was sliced up on a different grid pattern—one that cut through the middle of the previous blocks.

The alternative, *second-order rule* was even more ingenious. In this the next state of each cell was determined not only by the neighborhood but also by the neighborhood's previous state. Future states were therefore determined, in effect, by snapshots taken at two different times. Reversing such an automaton was elegantly simple because all it involved was calculating what should happen when the snapshots were taken in reverse.

In *Three Scientists and Their Gods,* Robert Wright evocatively summed up the significance of Margolus's billiard ball CA.

> This cellular automaton in action looks like a jazzed-up version of the original video game: Pong. It is an overhead view of endlessly energetic balls caroming through clusters of mirrors and off each other. In a way, it is the best illustration yet of Fredkin's theory: it shows how a very simple, binary cellular automaton could give rise to the seemingly more complex behavior of microscopic particles bouncing off each other. And, as a kind of bonus, these particles themselves amount to a computer.[19]

Margolus and Toffoli experimented with many other rules for their cellular automata and increasingly saw them as a way of modeling physics in artificial universes. By changing the rules, they were able to see what happens when you change the laws of physics in these synthetic worlds.

However, they became frustrated by the limitations of their early 1980s workstations in running the simulations, some of which were very large. With a million-cell universe, for example, these machines could update the screen only once every minute. What Margolus and Toffoli wanted was hardware that could run cellular automata at the speed of a movie rather than that of a slide show.

Using a few simple logic chips, Toffoli built a prototype machine at home that could do just that. This was the first of a new generation of "cellular automata machines." Toffoli and Margolus collaborated over the next few years to refine the hardware. The first commercial version of the machine was CAM-6. For around $1,500 you could turbocharge a PC's performance with their plug-in module and run state-of-the-art CA simulations. The results were impressive, including simulations of complex phenomena such as biological growth, fluid turbulence, chemical reactions, and crystal formation.

"On this machine we see a window on a synthetic universe," Toffoli told me. "We can see the show moving in real time. We can disturb it, we can play miracles, we can change the initial conditions and see what happens as a consequence. We can stop it and we can analyze it and then resume the situation. Somebody called this machine the God Game because, as in chess, we play a form of war—it's a tame, simplified war, but it's a war nevertheless. Here we play God with the universe—it's a tame, simplified universe, but it's a universe nevertheless."

In the 1990s Margolus continued development, and he was largely responsible for the machine's latest incarnation, CAM-8. This new and improved model is able to outperform supercomputers in generating patterns that model phenomena such as the evaporation of a liquid droplet, the spread of a forest fire, and the expansion of a ripple of sound-wave energy.

When a conventional computer runs a CA simulation, one processor has to grind its way through the process, cell by cell. The CAM machines divide up the task among several processors and also use fast lookup tables stored in memory chips rather than relatively slow microprocessors to implement the CA rules.

Although Toffoli and Margolus have devoted considerable efforts to the practicalities of designing and building efficient cellular automata machines, both of them have been driven by the desire to understand the link between physics and computation. While computers and CAs can model aspects of physics, Toffoli and Margolus see the connection as a two-way street. Though they both refrain from endorsing Fredkin's bolder claims that the universe is a computer, they preach the gospel by describing *physics* as a computational process. Toffoli, for example, wrote in one of his papers that "nature has been continually computing the 'next state' of the universe for billions of years. All we have to do is 'hitch a ride' on this huge ongoing computation and try to discover which parts of it happen to go near to where we want."[20]

Margolus has also made calculations of how much computing is going on in a lump of matter. The answer turns out to depend on the amount of energy involved, but even for very modest figures, the information processing rate is extraordinarily high. For each joule of energy, Margolus calculates a maximum processing rate of around 10^{33} operations per second—a million billion billion times faster than present-day computers.[21] That's hard to imagine when you look at an inert lump of matter like a brick, but there is no reason to assume that all that processing power is always put to the most effective use.

Fredkin has gone one better by estimating the total amount of computation going on in the universe, producing, ironically, a figure that seems puzzlingly low. He calls this the problem of the missing workload. Essentially what he has done is to calculate how large a cellular automaton would need to be to simulate the entire universe in all its details. The answer, he argues, is that a CA that operated at the tiniest quantum scales known as the Planck length and Planck time would only need to be not much larger than a biggish star to faithfully simulate the entire macroscopic evolution of our universe from the Big Bang to the present in about four hours. The difference in space-time volume between the universe and such a system is a factor of some 10^{63}. This figure is Fredkin's "missing workload," which he contrasts with two other great mysteries of the cosmos: the missing neutrinos from the sun (a factor of around 3) and the missing mass in the heavens (perhaps a factor of 50). So what explanation does he have for this missing workload? "Either something else is going on in the universe that we don't know about," he says, "or God was incompetent on a scale that boggles the mind."[22]

Low-Energy Computing

Cellular automata are undoubtedly powerful machines for simulating physics. However, to map computation onto physics even more closely, we would ideally want a 3-D cellular automaton—and one that contained a processor in every cell. Such a machine would require massive interconnection of sheets of cells in the third dimension. Though well beyond current technology, there is no reason in principle to think that such machines could not be built in the future.

Such a hypothetical leap presents a new problem: heat crisis. This is because of a cube/square relationship whereby the heat produced by the machine would be proportional to the number of cells and thus to the volume of the array, while the heat removed would all have to pass through the surface of this volume. The heat produced would therefore increase as the cube of the machine's size—outstripping the surface area, which would increase only as the square. The running temperature of the machine would thus rise inexorably the larger it became.

Manufacturers of supercomputers have long faced problems with getting rid of excess heat. Some Cray computers, for example, use liquid cooling. But current technology is still a million times more wasteful than Landauer's limit. So it would appear that there's still plenty of room for improvement before having to worry about reversibility. Nevertheless, over the last few decades, energy demands per gate have dropped remarkably fast, and if the trend continues at the same rate, we are likely to hit the thermal energy realm around the year 2010. How significant would the heat problem be then?

Ralph Merkle, a computational nanotechnologist at Xerox PARC in Palo Alto, California, anticipated the question in 1992. If it were possible to develop irreversible devices that approached the *kT* threshold, he argued, a computer operating at room temperature at a frequency of 1 gigahertz with 10^{18} logic gates packed into a volume not much larger than a sugar cube would dissipate more than 3 megawatts—approximately the same power output as the kind of nuclear reactors used for research. Such power in such a small volume would cause the device to explode like a bomb.

On the face of it, one solution might be to lower the operating temperature of the computer because the minimum energy is proportional to the temperature of the gates. By cooling a compact supercomputer in liquid helium at 4 degrees Kelvin, you could reduce the heat dissipated by each gate by a factor of nearly 100. However, it turns out that the inefficiencies involved in cooling more or less wipe out these savings in energy. Furthermore, refrigeration would make these computers very expensive to run and inconvenient to transport.

The obvious solution instead will be to exploit reversible technology. The question is, can semiconductor logic gates be made to work reversibly? Twenty years ago, when Fredkin was having a hard time persuading anybody to take his ideas of conservative logic seriously, few people would even have thought of asking. With the advent of the billiard ball computer, which proved that, in principle, reversible computing should be possible, most people still imagined that such ideas were merely the stuff of theorists' daydreams.

However, as early as 1978 Fredkin and Toffoli published a paper offering a semipractical implementation of reversible logic using simple electronic components—namely switches, capacitors, and inductors. Capacitors act as storage devices and form the basis of conventional memory chips. However, when new data are stored in memory, some of the capacitors inevitably have to be discharged to overwrite the old memory. Normally this is achieved by shorting the capacitor via a transistor, and the result is that a small amount of energy is wasted in the form of heat.

With Fredkin and Toffoli's scheme, the energy is recovered by using inductors. An inductor typically consists of a coil of wire with a ferromagnetic core. When a current flows through the coil, the resulting magnetic fields can store energy temporarily. It turns out that if you try to discharge a capacitor via an inductor (by connecting the two in parallel), the electrical energy stored in the capacitor is initially transferred to the magnetic fields of the inductor. With a suitable arrangement of switches it is then possible to transfer the energy from the inductor into another capacitor for safekeeping. The net result is to discharge the first capacitor but without losing the energy.

Unfortunately, Fredkin and Toffoli's scheme was not practical for microchip technology because it required a separate inductor for each logic gate. Unlike capacitors, inductors are too large to be accommodated on a silicon substrate. They cannot easily be miniaturized because the energy saving becomes less efficient the smaller the size of the inductance. Nevertheless, the idea led the way to a number of more promising schemes, known as *adiabatic switching,* in which a single inductor could be used in the power supply driving the whole chip, instead of separate inductors for each gate.

The term "adiabatic" is usually found in thermodynamics and means "without transference of heat." When a gas is allowed to expand gradually in a cylinder, if the system is maintained at a constant temperature, the expansion is said to be *isothermal.* But if the cylinder is thermally isolated from the surroundings, the expansion is adiabatic and the gas cools. You may notice this effect when you let air rush out of the valve of an automobile tire—the expanding air feels quite cool. The expansion is approximately adiabatic because the rushing air, although not isolated from its surroundings, expands so quickly that it does not have time to exchange heat with the atmosphere.

An *adiabatic circuit,* by the same token, is one that does not produce heat. One way this can be achieved is to ensure that circuit elements are switched only when there is no voltage across them and no current flowing through them. A capacitor that is charged or discharged instantly using a switch inevitably dissipates a certain amount of energy as heat, an amount given by the formula $\frac{1}{2}CV^2$, where C is the capacitance and V is the voltage. However, if the voltage is allowed to change relatively slowly, the energy dissipated is greatly reduced. It's a little like dropping a block of stone from a tall building. Its potential energy is converted to kinetic energy, which is then dissipated in the form of heat when the block crashes to the ground. But if you attach the block to a pulley with a counterweight, you can gently lower it and transfer most of its potential energy to the counterweight. The role of the counterweight in adiabatic circuits is typically played by the inductor.

This requirement for changing things slowly also applies, as we saw, to reversible thermodynamic systems. In the Carnot engine, for example, processes have to be carried out slowly to be reversible. If gas expands too fast, useful energy is irretrievably lost. Theoretically, to achieve perfect efficiency, adiabatic circuits need to be switched infinitely slowly. However, such a restriction in a computer would obviously make them of little practical use. There is thus a trade-off between switching speed and energy efficiency.

Some of the elements of adiabatic switching were first set out by Charles Seitz of Caltech in the mid-1980s. He and his colleagues devised a system known as *hot-clocking,* in which they sought to save energy by varying the power supply voltages. Normally the power rails for computer chips are held at constant DC voltages. Seitz showed that by varying the power voltages, like an alternating current or AC supply, it was possible to construct circuits in which the gates switched logic states with smoothly varying voltages rather than sharp-edged, energy-wasting pulses. Although this idea was a step forward in saving energy, it didn't exploit the principles of reversible logic and so couldn't recover all of the energy. It potentially reduced the energy wasted charging capacitors up but didn't recoup the energy lost whenever a capacitor was discharged.

It was not until 1992 that scientists working independently at three institutions announced *reversible* adiabatic-switching logic schemes. These were Ralph Merkle of Xerox PARC, J. Storrs Hall at Rutgers University, and Jeffrey Koller and William Athas at the Information Sciences Institute at the University of Southern California. Athas's group was the first to demonstrate their ideas, with a working CMOS chip. CMOS stands for *complementary metal oxide semiconductor,* which is the standard semiconductor technology used by chip manufacturers. Further work at MIT by Thomas Knight and Saed Younis led to the idea of a *reversible pipeline,* which they demonstrated with an 8-bit by 8-bit multiplier chip.

All these approaches exploited the idea of varying the power voltages, but with reversible logic it was possible to arrange for charges stored on capacitors to be returned to the power supply instead of being dissipated. This was typically done by switching a DC source electronically and feeding the current through a large inductor to generate an approximately sinusoidal waveform. So the inductor's role was both to smooth the voltage variations and to recover the energy stored in the capacitors.

Knight and Younis's reversible pipeline consists of a chain of circuits in which data are passed from one end to the other like the baton in a relay race. At intermediate points along this pipeline, signals are fed into a reverse pipeline, which consists of logic circuits that undo the work of the forward-facing circuits. It is as if whenever a runner passes the baton, she splits it into two, giving one half to the forward-facing runner and the other to a runner going in the opposite direction. The effect of the reverse circuits is to reset the inputs to each stage of the forward pipeline in a reversible and hence nondissipative way. The reverse circuits do this by returning the charges stored in the outputs of the forward circuits to the power supply. Needless to say, the precise way in which this is achieved has to be very carefully timed to avoid wiping out the forward-running signals completely.

Although these techniques have demonstrated substantial energy savings with relatively simple chips, it is far from clear at what point such technology could become sufficiently attractive for microprocessor manufacturers to adopt it. Seven years down the road from the discovery of adiabatic switching at the University of Southern California, Athas thought significant progress still had to be made to transform these ideas into a practical technology.

"It doesn't look too good for CMOS," Athas told me. "The overhead is a killer. The extra circuitry you need to support the reversible process is prohibitive, and you often end up expending nano-joules to rescue pico-joules." One problem is that, as with the purely theoretical schemes, reversible computing requires more than twice as much processing to perform a computation. Significant losses also arise elsewhere, particularly in generating the varying power supply voltages. So at the moment the gains made by recovering energy are being wiped out by the remaining inefficiencies.

Nevertheless, Athas is working on a number of nonreversible techniques for energy recovery that look promising. Commercial interest is also growing rapidly: researchers at IBM, AT&T, Intel, Motorola, Texas Instruments, Honeywell, and elsewhere have been filing patents on some of these ideas and are expected to make use of energy recovery techniques in their chips before long. Tom Knight at MIT has been advising some of them. "Power consumption is now the limiting factor," Knight told me. "When you start dissipating 50 watts and you talk about doubling the speed of the processor, it's simply impossible to dissipate 100 watts in the packages manufacturers are using. The only thing you can do is to be clever about how you use that power. So this is not an academic discussion anymore. This is about dollars and cents and performance."

The thrust of Knight's research at MIT, though, has been to look to the long-term future, where he thinks reversible techniques will be essential. Quite apart from the problems of getting rid of waste heat, he also foresees applications in the area of computer security. When computer programs crash, it's often difficult to find out what precisely went wrong. However, it would be a simple matter to get a reversible computer to backtrack from a crash, enabling the user to recover from and diagnose the program error more easily.

Also of particular interest to him and to us, of course, is the connection between reversible computation and quantum computation. Quantum computation, by definition, has to be reversible because the laws of quantum mechanics are reversible. The techniques of energy recovery such as the reversible pipeline won't be needed for quantum computation because each quantum operation has to be intrinsically reversible anyway. That's not to say that a quantum computer won't use any power. Reversibility doesn't guarantee nondissipation of energy. It is simply a requirement to minimize energy consumption. Nevertheless, we'll see in the next chapters how the ideas about reversible computation developed by Bennett, Fredkin, Toffoli, and others have laid some of the foundations for quantum computation.

In the meantime, I've noticed that the solution to the problem of using a computer on long-haul flights has arrived. It's called a *palmtop* computer. And it doesn't use any of the technologies we've talked about but instead relies on simpler chips, simpler software, and a stripped-down operating system. And its battery lasts for six hours plus.

Still, for flights from New York to Sydney, a reversible computer sure would come in handy. Otherwise, I'll simply have to learn to enjoy watching terrible airline movies. Or better yet, I'll return to another technology that requires very little external energy consumption apart from natural light: reading a book.

3

The Logic of the Quantum Conspiracy

. . . because nature isn't classical, dammit.

RICHARD FEYNMAN

Feynman's U-turn

Richard Feynman was not the first person to think about quantum computers or the first person to publish anything about them. But he probably *was* the first to generate widespread interest in them. The year was 1981. The event was a meeting at MIT, the first major conference ever on physics and computation. And Feynman was invited to give the keynote speech.

Feynman was by then a hugely influential figure in physics. He had worked on the atomic bomb project. He'd won the Nobel Prize in 1965 for his work on quantum electrodynamics, the theory that explains the behavior of light. His lectures on physics, recorded in three substantial volumes, had become a classic text, indispensable for physics undergraduates. He had pioneered the subject of nanotechnology decades before others seized on the subject, giving it the cult status it has today. He was also a wonderfully charismatic figure, a showman who could delight audiences with his hilarious and amazing tales, which he later recorded in two books: *Surely*

You're Joking, Mr. Feynman! and *What Do You Care What Other People Think?*

The MIT meeting was organized by three of the leading figures in reversible computation: Ed Fredkin, Tom Toffoli, and Rolf Landauer. During his year at Caltech in 1974, Fredkin had attempted to instill in Feynman his conviction that physics was essentially a form of computation. But Feynman was reluctant to accept Fredkin's vision. Despite their collaboration, Feynman seemed to distance himself from the computational world view for fear of being labeled a crank. Such caution was a little uncharacteristic for someone admired as the smartest physicist since Einstein, especially as Feynman normally reveled in breaking the rules. The issue came to a head after Fredkin suggested to Mike Dertouzos, his successor as director of the MIT Laboratory for Computer Science, that they invite Feynman to give the keynote lecture at the conference.

"We had as the title of the meeting Computational Models of Physics, or something like that," recalled Fredkin. "Feynman said to me, 'If you have a title like that, I'm not coming.' I thought that was odd, but we wanted him to come, so I came up with some other title for the meeting and he was happy with that. He explained that he didn't want there to be any implication that he believed there were actually true computational models of physics—that is, ones that take physics as a computational process.

"I told that to Dertouzos as a joke, but Dertouzos happened to be in a mood those days to tweak my tail. So he introduced the keynote speaker and started out saying the reason this meeting has been called some abstract thing like Physics and Computation is because of this story about Feynman refusing to come if the title implies there are computational models of physics. So Feynman gets up and says, 'Well, since that time I've changed my mind, and my talk *is* on computational models of physics.' So he moved in that direction but he was always hesitant about admitting it."

The title of Feynman's talk was "Simulating Physics With Computers" but, as he quickly explained, his purpose was not to explore how computers can model physics in an *approximate* way.[1] Weather forecasting, aerodynamic modeling, and predicting the orbits of planets and comets are examples of such approximate simulations. There was nothing particularly new in them. Feynman was interested instead in the possibility that there were *exact* simulations, such that a computer could do *exactly* the same as nature. That small difference in nuance indicated a conceptual leap as significant for a physicist as the difference for an astronaut between a ride in a simulator and a ride in a space rocket. It showed that Feynman really was beginning to entertain the idea that physics was, deep down, computational.

In the work on reversible computers and cellular automata explored in the last chapter, we saw how the addition of reversibility put computation on a more equal footing with physics. But nature, it seems, is not mimicked so easily. Something else was needed for the computational approach to measure up to the reality of our universe. The missing factor was quantum theory, and it was this Feynman identified as crucial to making computational models of physics.

Journey Into the Quantum Realm

Quantum theory is often described as the theory that governs the behavior of matter on the atomic scale. In fact, the theory governs the behavior of matter on *all* scales. It has to because all larger objects are made out of atoms and other quantum particles. It simply wouldn't make sense if the theory no longer applied when one looked at collections of these particles.

The reason we do not normally worry about quantum theory on macroscopic scales is that its more peculiar aspects usually become less and less noticeable the larger the system.[2] In fact, the laws of quantum mechanics appear to merge with the classical laws of everyday life on scales usually not much larger than a few atoms. This ability for the classical and quantum laws to connect with one another is known as the correspondence principle, which was first articulated by the great Danish physicist Niels Bohr back in the early days of quantum physics. Deep down, though, everyday objects such as rocks, trees, and people are ultimately quantum mechanical because they are all constructed from quantum building blocks.

The fact that everything in the world is ultimately quantum mechanical raises a problem for the computational view of physics. Can a computer simulate quantum mechanics? This was the question Feynman addressed in his 1981 MIT talk.

To understand the answer, one needs to know a little about what quantum theory tells us of the behavior of atoms and subatomic particles.

The development of quantum theory at the beginning of the twentieth century certainly marked one of the most extraordinary and revolutionary periods in science. With poetic timing, the first step actually took place in the very first year of the new century. In 1900 the German physicist Max

Planck solved a key problem in physics. Classical calculations of the spectrum of heat radiation emitted from a hot body suggested that the intensity of radiation would increase without limit as you looked at higher and higher frequencies. This clearly did not make sense because it implied that a hot body, if it was in equilibrium with its surroundings, ought to radiate an infinite amount of energy per unit time—and, since that was presumably impossible, it implied that thermodynamic equilibrium itself was impossible, which contradicted experiment as well as the whole of thermodynamic theory. The problem, known as the ultraviolet catastrophe, arose because the classical calculations treated the radiation as a continuous phenomenon. Planck hypothesized that the radiation was emitted from the body in discrete packets, which he called *quanta.* His new calculation of the so-called blackbody spectrum not only avoided the infinite energies but also agreed perfectly with experiment.

In 1905 Albert Einstein revealed how Planck's quantum hypothesis could explain another puzzling experimental phenomenon: the *photoelectric effect,* which occurs when light is shone onto a negatively charged metal plate. If the experiment is performed using ultraviolet light, the charge tends to leak away, but with visible light the charge is unaffected. Einstein argued that light must consist of a stream of discrete particles, which could knock electrons out of the metal. According to the quantum hypothesis, the energy of these particles—later called photons—depended on their frequency, so that only the ultraviolet photons had sufficient energy to displace the electrons.

In 1913 Niels Bohr proposed that electrons in atoms were also quantized in that they could only occupy certain fixed energy levels. If an electron was given a small kick (typically by a photon), it could absorb the energy only if it were able to jump from one level to another exactly; it could not settle in some intermediate energy state. Thus was born the notion of the quantum leap.

An important feature of these developments from a modern-day computational perspective was that they made atomic physics look much more digital in character. Classical physics was predicated on the assumption that properties of matter such as energy were continuous. On the face of it, the quantum revolution therefore helped to bring computation and physics closer together. But quantum physics also has attributes that are altogether more difficult to interpret.

One of the most important and also most puzzling features is that particles such as photons appear to exist in the form of both particles and waves. The wave properties of light had been the subject of intense controversy ever since Newton's time, in the seventeenth century. Newton actually concluded that light consisted of particles, but this view was eventually overturned in 1801 when Thomas Young performed his famous two-slit experiment. In this experiment, light is projected onto a screen through a mask pierced with two closely spaced slits. Unlike modern-day variants, which use lasers, photodetectors, special crystals, quarter wave-plates, and all the other paraphernalia of modern optics, Young's experiment relied on simple screens and candlelight. Yet even with such rudimentary equipment, Young noticed an amazing thing. Instead of seeing two bright lines on the screen, he saw a pattern of fringes. The image was a classic example of wave interference whereby waves radiating from the two slits overlapped with one another on the screen, producing a pattern of cancellations and reinforcements (see Figure 3.1).

With Einstein's explanation of the photoelectric effect in terms of particles, scientists were faced once again with the wave-particle dilemma. Gradually the view developed that photons—and, by the 1920s *all* particles—must exist in some form of fusion of waves and particles.[3]

Figure 3.1. Young's two-slit experiment

When light is shone onto a screen through a mask pierced with two closely spaced slits, the result is a series of fringes characteristic of wavelike interference.

The developments up to the mid-1920s are often described as the old quantum theory because they were still strongly based on classical notions of physics. But in 1925 Werner Heisenberg, at the age of twenty-four, published a revolutionary paper that cut a swath through the old-style thinking and proposed in its place a new system of mechanics: *quantum mechanics.* Pictorial models of the atom and concepts such as the position of an electron in an atom were abandoned as meaningless. Instead Heisenberg built a mathematical theory based entirely on experimentally observable quantities such as frequencies, amplitudes, and polarizations. The ideas were developed further with the collaboration of Max Born and Pascual Jordan. The same year, they jointly published a definitive paper on the new quantum mechanics that showed how observable quantities could be calculated using matrices, mathematical objects consisting of grids of numbers.

In 1926 the Austrian physicist Erwin Schrödinger published a very different analysis of the atom with the introduction of his famous wave equation. In this analysis, particles were treated as waves using differential equations, an entirely different type of mathematics from Heisenberg's matrix mechanics. To many people's surprise, though, both the wave equation and the matrix techniques gave the correct answers when applied to such problems as calculating the energy levels of the hydrogen atom.

The two approaches were like chalk and cheese, yet both apparently worked. Many physicists felt more comfortable with Schrödinger's analysis initially because it was, at least, rooted in more familiar territory. The wave equation, after all, came from an understanding of other wave phenomena. But there was a big question over the precise meaning of the wave equation. A particle could be represented by a mathematical object called a wave function, but what did this actually mean? For a water wave or a sound wave, the wave function simply represented the size of the displacement of the water or medium through which the wave traveled. But in the case of a particle in the void, what was the medium?

Schrödinger suggested that no meaning could be attached to the position or path of an electron in an atom and that the only reality was in the wave function. He then proposed that the wave function of an electron was related to the distribution of electric charge in space.[4]

Although Schrödinger's interpretation was almost correct, it lacked one vital ingredient. This came soon afterward, in 1926, when Max Born introduced *probability* into the quantum interpretation—an innovation that, in the words of Abraham Pais, "may well be the most drastic scientific change yet effected in the twentieth century."[5]

Born's probability interpretation marked the end of the old quantum theory and the beginning of the new. In his paper he examined what happened when particles were scattered by collision with a repulsive field. He argued that the wave function of a particle (often written mathematically using the Greek letter ψ, pronounced *psi*) represented the probability of finding a particle in any particular region of space. In fact, as he noted in a footnote, the probability was proportional to the *square* of ψ. Though to be accurate he should have said "absolute square,"[6] as Pais noted:[7] "He clearly had got the point, and so that great novelty, the . . . probability concept, entered physics by way of a footnote."

The significance of Born's interpretation was that it brought to an end the classical idea of a deterministic universe in which the outcomes of all measurements were dictated by exact laws. The Newtonian rules of billiards were, it seemed, suddenly replaced by the quantum rules of roulette. Particles such as electrons were described by wave functions, but these offered only probabilities when it came to predicting the particles' behavior in the real world. This idea became clearer the following year when Heisenberg introduced his famous uncertainty principle. According to this principle, you cannot know both the position and momentum (or speed) of a particle with arbitrary precision. The laws of quantum mechanics show that the more precisely you know or measure its position, the less certainty you will have about its momentum and vice versa. This rule can be generalized to many other properties of particles, such as energy and time, horizontal and vertical polarizations, right-handed and left-handed circular polarizations, and so on. These pairs of properties are known as *conjugate* variables: Whenever a precise measurement is made of one, information is lost about its conjugate.

The strange properties of quantum variables are, perhaps, most clearly seen in the case of *spin.* Many particles, such as the electron and the proton, behave as if they spin around their own axes like miniature spinning tops. The amount of spin is always the same for the same type of particle. For an electron, for example, the amount of spin is always $1/2$, in the units of angular momentum physicists like to use. The direction of spin is defined by the axis of rotation and whether the spin is clockwise or counterclockwise. An electron that appears to be rotating clockwise around a vertical axis, for example, is described as having "spin up," while a counterclockwise rotation is called spin down.

Now, an electron's spin can point in any direction, but what makes it distinctly different from classical spin is that no matter what direction we choose to measure the spin, we always get one of two answers: $+1/2$ or $-1/2$. The amount of spin is therefore strictly quantized. So even if an electron spin were known to be oriented along the horizontal, a measurement of its vertical spin would always yield either $+1/2$ or $-1/2$ and never 0. It's tempting to ascribe the inability to measure pairs of conjugate properties with arbitrary precision to the unavoidable clumsiness of the experimenter. The very act of measuring a particle's momentum, for example, inevitably disturbs the particle's position, thus limiting our knowledge of both. This picture is only partly correct. Although measurement usually does disturb particles, the conflict between knowing the momentum and position of particles is intrinsic to the laws of physics and not simply a matter of inadequate equipment.

Also in recent years, there have been intriguing *quantum eraser* experiments in which physicists have shown that in special circumstances they can actually undo the disturbances caused by measurement.

One of the consequences of the probabilistic interpretation of quantum mechanics is that while ensembles of particles obey strict mathematical rules, the outcomes of individual measurements in certain situations are inherently unpredictable. This is seen, for example, in modern-day versions of Young's two-slit experiment in which it is possible to observe the arrival of light on the screen photon by photon. Each photon is observed as a tiny point of light on the final image. Furthermore, the arrival of any particular photon looks random. But there is method in this madness because as more and more photons are observed, a pattern of fringes is seen to build up. Each photon behaves as if it were a tiny particle that is influenced by a kind of probability wave that spreads out in space and passes through the two slits. The emerging probability wavelets from the two slits reach the screen and interfere with one another. So the key is that whenever a single photon is actually observed on the final screen, it appears only as a sharp point, not a fuzzy blur, but the probability of arriving in any particular place is governed by the probability waves.

After he had published his famous wave equation paper in 1926, Schrödinger quickly solved the puzzle over the dichotomy between wave mechanics and Heisenberg's matrix mechanics.[8] Within two months he had shown that the two approaches, despite their totally different formalisms, were mathematically equivalent. The introduction of probability into quantum mechanics, on the other hand, caused Schrödinger considerably more difficulty. He had hoped to find some classical backdrop to the quantum facade. He was in good company because Einstein too strongly doubted the probabilistic nature of quantum mechanics, as exemplified by his famous phrase, "God does not play dice."

As we shall see, the understanding of the origin of the probabilities within quantum mechanics and the problem of making sense of the theory remain to this day among the great controversies and challenges of physics.

Strange Correlations

A question Feynman raised in his 1981 MIT lecture was whether the probabilistic features of quantum mechanics could be imitated exactly on a classical computer. His conclusion was no, they could not. He offered two main arguments.

One concerned the amount of computation you would need to do to simulate a quantum system consisting of a number of particles. It turns out that to describe a probabilistic system with only finite accuracy, the size of the calculation grows alarmingly fast as you add more particles. The growth is, in fact, superexponential, which means that for each extra particle, the problem, and therefore the running time of the simulation, inflates in size by a large factor. Feynman considered that to be "against the rules," because he wanted a simulation that could accomplish its task in no more time than the physical system itself. He wanted the simulation in effect to happen in real time or faster.

His other argument focused not on efficiency but on something altogether more unworldly. I am going to present his argument here first in a fictitious setting to highlight the essential details without worrying about all of the experimental details.

Imagine you wanted to play Monopoly or some other board game that involved throwing dice via the Internet with a friend who lived in a distant country. How are you going to handle the dice throws to avoid cheating? You could both throw dice and, after reporting the results over the Internet, combine the numbers in some complicated way that scrambled the original numbers to generate a new number. The scramble would help to avoid either party rigging the outcome. But the system wouldn't be foolproof. If you waited until you received your opponent's result before sending your number, for example, you might be able to change it at the last minute to improve the outcome in your favor.[9]

A more satisfactory solution could be an electronic gadget from the Morphic Resonance Corporation. Unfortunately, this device only works as a substitute for random coin tosses, but you could operate it several times in a row to simulate throwing dice. The gadget is a handheld electronic black box that features two lamps, one red and one green, a rotary switch that can be set to one of six positions on a 360-degree dial, an on-off switch, and a counter that tells you how many times the gadget has been used. The manufacturer sells these devices in matched pairs, and its manual says the following:

> Congratulations on your purchase of the Morphic Resonator Kit. Before use in remote game playing, please check the following features of each Morphic Resonator (MR).
>
> Randomly select one of the dial settings and switch on the device briefly. Do this several times. Each time you switch on, one of the lamps will light up. The behavior is guaranteed to be random, with a 50-50 chance of either the red or green lamp coming on. We also guarantee random behavior even if you do not adjust the dial setting.
>
> To use the MRs as a pair, you must first ensure that both have been switched on and off an equal number of times. Check that the counters on both MRs agree before proceeding with the next test. Your MRs will now give exactly the same response provided their dial settings are the same.
>
> Please note that your MRs will need to be serviced when the counters reach 1 million and that these devices cannot be interchanged with any other Morphic Resonators or similar devices.

At this stage one might consider how these devices actually work. One way might be that they use a noise generator to produce random numbers, which would determine the choice of lamp illuminated each time the device was switched on. However, noise generators are so unpredictable, you could never get the two gadgets in a matched pair to produce the same random output.

Instead we could imagine that inside each device was a nonvolatile memory chip preprogrammed with a large table of random 0s and 1s that determined the behavior of the gadget for any particular setting. The table would have six columns, one for each switch setting, and a million rows, one for each time the device was switched on. The device would then read off the correct number from the table by looking at the appropriate row and column. If the number was 0, the red lamp would light up, and if it was 1, the green lamp would light up. Both devices would then behave exactly the same with the same settings, provided they both carried the *same* table of random numbers.

However, there are some further notes in the manual.

> If your opponent selects the opposite position for her rotary switch, she will always see a different color lamp from you.

So if my switch points at twelve o'clock and I see a green lamp, and my opponent's switch points at six o'clock, then I know she will see red. The same if my switch points at two o'clock, and hers points at eight o'clock, and so on.

How difficult is that to accommodate? Well, again it's no problem, because all we have to do is modify the table of random numbers so that the fourth column reads the opposite to the first, the fifth column reads the opposite to the second, and the sixth is opposite to the third. By "opposite" here, I mean 1s are substituted for 0s, and 0s for 1s.

But now we see another entry in the manual.

> If you and your opponent choose adjacent switch settings, the lamps will agree on average three out of four times.

So if I choose twelve o'clock and my opponent chooses two o'clock, or I choose four o'clock and she chooses six o'clock, the lamps will show some variation from each other but remain in agreement three quarters of the time. Can we simulate that with our information table? If we chart every possible combination, there are three distinct patterns of behavior that a random table could generate (see figure 3.2). Other possible patterns are either rotations or mirror images of these.

If we examine each diagram, we see that adjacent settings will illuminate the same color lamp (a) in four out of six cases, (b) in two out of six cases, and (c) never.

So the best possible match we can get if we fiddle with our "random"

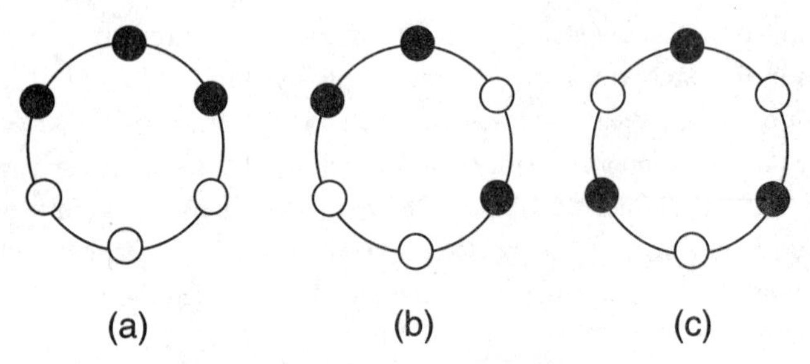

Figure 3.2. The Morphic Resonator's patterns of activity.

table is four out of six (that is, two thirds), which is less than the manufacturer's specification of three out of four.

So it turns out that no matter how much we adjust the numbers in the table, there is simply no consistent way that we can arrange to get a three-out-of-four agreement between all of the columns. Therefore we conclude that Morphic Resonators cannot rely on stored information. It is as if these gadgets were able to generate numbers randomly but then communicate some information to one another to ensure the close match in performance. But the manufacturer claims that these devices use no signaling of any kind. So how do they work?

The answer is quantum entanglement. Such experiments have actually been performed and confirm that quantum mechanics can give rise to correlations that defy anything we can do classically. The real experimental apparatus is very different from the gadgets I described above, which as I said are fictitious,[10] but the principle is the same.

The EPR Puzzle

The first practical demonstrations of the quantum "conspiracy" we've just been examining date back to the early 1970s and involved measuring the polarization of entangled pairs of photons. These are photons that because they are born together share many of the same quantum properties. Entanglement, as we saw in Chapter 1, is the essential feature that makes quantum computation potentially much more powerful than classical computation.

There are now several standard laboratory techniques for generating entangled photons. The method used in some of the early experiments was to excite a beam of calcium atoms with laser light. When excited, an electron around an atom occupies a higher energy level than normal. These excited states tend to be unstable, so that the atoms subsequently return to their natural ground state, giving up energy in the form of photons. Calcium has a particular excited state that leads to a two-stage cascade in which two photons are emitted. Because the pair of photons are born together from a symmetrical atom, the laws of conservation dictate that they must fly off in opposite directions, carrying many of the same properties, including polarization.

In more recent experiments, the favored method of generating entangled particles has been to use a special kind of crystal known as a *down conversion crystal.* When a beam of green laser light is directed toward a crystal of potassium dihydrogen phosphate (known for short as KDP), for example, some of the green photons are destroyed, and in their place pairs of lower-energy red photons emerge along divergent paths. This process, known as *parametric down conversion,* is very efficient and can be used to produce polarization-entangled photons. Alternatively, it can be used to generate photons whose phases, or timings, are entangled, a feature exploited by many experiments these days because it is possible to measure phase very accurately using a device known as an interferometer.

We'll stick to polarization entanglement, and we'll imagine we're using a calcium source because it's a little easier to explain what's going on. If we measure the polarization of any one photon from the calcium source, it is seen to be randomly oriented. However, whenever one photon is seen to pass through a polarizer, its twin will also pass through a similar polarizer, provided the two polarizers are oriented in the *same* direction. (With other sources, including most down conversion crystals, the polarizations always come out *opposite*—that is, orthogonal, such as horizontal/vertical. This makes no real difference to the physics apart from inverting some of the measurements.)

Although light can be polarized in any direction—horizontal, vertical, or anything in between—polarization is a quantum property and is analogous to quantum spin. When subject to measurement, the answer is always a yes/no decision by any single photon. If a pure beam of vertically polarized photons is measured using a polarizing filter set horizontally, none pass through. If the filter is set at 45 degrees to the vertical, 50 percent of them pass through, but the decision as to whether any *individual* photon passes through is completely random. If you ask what decides which photons pass through, the answer according to quantum mechanics is that there is no answer. It's just a matter of probability, no different from throwing a coin.

Leaving it all to chance seems reasonable until you examine the situation with entangled particles. If you place polarizing filters on both sides of our source, you find that when they are aligned at the same angle, each photon passes through *only if its twin does the same.* The entangled particles seem to share some information about what to do when confronted by polarizing filters.

This, in itself, is not particularly strange. The late Irish physicist John Bell, who did much to lay the theoretical foundations for these experiments, offered the example of Bertlmann's socks in one of his papers on the subject. Dr. Bertlmann was a mathematician who liked to wear socks of different colors. Whenever he wore a pink sock on one foot, he would have a green sock or some other color on the other foot. If you saw just one of his feet and spotted a pink sock, you knew immediately that his other foot would *not* be pink. The point of the story was that there was nothing particularly puzzling about this form of correlated behavior (except perhaps Dr. Bertlmann's color sense).

The puzzling aspect of the polarization experiment becomes apparent when we examine it in a little more detail. First of all, to make our measurements more informative, we'll use crystals of calcite, a naturally occurring polarizing material composed of calcium carbonate. A calcite crystal can separate a beam of light into photons polarized horizontally and vertically, or in any two orthogonal (that is, perpendicular) directions, depending on the orientation of the crystal. These beams are known as the ordinary ray and the extraordinary ray. So just as an electron when measured has a spin pointing either up or down, a photon when measured with a calcite crystal will always appear to be in just one of two orthogonal polarization states.

Now, if we measure the photons emerging from one side of our source, the polarization of any one photon is random, making it impossible to predict the outcome of any individual measurement in the stream. However, if we measure the polarizations of both photons in a pair using a pair of calcite crystals both oriented at the same angle, we discover that each photon always does the same as its twin. If one photon emerges along the path of the ordinary ray in the first crystal, then we can confidently predict its twin will emerge along the path of the ordinary ray in the second crystal. The results are said to be perfectly *correlated.*

A question arises over what happens when the calcite crystals are oriented at different angles. If they are set at right angles, each photon does the opposite of its twin. If one photon emerges along the path of the ordinary ray, the other emerges along the path of the extraordinary ray. The results are perfectly *anticorrelated.*

What happens if the angle between the crystals is set at the midpoint—that is, at 45 degrees? According to theory and experiment, there is *no* correlation in behavior. In other words, the behavior of each photon is random and independent of the others. Note that this means they will do the same thing 50 percent of the time, just as flipping two coins will produce agreement 50 percent of the time.

The interesting situation arises when the crystals are set at intermediate angles between full and no correlation. At 30 degrees, for example, the experiments show the photons do the same thing, on average, 75 percent of the time—that is, in three out of four cases. The remaining quarter of the time they do the opposite of one another. So this experiment is the real-world version of the situation we saw when the gadget boxes were switched to adjacent settings.

There is no way to explain the correlation strength of the photon polarizations in terms of classical information. When pairs of photons do the same thing in correlated measurements, it is tempting to think of Bertlmann's socks or, a more elaborate analogy, the idea that each photon carries its own table of information that tells it how to behave for each particular orientation of the filter. However, such a system cannot account for the strength of the correlations when measurements are made at certain ranges of intermediate angles.[11] That is why the photons are said to be entangled: because it seems that making a measurement of one instantly "influences" the outcome of a measurement of the other, no matter how far apart they are.

This idea of particles carrying information (like the color code of Bertlmann's socks) is known as a *hidden-variable theory,* a concept that has been advanced in different forms over the years to try to explain the strange features of quantum theory in purely classical terms. The scenario of the inexplicable correlations is better known as the EPR experiment, or EPR paradox, after its originators—Einstein and his two young colleagues Boris Podolsky and Nathan Rosen, who were trying to find a way of proving that quantum theory was flawed. Einstein, who was deeply dissatisfied with the implications of quantum theory, referred disparagingly to the apparent quantum conspiracy that enabled particles to behave in this supercorrelated way as "spooky actions at a distance" (*spukhafte Fernwirkungen*).[12]

Einstein, Podolsky, and Rosen's original proposal in 1935 was a pure thought experiment that imagined simultaneous measurements of position and momentum on correlated pairs of particles. In their paper they argued, "If, without in any way disturbing a system, we can predict with certainty the value of a physical quantity, then there exists an element of physical reality corresponding to this physical quantity." In other words, given that we can deduce the position or the momentum of one particle by measuring only its twin, both of these physical quantities must already exist as "elements of reality." Yet according to the Heisenberg uncertainty principle, both position and momentum cannot be well defined simultaneously. Thus, if the position of one entangled particle is precisely measured, we therefore know the position of its twin; measuring the twin's momentum would give both measurements and, therefore, a paradox. Bohr and his followers countered by arguing that the two particles formed an inseparable quantum system and that in discussing the circumstances of one, you cannot ignore measurements carried out on the other. The two particles are, to use the modern parlance, entangled.

In 1951 David Bohm introduced the idea of measuring correlated spins, a modification that made the EPR experiment easier to explore theoretically and potentially in the laboratory. But it was John Bell who showed in 1964 in a famous mathematical expression known as Bell's inequality exactly how classical and quantum correlations ought to diverge from one another. The quantum correlation of ¾ in the case we looked at is in excess of the maximum classical correlation of ⅔. Bell's inequality was a generalization of the limit on the strength of correlations we can expect to see classically—that is, if hidden variables are involved. Bell showed that quantum mechanics *ought* to violate this inequality, thereby ruling out hidden-variable theories and with them Einstein's "elements of reality."

In the 1970s and 1980s the idea was put to the test in a number of experiments, culminating in the most famous, which was carried out by Alain Aspect and his colleagues in Paris. They measured the correlations in polarizations between entangled photons separated by some 15 meters. Their results confirmed the violation of Bell's inequality and agreed perfectly with the predictions of quantum mechanics. The results, although not unexpected, convincingly demonstrated that hidden-variable theories, at least in their most simple form, could not work. An important feature of Aspect's experiments was that the detectors for each photon in a pair were randomly oriented at very high speed (using electronics) to eliminate any possibility of signals traveling between them. The idea is based on the notion of *delayed choice.* If the orientation of the second detector is randomly chosen *after* the first photon has been measured, there is no obvious way any interaction between the detectors could explain the superstrong correlations unless it somehow traveled faster than the speed of light.

The apparent ability of quantum systems to act at a distance is known as *nonlocality.* Some theorists, notably David Bohm and his followers, have attempted to explain these nonlocal quantum effects in terms of hidden variables using a "potential" that travels across space instantly, but such ideas run into trouble with Einstein's special theory of relativity, which forbids faster-than-light signaling. Although the EPR phenomenon seems to gives rise to an instantaneous interaction between two objects no matter how far apart, it offers no way to send *signals* faster than light.

Imagine you set off on a journey to the nearest star, 10 light-years away, carrying a box of entangled particles that were emitted from one side of an EPR source, while a friend on Earth kept the entangled particles from the other side of the source. Suppose on arrival you wanted to send a message using your entangled photons. What would you do? It doesn't matter what measurements you make on your entangled particles—they will not reveal any pattern in the results of measurements made on Earth because the outcome for each is random. It's only when we can compare the results for *both* sets of measurements that we see any pattern.

In his 1981 MIT talk, Feynman pointed to the EPR effect as evidence that there were things in quantum theory that could not be imitated exactly on a classical computer, if it was to behave entirely locally. In one fascinating and highly revealing section, he made the following comment, which although he didn't mention it by name was clearly about Bell's inequality. "I've entertained myself always by squeezing the difficulty of quantum mechanics into a smaller and smaller place, so as to get more and more worried about this particular item. It seems to be almost ridiculous that you can squeeze it to a numerical question that one thing is bigger than another. But there you are—it is bigger than any logical argument can produce. . . .

"There are all kinds of questions like this, and what I'm trying to do is to get you people who think about computer-simulation possibilities to pay a great deal of attention to this, to digest as well as possible the real answers of quantum mechanics, and see if you can't invent a different point of view than the physicists have had to invent to describe this. In fact, the physicists have no good point of view. . . ."

Here then was the call for computer scientists and physicists to start thinking seriously about the connection between computation and quantum mechanics. Although Feynman devoted most of his talk to looking at why conventional computers could not simulate quantum mechanics to arbitrary accuracy, he added as an aside an idea for how a computer might be modified so that it could. "Can you do it with a new kind of computer—a quantum computer?" he asked. Feynman then went on to describe his idea for a *universal quantum simulator,* a machine that could "imitate any quantum system, including the physical world."

At first blush such an idea might sound unbelievable. Such a machine would have to be able to simulate, in principle, any system that you could possibly think of, ranging from fundamental things like atoms to complex entities such as human beings, societies, and entire virtual-reality environments in as much detail as the original. However, Feynman's proposal was in some senses an extension of the idea of *universality* (see Chapter 2, page 62). Alan Turing showed that all computers were universal in the sense that given enough time and memory, one computer could do the job of any other computer. If, therefore, there was a way to enhance a computer to accommodate the behavior of quantum systems, why should it not be truly universal? It was a bold and inspirational idea but sketched only in minimal detail.

Feynman, in his inimitable style, concluded with, "And I'm not happy with all the analyses that go with just the classical theory, because nature isn't classical, dammit, and if you want to make a simulation of nature, you'd better make it quantum mechanical, and by golly it's a wonderful problem, because it doesn't look so easy. Thank you."

Designer Hamiltonians

"Everybody has their own sloppy version of history," Rolf Landauer told me, complaining that all too often, accounts of the history of quantum computation attributed the beginnings of the field to the cult-hero figure of Richard Feynman. Landauer was concerned about giving due credit to Paul Benioff of the Argonne National Laboratory in Illinois, the man who actually published the first model for a computer based on quantum mechanical components. His paper on the subject was submitted in 1979 and published the following year, but it was still a year ahead of anyone else.[13]

Unlike Feynman, who began by trying to simulate quantum mechanics on classical machines, Benioff did the exact opposite: he tried to use quantum physics to simulate a classical computer. Though this ambition seems far less exciting now, it was a worthwhile issue to investigate because there was a widespread belief that Heisenberg's uncertainty principle might cause problems for a computer that worked at the quantum level. The question then was whether quantum computers would be *less* capable than their classical counterparts.

Benioff envisaged a sequence of quantum operations that would choreograph a "lattice of quantum spins" such as a string of atoms into performing the operations of classical logic gates. He modeled his quantum machine on a *Turing machine,* a mathematical idealization of a computer proposed by Alan Turing before the first computers were ever built. A Turing machine is an abstract device that carries out its activities step-by-step according to instructions stored on a tape. The machine reads one symbol at a time on the tape and moves the tape backward or forward according to these instructions. The machine is also able to erase symbols and write new ones in their place. Turing showed that such a machine could be computationally universal. Benioff's model was pretty abstract and did not go into practical details. But the implication was that it would emulate a classical Turing machine by reading the spin states of individual atoms on the lattice and manipulating them by means of quantum operations built into the machine's hardware.

These operations are described mathematically by what is known as a *Hamiltonian.* Hamiltonians are mathematical rules that govern how systems change from one state to another and are the sort of thing quantum physicists live and breathe by. However, they were actually discovered in the analysis of classical systems by the British mathematician and astronomer Sir William Hamilton in the nineteenth century.

Traditionally, if you wanted to calculate the behavior of billiard balls bouncing around a table, or planets orbiting the sun, you would apply Newton's laws and derive equations governing the motion of each individual object. Hamilton showed that it was possible to recast the equations in terms of the total energy of the system, which, assuming no external influences, is a constant because of the law of the conservation of energy. The Hamiltonian is, in fact, a mathematical function that represents the total energy of a system in terms of the motion of all of its components. The beauty of this description is that it gives a God's-eye view of a system's dynamics rather than the on-the-spot view offered by Newton's laws.

Classically, both perspectives have their uses, but in quantum mechanics local Newtonian rules no longer hold true, while global laws such as the conservation of energy do. The Hamiltonian approach therefore reigns supreme in describing the dynamics of quantum systems.

Every quantum system has an associated Hamiltonian, or rule, that defines how it will evolve. So if you want a quantum machine that will manipulate lattice spins to perform computations, mathematically all you need to do is work out the necessary Hamiltonian for the job.

Benioff's model was based on a Turing machine that had several components, including the tape, record and read heads, and the logic and control center. The operation of such a machine can be broken down into three main steps.

1. Reading an instruction on the tape.
2. Performing the computation in the logic center.
3. Moving the tape to the next instruction under the direction of the control center.

Benioff required his model to switch between three different Hamiltonians, each of which was designed to execute the appropriate step quantum mechanically on the Turing machine. The agency that actually did the switching was left unspecified, but the implication was that the machine would need to be externally controlled, rather like the way microprocessors are driven by external clocks.

His system was complicated, and although Benioff pointed out that the machine would be reversible, there was no hint that such a machine could go beyond the classical repertoire of computation. Nevertheless, Benioff's paper "was a remarkable accomplishment," according to Landauer. "I had asked myself somewhat related questions, but I was too much of an engineer to make Benioff's idealization."

Benioff didn't receive immediate recognition for his work. He, like Feynman, presented his work at the 1981 MIT meeting, but his audience had difficulty with his message. "Benioff's papers were hard to understand," said Landauer. "A lot of us were trying to understand him. What's he talking about? Do we believe it? Feynman took part in discussions at that conference. I think he caught the question and ran with the ball."

The result was that three years later, in 1984, Feynman presented his own version of a fully quantum mechanical computer in a plenary lecture delivered to a meeting held in Anaheim, California. "Unfortunately Benioff is never mentioned in that paper," Landauer reflected. "Feynman's paper made the whole thing clearer and more convincing. But it's unfortunate because for Feynman, in his history of great intellectual accomplishments, this was number 29 in the list down from the top. For Benioff, it was probably the accomplishment of his career. And I think he's had a hard struggle getting credit for it."

In recent years, though, researchers have acknowledged Benioff's contribution more readily. His papers are now, at least, given due mention in many articles published in the scientific journals.

The significance of Benioff's quantum version of a Turing machine was that it showed that there was sufficient discipline within the quantum mechanical regime to construct an essentially classical computer. The practical issue of whether quantum mechanics would cause problems for the construction of molecular computers also appears to have been a motivation for Feynman's 1984 Anaheim talk, entitled "Quantum Mechanical Computers." In it he described a quantum mechanical computer based on a model of reversible computation developed by Tom Toffoli from earlier work with Ed Fredkin.[14] Like Benioff, Feynman chose to write a Hamiltonian that would do the job he wanted, except that Feynman used only one Hamiltonian as opposed to three. Seth Lloyd, a leading researcher in quantum computation at MIT, describes this time in the 1980s as the period of "designer Hamiltonians" because researchers gave themselves carte blanche to write more or less whatever Hamiltonian they wanted, regardless of how difficult it would be to implement practically. This was something Landauer tried to put a stop to, complaining that such ideas were physically unrealistic. Nevertheless, this was pioneer country and its interlopers wanted to see how far they could go before worrying about the small question of whether any of these wild ideas could ever work.[15]

Feynman's model was less like a conventional computer than Benioff's because it didn't need to be clocked. Also, its program had to be incorporated into the machine's Hamiltonian instead of being read from tape. Once the Hamiltonian was set up, you would, in effect, press the button and the machine would evolve through its computational steps, rather the way a ball thrown in the air automatically computes a parabolic arc. In many respects Feynman's model was therefore more like an analog than a digital computer. Nevertheless, Feynman showed how such a machine could do various logical operations, such as act as a simple binary adder. He also argued that with a sufficiently complex Hamiltonian, the machine would be able to mimic the operations of a general purpose computer.

Although the Feynman model was an advance, he made no further references to simulating quantum mechanics. So once again it appeared that these quantum computers were no more capable than their classical cousins. Feynman's final sentence summed up the general conclusion of the time: "At any rate, it seems that the laws of physics present no barrier to reducing the size of computers until bits are the size of atoms, and quantum behavior holds dominant sway."

In retrospect, this whole period between Benioff's first paper in 1980 and Feynman's contribution in 1984 was somewhat reminiscent of the phase quantum theory went through in its early years, up to the mid-1920s. During the years of the old quantum theory, many scientists clung to classical notions. Although Benioff and Feynman had adopted a quantum perspective, they were still thinking of classical computation. Just as Heisenberg, Schrödinger, Born, and others transformed quantum theory, the study of quantum computation awaited its next conceptual leap. It wasn't long in coming.

A Matter of Interpretation

Quantum theory has been astonishingly successful. It explains a vast range of phenomena including the behavior of subatomic particles, the nature of chemical bonds, the flow of electricity through metals and semiconductors, the properties of light, and even certain features of black holes. It has been tested in all manner of circumstances and proven itself accurate to a level of accuracy unequaled by any theory apart from general relativity.[16] Yet from its earliest days it has been bedeviled by conceptual and philosophical problems.

After the revolutionary developments that established the new quantum theory in the mid-1920s, Niels Bohr, founder of the Institute for Theoretical Physics in Copenhagen, pioneered an interpretation that became for many years the standard view among many physicists. Today the *Copenhagen interpretation,* as it is known, seems to be less widely touted than it used to be, although aspects of it still prevail. According to this view it doesn't make sense to ascribe intrinsic properties such as position and velocity to isolated quantum entities such as electrons. The properties of quantum systems only make sense in light of the measurements we make. This idea was summed up more graphically by Bohr's onetime collaborator John Wheeler. "No elementary quantum phenomenon is a phenomenon until it is brought to a close by an irreversible act of amplification by a detection such as the click of a Geiger counter or the blackening of a grain of a photographic emulsion."[17]

The Copenhagen interpretation thus more or less denies the reality of an individual photon or electron until it has been observed. It's not hard to see why this view might have seemed quite reasonable back in the 1930s, when nobody had ever directly seen *individual* atoms, let alone individual particles like electrons and photons. Today, though, with modern quantum technology, it is almost routine to observe individual atoms, measure the presence of single electrons, and record the arrival of individual photons. Denying the reality of these entities (even if it is in the absence of measurements) begins to look a little perverse when one is playing and experimenting with them on a daily basis. But the Copenhagen interpretation was always odd in that what happens *between* measurements of a quantum system is actually perfectly described by the equations of quantum mechanics. It's only when we make measurements that determinism goes out the window.

The distinction between these two types of quantum behavior was first made explicit in 1932 by John von Neumann in his book *Mathematical Foundations of Quantum Mechanics.* It was his analysis that brought out into the open what is known as the *measurement problem*—to explain why measurement seems to give rise to a fundamentally different form of quantum evolution from that between measurements. In modern treatments the measurement problem is often circumvented by simply writing the different behaviors into the laws of quantum mechanics, which can actually be summed up in just four mathematical rules, or postulates.[18]

Consider these two types of behavior for a moment. Between measurements, so long as a quantum system is isolated from its environment, its progress is described by what mathematicians call a *unitary* operation, or transformation. Unitary operations are fully reversible because they preserve certain important mathematical properties of quantum states. Of interest to us here is that unitary operations are the link between classical reversible computation and quantum computation because, as we will see, they are the stuff of which quantum logic is made.

In contrast, there's the kind of evolution that happens when a quantum system is subjected to measurement. If the quantum system starts off in a *superposition* of states—a linear combination of different states—then the act of measurement causes it to undergo a dramatic change, collapsing onto just one state. This collapse is generally irreversible[19] and nonunitary. Measurement plays a fundamental role in that it enables us to see the results of *interference* between different quantum states. The power of quantum computation comes from the ability to exploit entangled states and to perform interference experiments on them. But that's running a little ahead of the story.

From a philosophical point of view, the most glaring problem with the Copenhagen interpretation is that it gives observers a special status in physics. This emphasis on the role of the observer led to a huge amount of controversy ever since the 1930s because it begged the question of what exactly constituted an observation. Einstein's skepticism over this issue was aptly captured in a now famous remark noted by Abraham Pais in the opening of his biography of Einstein, *Subtle Is the Lord:* "It must have been around 1950. I was accompanying Einstein on a walk from the Institute for Advanced Study to his home, when he suddenly stopped, turned to me, and asked me if I really believed that the moon exists only if I look at it."[20]

Although Einstein, Podolsky, and Rosen failed to prove that there was anything wrong with quantum theory with their thought experiment of 1934, it inspired Schrödinger, who maintained a long friendship with Einstein, to publish the following year his own thought experiment, now well known as Schrödinger's cat paradox. The experiment has in recent years received so much popular attention that Stephen Hawking reputedly said, "When I hear about Schrödinger's cat, I reach for my gun." But in case you're not familiar with the setup, here's a brief description.

A cat is placed inside a box along with what Schrödinger described as a "diabolical device." The device contains a tiny mass of radioactive material that has a half-life chosen so that in the course of one minute there is a 50-50 chance of one atom decaying. The diabolical aspect is that when a decay happens, it is detected by a Geiger counter wired up to release a small hammer, which smashes a flask of poison, killing the cat. In a more politically correct version of this experiment—and it is *only* a thought experiment—the animal and flask are replaced by a porcelain cat, which is smashed by the hammer instead.

The point of the exercise is to ask what state the cat is in after the system has been left for one minute. Radioactivity is ultimately a quantum process, which means that the possibility of a decay within one minute is a purely random event dictated by the laws of probability. According to quantum mechanics, though, the radioactive atoms evolve into a *superposition* in which each atom exists in both its intact and decayed states simultaneously. In the language of Schrödinger's wave mechanics, an atom would be described by a wave function in much the same way that we can describe the motion of a water wave using mathematics. But though it's easy to see how water waves can superimpose themselves on one another, it's harder to envisage how different physical states of an atom can go into a superposition. This strange state of affairs is analogous to the situation in Young's two-slit experiment where individual photons are apparently able to pass (as waves) through both slits simultaneously.

But in the cat experiment, the result of a decay is amplified by the apparatus linking the microscopic quantum world directly to the macroscopic state of the cat. When you look inside the box you obviously see only one state or the other—not some crazy superposition of dead and live cats. But at what point in this process does the system make up its mind? Somewhere between the atom and our observation, the wave function is said to "collapse" onto one state or the other. The precise point at which this happens is the essential feature that distinguishes many of the different interpretations of quantum mechanics.

According to the Copenhagen interpretation, "No phenomenon is a phenomenon until brought to a close by an irreversible act of amplification." From this one might conclude that the process of amplification settles the question and collapses the wave function. But therein lies a problem. The apparatus is made of atoms, which are subject to the same quantum rules as the radioactive material, so in principle they should go into a superposition of states as well. The same could also be said of the cat. So we are still left with the questions of where and when the superposition collapses onto one outcome. This is the measurement problem writ large.

No one to this day has provided a totally satisfactory resolution to this problem. If you ask a bunch of physicists, you will get half a dozen different answers. That is precisely what Paul Davies and I did in 1982 when we made a radio documentary for the BBC in London about the problems in understanding quantum theory. We talked to eight leading scientists about these issues and got an extraordinary range of responses.

One of our interviewees was the late Sir Rudolf Peierls, an eminent physicist who had studied under Heisenberg and Wolfgang Pauli. He supported Bohr's view of quantum theory, although he objected to the phrase "Copenhagen interpretation" on the grounds that it implied there were other viable interpretations. "There is only one way in which you can understand quantum mechanics," he said with resolute confidence. He then went on to articulate a point of view that went beyond anything Bohr ever said but, according to David Deutsch, was probably the only consistent position *if* you subscribed to the standard interpretation. Peierls suggested that the outcome of a quantum experiment depended on an observer becoming *conscious* of the result.

This view that consciousness was crucial to turning the indeterminacies of quantum mechanics into concrete reality was first advocated by Von Neumann in 1932 and again by Eugene Wigner in the 1950s. But as Walter Moore points out in his biography of Schrödinger,[21] "Only a few commentators on the cat paradox . . . have defended the uncompromising idealist position that the cat is neither alive nor dead until a human observer has looked into the box and recorded the fact in a human consciousness."

Of course, if we *are* going to invoke consciousness, it would surely not be unreasonable to assume that the poor cat was sufficiently conscious to bring about its own wave function collapse. But this raises the question of whereabouts in the animal kingdom consciousness becomes strong enough to effect the collapse. And given that brains are made of atoms just like everything else, why aren't they too subject to the laws of quantum mechanics and therefore able to go into a superposition as well?

Faced with such difficult questions, advocates of consciousness as the answer to the measurement problem are decidedly thin on the ground these days. Instead, many physicists duck out of answering these conceptual problems on the basis that they are irrelevant to virtually all practical problems of physics.

While commendably pragmatic, this attitude is philosophically unsatisfying. It is, in my view, somewhat reminiscent of the mind-set of the behaviorists who dominated American psychology from the 1930s to the 1960s. Their attitude, at its most extreme, was informed by a belief that only phenomena that could be objectively measured had any claim on reality; they thus treated mental experience with disdain and regarded the brain as little more than an automaton that responded to the outside world via a set of conditioned reflexes.

At any rate, the inadequacy of the physicists' position was amusingly exposed in the following exchange, broadcast on BBC radio in 1996, between the British particle physicist Frank Close and Melvyn Bragg, a doyen, surprisingly enough, of the arts and media establishment in Britain.

> CLOSE: *What intrigues me about this debate on the meaning of quantum mechanics is that I have worked for the whole of my career using quantum mechanics. I have never had to worry about these questions. They never come up. They are irrelevant to the day-to-day practicalities.*
>
> BRAGG: *I never worry about what happens inside an airplane, but I still fly in one, and I'm very pleased that someone knows!*

To be fair, though, quantum theory has been so thoroughly battle tested, one might forgive Close his lack of concern. It's not as if anybody's life is depending on it. Yet it was this very issue that led David Deutsch toward his first ideas about quantum computation. He had a big incentive to find a method of testing different quantum interpretation because the interpretation he favored was so strange, it was almost unbelievable. It was the idea that quantum mechanics depended on the existence of an infinite number of parallel universes.

The Case for Many Universes

The many-universes interpretation of quantum mechanics, sometimes also known as the many-worlds interpretation, was first proposed by Hugh Everett III in 1957. It was a complete departure from the ideas enshrined in the Copenhagen interpretation because it attempted to remove completely the measurement problem, and with it all talk of wave function collapse, by eliminating any references to observers.

According to the modern version of the many-universes idea, our universe is embedded in an infinitely larger and more complex structure, called the *multiverse,* which to a good approximation can be regarded as a system of parallel universes. In the multiverse there are endless copies of our universe—some slightly different, some completely different. Furthermore, every time there is an event at the quantum level, such as a radioactive atom decaying or a particle of light impinging on your retina, the assemblage of universes differentiates along different paths.

So in the case of Schrödinger's cat what happens is that the decay of the radioactive atom does indeed lead to a superposition of dead cat and living cat. The reason we don't notice this strange coalition of states is that the different states are in different universes. If you are the observer, then you see a live cat in one universe, and a copy of you sees a dead cat in the other universe. According to this interpretation, then, you have to accept that there are endless copies of yourself doing different things in the different universes. In some universes things will be so different there won't even be any copies of you, and there may not even be any human beings either. The multiverse is unlimited in its diversity.

Hugh Everett developed the idea when he was a Princeton graduate student working under John Wheeler, who was then closely allied to the Copenhagen interpretation. However, Wheeler was sufficiently impressed by the idea to write a paper to accompany Everett's in an effort to increase attention to what otherwise might have been regarded as too way out for serious consideration. Wheeler later commented that he had wanted to alert other physicists to the idea because "it seemed to represent the logical follow-up of the formalism of quantum theory."

Not surprisingly, though, most people found the idea utterly bewildering. The physicist Bryce de Witt, who later became an ardent supporter of the many-universes interpretation, remarked in an early critique of Everett's idea that he couldn't feel himself "split" into multiple, distinct copies every time a decision was made. Everett responded by echoing Galileo's answer to the Inquisition: "Do you feel the Earth move?"

Despite Wheeler's intervention, the theory did not gain currency for many years, and Wheeler himself decided that many universes carried too much "metaphysical baggage." Nonetheless, the theory enjoyed renewed interest in the 1970s as research grew in the subject of quantum cosmology, which attempts to apply quantum mechanics to the universe as a whole. It quickly became apparent that the concept of a wave function for the entire universe makes no sense in the standard interpretations of quantum mechanics, because there is by definition nothing external to the universe to make measurements. Because the many-universes idea did away with the need for external observers, post-Everett interpretations became the only serious option for the new breed of quantum cosmologists.

David Deutsch saw this, but he was not happy to accept Everett's theory merely as a convenience for cosmological calculations. He wanted to understand the theory and acclimatize himself to it, in the same way that most of us today, I hope, feel comfortable with the notion that the Earth is round rather than flat. So he began to ponder whether there might be a way of testing the theory experimentally. That, in itself, was an unusual line of thought, because most physicists have regarded the different interpretations of quantum theory as experimentally indistinguishable.[22] That, indeed, is why they are called "interpretations" rather than "alternatives" to quantum theory.

Around 1977 Deutsch developed an idea, which, as we saw in Chapter 1, more or less called for a quantum computer endowed with consciousness. The idea remained unpublished until 1985, after the more down-to-earth work of Benioff, Feynman, and others had established some first principles of quantum computation. What's more, a paper by David Albert, then at Tel Aviv University, appeared in 1983, describing how a quantum "automaton" could make measurements on itself that would be quite unlike anything that could be done classically. Other people, it seems, were catching up with Deutsch's thinking.

Nevertheless, the idea that there might actually be a way to test the many-universes interpretation of quantum mechanics remained uncharted territory. In 1982 Deutsch discussed his idea with Paul Davies and me for our BBC documentary. His account of the proposed experiment, while hugely entertaining, sounded like pure science fiction. The experiment required the participation of an advanced artificial intelligence machine with a quantum-based memory and sense organs capable of making quantum mechanical measurements. Furthermore, to satisfy the requirements of what Deutsch regarded as the clearest version of the Copenhagen interpretation, the artificial intelligence would need to be conscious. The reason for that was if consciousness was required to collapse wave functions, then it would be necessary to invoke consciousness in the experiment. The clever part of Deutsch's proposal was in seeing how to control the intrusion of the conscious observer so that it wouldn't disturb the quantum state *when the situation was viewed from a many-universes perspective.*

Even today the experiment is still wildly futuristic. No one really has a clue what will be required to build a conscious computer using conventional logic circuits, let alone their quantum versions. But other aspects of the experiment are now easier to envisage in the light of current knowledge about quantum computing. And the key point of the experiment was that it suggested there was indeed a way to distinguish between different quantum interpretations, contrary to what many physicists had thought. (See Appendix C for a description of the experiment, though to understand it you should read the next chapter first.)

The Universal Quantum Computer

Although Deutsch set out to find a way to test the many-universes idea, he had already become convinced that it had to be the correct way of looking at quantum theory. The Copenhagen interpretation, for him, seemed too ad hoc and contrived to be philosophically acceptable. Nevertheless, in its place he had to accept the extraordinary concept of the multiverse. As Paul Davies put it, the many-universes idea was "cheap on assumptions but expensive on universes."

Having accepted this Faustian bargain, Deutsch began to ponder the further implications of quantum machines. What else might they be capable of? In 1985 Deutsch published another paper, which proved to be a turning point in the development of this subject. Entitled "Quantum theory, the Church-Turing principle and the universal quantum computer," the paper had wide-ranging implications for computing and physics. Strangely, though, it took quite a while for its implications to sink in with other researchers.

The paper outlined Deutsch's scheme for a *universal* quantum computer, which he showed could do three important things. First, it would act as Feynman's universal simulator. Second, it would do everything a general purpose computer could do. Third, and perhaps most important, it could take advantage of *quantum parallelism* to do things a general purpose computer could not do.

These advances opened the way to some fundamentally new thinking. The work of Benioff and Feynman had established that quantum mechanics would not represent an obstacle to computing with components at the atomic level. The quantum simulator had shown that special-purpose quantum hardware could emulate other quantum systems much more efficiently than conventional computers. Deutsch brought the two together to propose a new kind of universal computer with profoundly novel powers.

Like Benioff's quantum computer, Deutsch's idea for a universal quantum computer was based on a Turing machine. It called for a quantum mechanical processor and an infinitely long tape, which the machine used for its memory. The program would be stored on the tape along with the input data. The processor would then execute the program by repeated application of a Hamiltonian operator—that is, a set of quantum rules. The rules would be like instructions in a microprocessor: they would manipulate individual "bits" on the memory tape. Again like the work of Benioff and Feynman, though, this paper said very little about the practicalities of building such a computer. Its focus was to explore the implications of such machines for physics and computation.

Besides revealing the possibilities of quantum parallelism, Deutsch's additional breakthrough was to demonstrate the *universality* of his model. In particular he showed that it was possible to construct a single Hamiltonian that could perform *any* calculation—classical or quantum—to an arbitrary level of accuracy. This was a considerable advance beyond Benioff's model, which, despite needing to use a sequence of different Hamiltonians, could only mimic the operations of a classical machine, and from Feynman's quantum simulator, which had to be set up with a different Hamiltonian for each program.

The Turing Principle

Even more significant, from Deutsch's point of view, was that he saw the universal quantum computer as evidence for a hitherto unrecognized principle of physics that was on a par with such principles as the laws of thermodynamics. The new principle, Deutsch claimed, sprang from a famous conjecture in computation theory. In 1936 the American mathematician Alonzo Church and Alan Turing independently suggested that no matter how ingenious the technology and how sophisticated the mathematical techniques, there were universal limits on what could be computed. This view became enshrined in the so-called Church-Turing thesis, which in the form articulated by Turing said:

> Every function which would naturally be regarded as computable can be computed by the universal Turing machine.

At first sight this statement looks almost tautological. After all, you might think that universal Turing machines would not be *universal* if there were things they could not compute. This issue of computability first came to people's attention in 1900 when, as a kind of intellectual curtain-raiser for the twentieth century, the German mathematician David Hilbert posed twenty-three problems he wanted to see solved in mathematics. One of them was the question of whether mathematics was *complete:* Hilbert asked whether every mathematical statement could be proved or disproved. Another question was whether mathematics was *decidable:* Was there a definite method that could guarantee finding all such proofs or disproofs?

In 1930 neither of these questions had been answered, but Hilbert boldly proclaimed that he could not envisage any unsolvable problems in mathematics. The following year he was proved decidedly wrong by the brilliant twenty-five-year-old Austrian logician Kurt Gödel, who produced what is now one of the most famous theorems in the whole of mathematics. Gödel's theorem provided a method of constructing mathematical statements that can never be proved or disproved, thereby demonstrating that mathematics was incomplete.

But if you put Gödel's unprovable propositions to one side, this still left open another of Hilbert's problems—the *Entscheidungsproblem,* as it was known: Did there exist a general procedure, or "algorithm," that could decide all mathematical questions that *were* amenable to proof or disproof? Church and Turing demonstrated by different means that there was no such algorithm.

Turing later offered a concrete example of such an undecidable problem. Known as the *halting problem,* it referred to the difficulty in deciding those situations in which a Turing machine will halt or carry on running forever. This would be like programming a computer to examine other computer programs and decide which ones would produce an answer in a finite time and which ones would never finish. Take the following program written in old-fashioned BASIC, for example. It clearly loops forever:

```
10 PRINT "HELLO"
20 GOTO 10
```

The following, on the other hand, terminates after just two steps:

```
10 PRINT "HELLO"
20 STOP
```

The halting problem, then, is to ask whether there is a way of inspecting such programs and predicting their behavior without actually running them. This has modern-day relevance because compilers for computer languages such as C++, Java, and BASIC are designed to look at programs written in high-level languages and translate them into low-level machine code. Compilers generally include sophisticated error checking so that if you include nonsensical instructions in your program, the compiler warns you with error messages (which, alas, are all too often incomprehensible). If it were possible to check programs in advance for infinite loops, it's more than likely that software companies like Microsoft, Symantec, and Inprise would have done so. It would certainly make a very handy debugging tool for computer programmers. But although infinite loops are obvious in some programs (as in the first example above), Turing was able to prove that there were no surefire procedures for detecting them automatically.

Despite this proof, the Church-Turing thesis—that any computable function could be implemented on a Turing machine[23]—was only a *hypothesis,* because it depended on what exactly was meant by "computable." Even today, although it is intuitively obvious to computer programmers what is meant by the word, it's hard to pin it down mathematically. However, in his 1985 paper Deutsch put the Church-Turing thesis in a new light.

It was evident that the universal quantum computer could simulate not only the functions of a Turing machine but also the behavior of other quantum systems. Therefore the notion of computability, Deutsch reasoned, depended not on *mathematical* ideas about the nature of algorithms but on the computational capabilities of *physical* systems. Deutsch chose to reinterpret Turing's "functions which would naturally be regarded as computable" as "the functions which can in principle be computed by physical systems." This new interpretation thus replaced the somewhat vague mathematical notion of computability with a more concrete physical idea.

With its new, more precise definition, Deutsch elevated the Church-Turing thesis into a physical principle that, after a suggestion from his Oxford colleague Roger Penrose, he called the *Turing principle.*[24] In his book *The Fabric of Reality,* Deutsch simplified the language of his original version of the principle to articulate it in a number of different settings. The following form applies to abstract computers simulating physical objects:

> There exists an abstract universal computer whose repertoire includes any computation that any physical possible object can perform.[25]

The important message here according to Deutsch was that computing was not a purely mental construct belonging to some Platonic realm of mathematical ideals but a manifestation of the laws of physics. Reaction to Deutsch's extraordinary paper was slow in coming, but it clearly struck a chord with IBM's Rolf Landauer. "In a sense what David Deutsch had to say [on the Church-Turing principle] was related but perhaps not equivalent to the thing I keep stressing, which is that information is a physical quantity. It's not an abstract thing, it's not part of theology, it's not part of philosophy. Information is inevitably represented by real physical entities and is therefore tied to the laws of physics."

According to Artur Ekert, one of Deutsch's collaborators at Oxford, it took a long time for many people to react to the paper partly because it appeared in a journal that, although highly respected, did not have a wide circulation and also because of its philosophical emphasis, which Ekert thinks might have turned off many physicists and computer scientists. Take, for example, this comment in a paper published by André Berthiaume and Gilles Brassard in 1992:

> Even though [Deutsch's paper] may come one day to be recognized as one of the most important papers of that decade in theoretical computer science, it was written primarily for the benefit of physicists, in a language somewhat foreign to computer scientists.[26]

Landauer also pointed to Deutsch's adherence to the many-universes interpretation. "Deutsch came into this arena with a particular set of metaphysical prejudices. The term "many worlds" makes some people bristle."

Yet ironically it was this prejudice that helped Deutsch develop the idea of quantum parallelism, the concept for which he now gets most credit. In his paper he showed how the quantum computer could carry out separate computations in parallel universes and produce a result that depended on all of the answers by getting the universes to interfere. But many physicists claim that you don't have to buy the many-universes interpretation to accept the idea of quantum parallelism. You can, for example, just say that a quantum computer follows many computational paths simultaneously, after which you can make measurements that depend on all of these paths coming together. Somehow, though, the many-universes view of the process, bizarre as it is, makes the process more concrete and therefore easier to imagine.

In the next chapter, we will see how Deutsch demonstrated the idea of quantum parallelism. His examples of quantum algorithms were not particularly spectacular, as he himself readily admits. But they were a straw in the wind for the revolutionary developments that were to come. If you are familiar with conventional computer programs these quantum programs will seem like they are from, well, another universe.

4

Quantum Parallelism

> The classical theory of computation, which was the unchallenged foundation of computing for half a century, is now obsolete. . . .[1]
>
> DAVID DEUTSCH

The New Paradigm

For physicists inured to the Alice-in-Wonderland logic of quantum mechanics, the idea of quantum parallelism now seems almost second nature. When photons, for example, pass through Young's two-slit apparatus, they appear to travel along both paths simultaneously. Everyone in physics these days lives with that, even if they're not absolutely sure how it happens. So the fact that quantum parallelism could be useful in computation now seems quite natural, at least as a theoretical proposition. However, that's not how things seemed in 1985, when David Deutsch first announced the possibility.

Norman Margolus of Boston University, long a convert to the idea, remembers the general mood and his own skepticism: "We have to give Deutsch and his colleagues credit. A lot of us didn't really share their intuition that you might be able to do better in computing with quantum mechanics. I certainly had this strong prejudice that you shouldn't be able to do any extra information processing using quantum mechanics, because, after all, the speed of information transmission isn't any faster because of quantum mechanics. A finite quantum system has only a finite amount of information in it, so you should only be able to do a finite amount of information processing. But that only argues to the point that they can't do any new types of computations. The thing that these people got interested in was, maybe you could do conventional computations faster."

What types of computations and how much faster? On the face of it, the most suitable tasks would be those that lend themselves to being broken up into numerous subtasks, each of which could be performed in parallel. André Berthiaume and Gilles Brassard of the University of Montreal aptly summed up this ambition in a paper published in 1994.

> What makes the quantum computer different from a classical machine is the possibility it offers for massive parallelism within a single piece of hardware.[2]

To give a crude example of what this means, suppose you wanted to calculate something such as how far a stone would travel away from you if you threw it in the air at 45 degrees to the horizontal at various different speeds. To compute these distances, you can use a simple formula: Ignoring air resistance and your height above the ground, the distance is given by v^2/g, where v is the velocity and g is the acceleration due to gravity. Taking into account the conversion of units, this gives a distance in feet of $0.0669 \times v^2$. So to get the answer for, say, 20 mph, you simply plug the number into the formula, perform the calculation, and get the approximate answer of 27 feet. Now, if you wanted the answer for a different speed, such as 30 mph, you would do the calculation again with the new value for v, getting this time 60 feet. Clearly, to do the calculation for two input values takes twice as long as for one. Of course, in this case the calculation is trivial. But imagine a much more complicated formula or, as mathematicians prefer to call it, a function f.

What Deutsch showed was that with a hypothetical quantum computer one could prepare an input that encoded two input values x and y (such as the two different speeds in our previous calculation) in a quantum superposition. Moreover, if you ran a program that computed f on that quantum computer, it would produce the quantum superposition of the two answers, symbolically represented by $f(x)$ and $f(y)$, in the time needed to compute the function *only once.* What is more, the program wouldn't need to be limited to just two input values. You could prepare a quantum superposition of a huge number of inputs, after which a single call to the program would compute the quantum superposition of all the corresponding outputs.

Berthiaume and Brassard expressed the implications of this idea as follows:

> If this sounds too good to be true, it is in a sense. The catch is that even though you have produced exponentially many outputs for the price of one, quantum mechanics severely restricts the way in which you can look at the result. Moreover, by virtue of Heisenberg's uncertainty principle, any measurement of the output state that yields information in a classical form induces an irrevocable destruction of some remaining information, making further measurements less informative or even completely useless.[3]

So although quantum computing offers seemingly unlimited amounts of massive parallelism without the need for extra hardware, there is also a price to be paid: It's impossible to read all the darned answers. The best we can hope to do is perform a computation that somehow depends on all of the many pathways coming together. From that, we can seek a result that depends on the totality of the answers and not on any one of them individually.

This is a crucial point because people sometimes ask why quantum computers won't be used for weather forecasting, aerodynamic modeling, and other tasks for which massively parallel computers are already in demand. Although it is not impossible that quantum computers could one day find applications in such tasks, it seems rather unlikely, at least at present. These modeling problems require computations at billions of grid points in 3-D space: on the surface of the Earth, on the surface of an airplane, and so on. The calculations typically involve working out air flows, temperatures, pressures, and so on at each point, plugging them into equations to predict how they interact with one another, and so on. In a conventional computer, programmers have complete flexibility on how they manipulate these quantities. In a quantum computer, however, you wouldn't get that kind of independent access to each computational pathway. Instead, what you get are some rather limited means of processing the numbers en masse.

How limited? As we shall see, computer scientists and mathematicians have become increasingly resourceful in squeezing more and more out of quantum parallelism.

The Meaning of Superposition

One of the intriguing facts about the discreteness of the quantum world is its similarity to the digital nature of the computer. However, the similarity with classical computers runs only so far. Quantum information comes in 0s and 1s just like the binary numbers inside computers. But unlike the bits in digital computers, these quantum values can be represented within states known as superpositions, which, as we saw in Chapter 1, behave somewhat like overlapping waves.

Because superpositions play a central role in quantum computation, it is helpful to examine them in a little more detail. Although their physical interpretation is a little hard to envisage in the case of Schrödinger's cat, they aren't particularly mysterious when we consider polarized light.

Light consists of electromagnetic waves in which electric and magnetic oscillations propagate together either through some medium or through empty space. The electric and magnetic fields are always perpendicular to one another, but as a pair these components can point in any direction. However, in some materials only certain directions are allowed. Some crystals, for example, may only allow photons with electric fields aligned in a particular direction to travel through. The emerging light is then said to be linearly polarized. Some materials allow only circularly polarized light to pass, photons whose electric and magnetic vectors continuously rotate in one direction as they travel through space.

If you have ever tried looking through two pairs of polaroid sunglasses, you will know that you see an interesting effect as you rotate one pair: The filters alternately blacken and then brighten again. The darkest point occurs when the polarizers are perpendicular to one another. Theoretically, at this point no light should pass through at all. However, filters are never perfect, so you will still see some light.

Now consider a beam of vertically polarized light passing through a polarizer set at 45 degrees to the vertical. At this angle half the light passes through. An interesting question is, what happens to individual photons? The photons are initially all polarized vertically, yet half of them still manage to pass through the polarizer. Furthermore, the ones that emerge are all polarized at 45 degrees to the vertical. The process of passing through the second polarizer, which is in effect a quantum measurement, actually changes the polarization of the photons.

So why do some photons make it through and others don't? The answer is that like all quantum phenomena, the rules are probabilistic. The outcome for any individual photon (at this angle of 45 degrees) is a 50-50 random chance.

But one thing that might seem odd about describing polarization as a quantum phenomenon is that it doesn't appear to be discrete. Light can, after all, be polarized in any of an infinite number of directions. The same applies to quantum spin. An electron spin can point in any direction. When we *measure* an electron's spin in the vertical direction, however, we only ever get the answer "up" or "down," even if the spin was pointing at some other angle before the measurement. The same two-valued behavior applies to photon polarization. When we measure a photon's polarization, we get the answer "yes, it's polarized in this direction" or "no, it isn't."

The idea of superposition becomes more concrete when we discover that a photon polarized at an angle to the horizontal is actually in a superposition of horizontal and vertical polarization states. From a multiverse perspective, a diagonally polarized photon can be regarded as being horizontally polarized in one universe and vertically polarized in another.

Quantum physicists sometimes talk about "mixed quantum states," but it's important not to confuse that idea with "superposition," because they mean very different things. If the photons in a beam are all in the same quantum state, regardless of whether the state is a superposition, they are said to be in a *pure,* or *coherent,* state. The reason a beam of light polarized at 45 degrees is regarded as pure is that such a beam is no different in character from a horizontally or vertically polarized beam. There's nothing special about the vertical or horizontal polarizations—the choice of reference frame is arbitrary. Indeed, from the point of view of someone looking at an angle to the horizontal, a horizontally or vertically polarized photon is in a superposition of states.

The light from an ordinary electric light bulb, however, is in a *mixed* state because the polarization varies randomly from photon to photon.

Counting on the Qubits

The ability of quantum properties like photon polarization and electron spin to take on these strange "intermediate" states—superpositions—

means that the two-valued logic of binary computing is inadequate for describing their behavior. Early in the 1990s Ben Schumacher of Kenyon College, Ohio, coined the term *qubit,* short for "quantum bit" and pronounced like the biblical measure "cubit," to describe each unit of quantum memory. A qubit is like a classical bit inasmuch as it can store either a 0 or a 1, but it also has the capacity to hold a superposition of states.

To see how these states enhance the capabilities of computers, it is helpful to introduce some notation. If you hate abstract-looking math, don't worry about the details too much, but the notation will make it possible to understand the ideas behind some of the quantum algorithms we're going to meet.

The symbols |0> and |1> are labels we attach to the quantum states of a single qubit and represent the numbers 0 and 1. The physical manifestation of these states depends on the underlying hardware: for polarized light it could mean horizontal and vertical polarization, for atomic spins it could mean down and up, or in the case of excited atoms it could mean the ground state and first excited states. I say "could mean" because the choice of representation is up to the observer.

Now, if we have a group, or register (see Chapter 1, page 35), of three qubits, all set to 0, we represent the complete quantum states as |000>, rather as we might represent the state of three conventional bits more simply as 000. Likewise, a register of four qubits all set to 1 would be written |1111>. A qubit register can represent in binary any number, in much the same way as does a conventional computer register. So the quantum state |1001>, for example, would represent the number 9 because 1001 is the binary for 9.

So far so good. How do we represent superpositions? For a single qubit, we write |0> + |1> to represent one particular superposition of |0> and |1>. This is *not* the same as the numerical addition of 0 + 1. For polarized light, |0> + |1> represents light polarized halfway—that is, 45 degrees—between the horizontal and the vertical. For light polarized at other angles, it turns out we need to add the two components in different amounts. This is achieved by multiplying each term by a weighting factor, or *amplitude.*[4] Think of the idea of mixing colors. If you mix red, blue, and green light in the right properties, you get white light. If you change the strength of any of the colors, the light takes on an infinite variety of different hues. Similarly for quantum states: By varying the amplitudes of the two states |0> and |1>, we can create an infinite palette of different superpositions.

The general state $a|0> + b|1>$ represents a superposition in which the amplitudes are a and b. If a beam of photons in the (pure) state of $a|0> + b|1>$ is passed through a horizontal filter, it would be tempting to suggest that the number of photons that pass through would be proportional to the amplitude a. But as we saw in chapter 3 (page 89), probability is related to the *square,* or more accurately, the *absolute square,* of the amplitude a. Likewise, the probability of passing through a vertical polarizer is given by the absolute square of b.

An important feature of quantum amplitudes is that they can be negative or more generally complex.[5] Furthermore, when two quantum states overlap, the rule is that you add amplitudes rather than probabilities. These two facts have profound consequences.

To understand why, it's helpful to know that whenever you calculate the probability of something happening in different independent ways, you always add probabilities. For example, if you bet on several horses in a race, the probability of winning anything is the sum of the probabilities for each horse winning individually. Likewise for roulette. Placing bets on several different numbers boosts your chances of winning because the probabilities for each number appearing all add up.

If the quantum rule of adding *amplitudes* instead of probabilities applied to horse racing or roulette, you'd certainly think twice before betting on different horses or numbers simultaneously, because it turns out that this rule enables probabilities to *subtract* from one another rather than always adding up. The rule, after all, explains how two wave fronts can overlap with one another to produce destructive interference. It enables "probabilities" along different paths, in effect, to negate one another rather than adding up. Without negative amplitudes, for example, there would be no dark areas on the interference patterns seen in the two-slit experiment. It is this quantum mechanical ability of different pathways to cancel one another out that is the cause, some say, of all the trouble with understanding quantum mechanics.[6]

Well, it's kind of hard to imagine what a horse with a negative amplitude might look like, but what meaning does a negative amplitude have in the case of polarization? Particles in the state $-|0>$ actually behave for most purposes exactly the same as particles in the state $|0>$. Both will pass through the same orientation of polarizer because when you square the amplitudes the minus sign disappears. However, the state $|1> - |0>$ is clearly different from $|0> + |1>$. $|1> - |0>$ has a polarization 45 degrees away

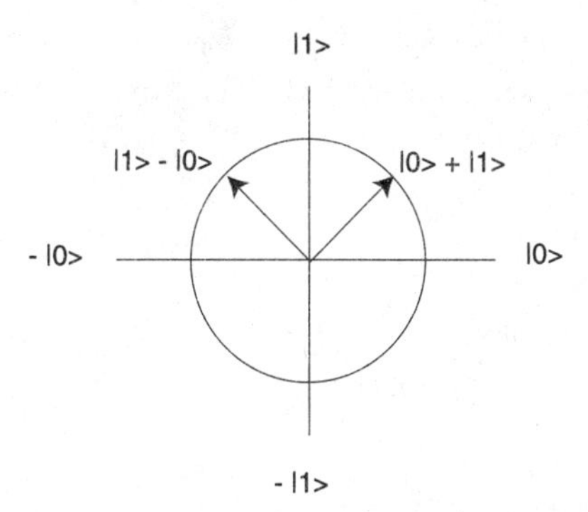

Figure 4.1. Vector representation of qubit states

The state of qubit, $a|0> + b|1>$, can be represented by a vector reaching any point on a circle if the amplitudes a and b are real numbers. For qubits based on polarization, the vector points in the same direction as the polarization. To represent complex amplitudes, it is necessary to use a three-dimensional representation (see page 130).

from the horizontal but facing in the opposite direction to that of |0> + |1> (see Figure 4.1). The negative sign in front of the |0> is sometimes referred to as the *phase.*

The relationship between this mathematical representation of quantum states and polarization (Figure 4.1) suggests a geometric interpretation using *vectors* (mathematical entities that specify size and direction) that is applicable regardless of the underlying physical system. Vector techniques, and their generalizations to higher dimensions called *tensors,* are widely used in the mathematical treatment of quantum systems. However, we won't need to worry about that here.

The Square Root of NOT

In a quantum computer one of the most useful operations is to be able to put qubits into superpositions of 0 and 1. The reason becomes clearer when you consider a quantum memory register. For example, let's take a register with just two qubits. Let's assume each qubit is initially set to 0, so in binary the register reads 00. If we somehow put both qubits independently into a superposition of 0 and 1 of the form |0> + |1>, what number is now represented by the total contents of this register? The answer is all possible numbers that can be held by two classical bits: 00, 01, 10, 11. Why? If you measure the content of each qubit individually, you will randomly see either 0 or 1. The result for each qubit is independent of the outcome for the other qubit. So the result is rather like tossing two coins. If you were able to repeat the experiment many times (by setting up the superpositions many times), you would see all possible combinations of 0 and 1. Seen from the many-universes perspective, the two qubits hold each of these numbers in four different universes.

We can write this more succinctly as follows. If two qubits are each independently set in the superposition $|0> + |1>$, we can write the whole state as:

$$\underbrace{(|0> + |1>)}_{\text{qubit 1}}\underbrace{(|0> + |1>)}_{\text{qubit 2}} \quad or \quad |00> + |01> + |10> + |11>$$

The second version makes it explicit that if we measured two photons polarized at 45 degrees using a horizontal or vertical polarizer, we would see at random any one of the four results:

↔↔ ↔↕ ↕↔ ↕↕

If we did the same with three qubits, putting each qubit in the superposition $|0> + |1>$, we would get all possible numbers 000, 001, . . . 111. There are eight possible combinations here (spanning, in effect, eight universes). Each time we add another qubit, the number of possibilities doubles. So 10 qubits can hold 1,024 different numbers simultaneously, and 100 qubits can hold approximately 10^{30}, or 10 followed by 30 zeros. On a classical computer 10^{30} numbers would occupy 100 million trillion gigabytes. Such calculations begin to hint at the awesome power of a quantum computer.

So how do we change the state of a qubit from 0 to a superposition of 0 and 1? The answer depends on the particular physical implementation of our qubit. If the qubit were represented by the polarization of light, it would simply require the polarization to be rotated by 45 degrees. Such a rotation can be brought about merely by passing the light through a solution of sugar. Sugar molecules can do this because, being biologically produced, they are right-handed. Although left-handed sugar molecules can exist, biology makes use only of left-handed amino acids and right-handed sugar molecules. Such handedness makes these molecules "optically active" in that they affect the way polarized light passes through.

For practical systems polarized photons would be fine for transmitting quantum information, but they're so fast moving it's hard to see how they could be stored for long enough to be of use in a memory register. Another possibility we saw briefly in Chapter 1 is to use excited atoms. The electrons around the nucleus of an atom naturally occupy fixed energy levels. Using photons of particular energy, it's possible to excite these electrons into higher levels. This works best for atoms in which there is a solitary electron in the outermost orbit of the electron cloud that surrounds the nucleus. The rubidium atom, for example, has a lone electron in its outermost orbit that can be excited into a higher orbit by shining a particular kind of laser light for a set period of time.

We noted before that if you shine the light on a ground state atom for only *half* the time, a very interesting thing happens: The electron goes into a superposition of both ground state and excited orbits. If you then apply another half pulse, the electron goes from the superposition to the excited state. So we now have an operation that, when it is applied twice in succession, is equivalent to a NOT function. In logic circuits, when a signal passes through two successive logic gates with logic functions X and Y, the overall effect is written as the product of X and Y—that is, X × Y. If, for example, you apply a NOT operation twice, you get NOT × NOT. Repeating a NOT operation actually returns you to where you started, so in logic it's possible to write NOT × NOT = 1 because multiplying your initial logic state by 1 clearly leaves it unchanged. By similar reasoning, the half-pulse operation is called the square root of NOT because when you apply it twice, you get a NOT operation. Symbolically this can be represented as:

$$\sqrt{\text{NOT}} \times \sqrt{\text{NOT}} = \text{NOT}$$

Initially the idea of $\sqrt{\text{NOT}}$ looks like something out of the same twilight zone inhabited by *i,* the square root of -1. Perhaps so, but mathematicians and scientists have lived with *i* long enough now for it to be regarded as a fully respectable number even in the cold light of day. Despite the unfortunate label "imaginary," *i* is just as much a real entity as numbers like 1, 2, and 3 (though granted it's hard to envisage the use of imaginary numbers in such tasks as counting sheep or money). Much the same can be said for the square root of NOT. Although in its incarnation as a 45-degree rotation of polarized light it doesn't appear to be particularly mysterious, it nevertheless has some odd properties. In a article published by the *American Scientist* magazine entitled "The Square Root of NOT,"[7] Brian Hayes pointed out that $\sqrt{\text{NOT}}$ has the ability to randomize a signal like 0 or 1 because the resulting superposition behaves randomly when subject to a measurement. What you get when you look at the superposition 0 and 1 is a definite answer of 0 or 1—not both or something in the middle. But the answer is completely random. However, when the $\sqrt{}$NOT operation is repeated, the signal is restored to a definite predictable state again. As Hayes says, "It is as if we had invented a machine that first scrambles eggs and then unscrambles them."

Having established the importance of this $\sqrt{}$NOT operation, we can imagine applying it to a string of rubidium atoms set up in their ground state. It would, in principle, be a fairly straightforward matter to put them all into a superposition by shining light of the right frequency and duration onto each of them in turn, or perhaps all of them together.

We will examine realistic hardware implementations in more detail later. The important point for now is that producing a superposition presents no particular problems. Where there are difficulties, as we will see, is in preventing the collapse of the superpositions, because they turn out to be exceedingly fragile.

Rotations in Quantum Space

Exotic though the square root of NOT is, we need some extra gadgets in our logic toolbox if we are to put together a useful quantum program. In terms of polarization, we saw that the effect of the square root of NOT was equivalent to a 45-degree rotation. There are, of course, many other possible rotation angles one might want, ranging from 0 degrees to 180 degrees (90 degrees is equivalent to a classical NOT operation, while 180 degrees brings us back to where we started). Such rotations could be achieved in an atom by varying the length of the laser pulses we use to excite the outer electron. The square root of NOT function and also the classical NOT function both then become special cases of quantum state rotation.

Now, it turns out that there are other types of rotations as well. To understand these, it's helpful to consider the underlying hardware implementation in terms of spin orientations instead of polarizations. When considering spin orientations, the angles are effectively doubled because spin axes pointing up and down are analogous to horizontal and vertical polarizations. So the square root of NOT rotates a spin axis by 90 degrees instead of 45 degrees (see Figure 4.2).

It becomes apparent that there are additional kinds of rotations when we consider spin orientations in three dimensions rather than just two. A

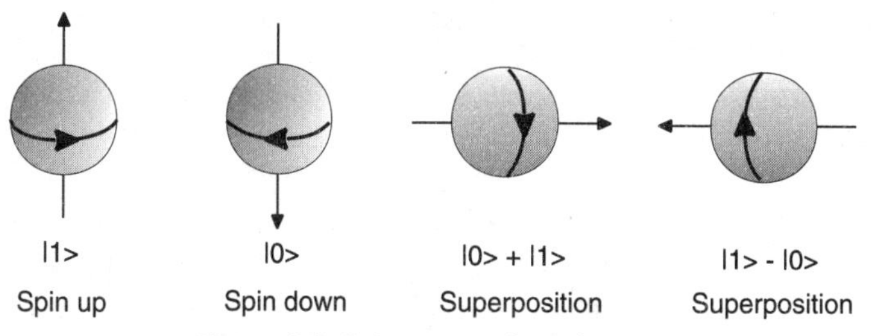

Figure 4.2. Spin states of an electron

When the spin of an electron is measured, no matter how the apparatus is oriented, it appears to point in either one of two directions—up or down with respect to the apparatus. If the spin of an upwardly polarized electron is measured horizontally, the electron will be seen to point at random one way or the other along the horizontal. A horizontal spin orientation can be considered as a superposition of up and down. In many situations an electron may not have *any* definite spin orientation until it is measured.

spin can, after all, point in any direction in three dimensions. We can represent the totality of directions by a sphere in which spin up points to the north pole and spin down points to the south pole (see Figure 4.3). Now when a spin down is rotated by 90 degrees, it ends up pointing toward the equator. The superpositions $|0> + |1>$ and $|1> - |0>$ are therefore on opposite points of the equator.

However, there are many other points on the equator, so if one is $|0> + |1>$ and its opposite is $|1> - |0>$, where are all the rest? This is where complex numbers come in. We can represent any point on the sphere by $a|0> + b|1>$, where a and b are complex numbers. So $|0> + |1>$, $|1> - |0>$, $i|0> + |1>$, $i|0> + |1>$, for example, are all points on the equator and are all valid superpositions. It is apparent from this that the operation the square root of NOT can take many forms: Each would transform a spin pointing north to different points on the equator. In every case, though, a second application of the same operation will rotate the spin to the same point—that is, the south pole.

Another variation on this theme is an important rotation called the *Hadamard transformation.* This is very similar to the square root of NOT in certain aspects but has a different geometric interpretation. Imagine an axis of rotation pointing at 45 degrees to the plane formed by the equator (see Figure 4.4). If a spin pointing upward is rotated around this axis by 180 degrees, it will point toward the equator. If it is rotated a further 180

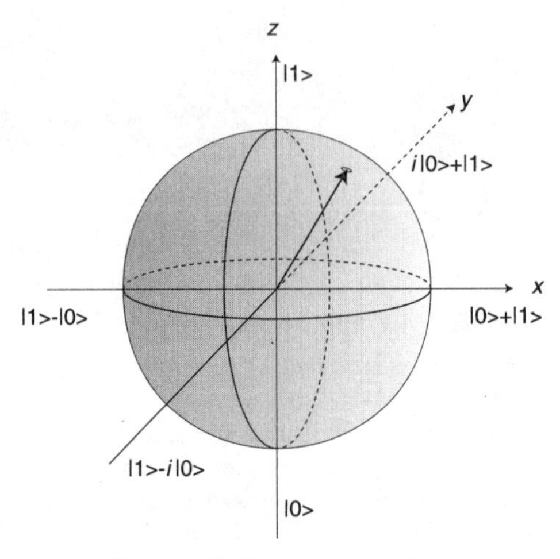

Figure 4.3. Quantum rotations

The spin of an electron can point in any direction within three-dimensional space. Imaginary numbers are used to represent spin vectors in the third dimension (that is, pointing toward or away from us). This explains why quantum amplitudes are generally represented by complex numbers. Quantum operations that change the spin orientations from one direction to another are described as rotations. Such rotations, which include the square root of NOT and NOT, are one of the basic building blocks of any quantum program.

degrees, it points upward again. This 180-degree rotation around a slanting axis is the Hadamard transformation.

The Hadamard transformation has the effect of rotating a spin up or spin down into a superposition, but when applied again, it transforms the superposition back into its original state:

$$\begin{array}{ccccc} & \textit{Hadamard} & & \textit{Hadamard} & \\ |0> & \rightarrow & |0> + |1> & \rightarrow & |0> \\ |1> & \rightarrow & |0> - |1> & \rightarrow & |1> \end{array}$$

Contrast that with the behavior of the square root of NOT:

$$\begin{array}{ccccc} & \sqrt{NOT} & & \sqrt{NOT} & \\ |0> & \rightarrow & |0> + |1> & \rightarrow & |1> \\ |1> & \rightarrow & |1> - |0> & \rightarrow & -|0> \end{array}$$

As we'll see shortly, the Hadamard transformation is widely used in quantum algorithms as a means of putting qubits into superpositions.

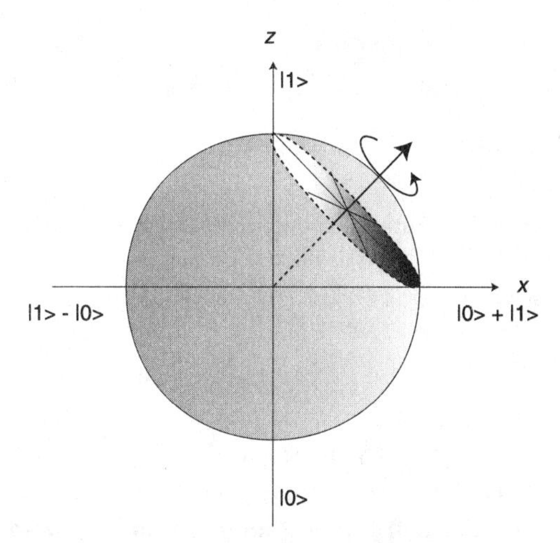

Figure 4.4. The Hadamard transformation

The Hadamard transformation involves a 180-degree rotation around a slanting axis. This rotation is widely used as a form of square root of NOT, though strictly speaking it isn't quite the same. Both have a similar effect, except that applying a Hadamard transformation twice restores the quantum state to its initial state, whereas two applications of the square root of NOT perform a NOT operation.

Controlled-NOT and the Toffoli Gate

So far we have examined quantum rotations, operations that affect only one qubit at a time. Such operations can be regarded as single-bit quantum gates, so the Hadamard transformation would typically be represented as a Hadamard gate, or H gate for short. Now, if you tried to build a conventional computer out of nothing but single-bit gates (such as NOT gates), you wouldn't get very far at all. Although single-bit quantum gates are more versatile than their classical versions, we still need some logic that operates on more than one qubit if we are to do something truly adventurous in the quantum world.

We saw in Chapter 2 how the two-input, single-output NAND gate was sufficient as a building block to construct a classical computer but that for a reversible computer we needed to upgrade to a three-input, three-output gate such as the Fredkin gate to achieve computational universality. Given that quantum mechanics is fundamentally reversible, we might expect that we would need to use quantum versions of reversible gates such as the Fredkin gate.

c	t	c'	t'
0	0	0	0
0	1	0	1
1	0	1	1
1	1	1	0

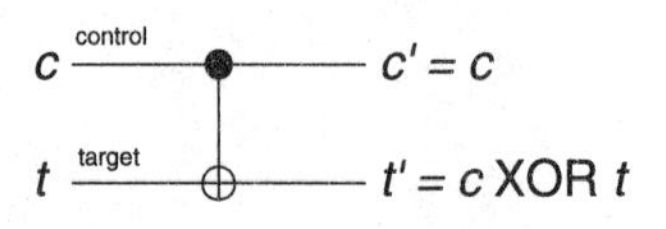

Figure 4.5. Truth table and circuit symbol for XOR, or controlled-NOT, gate

The truth table for the classical exclusive-OR, or controlled-NOT, gate shows that the control input is left unchanged by the gate whereas the target is conditionally flipped if the control bit is 1.

In his 1984 model of quantum computation, Richard Feynman took his cue not from the Fredkin gate but from reversible logic elements proposed by Fredkin's colleague Tom Toffoli. Toffoli had considered ways of implementing AND and XOR gates in a reversible form because together these gates are classically universal. XOR is the *exclusive-OR* gate and obeys the rule that the output is 1 only if the two inputs are different, that is, 0 and 1. If the inputs are both 0 or both 1, the output is 0. The reversible version of XOR simply extends one of the inputs to the output.

The XOR gate can be understood the following way: *c,* the control input, goes through unchanged, while *t,* the target input, is flipped if *c* is 1. So the XOR gate is also called *controlled-NOT* because the second output is *t* if *c* is 0, and NOT *t* if *c* is 1. Useful though this gate is, the classical version isn't universal.

For classical universality you also need Toffoli's version of reversible AND, which is officially known as the Toffoli gate but is sometimes also referred to as *controlled-controlled-NOT* (see Figure 4.6).

With the Toffoli gate, the first two inputs go through unchanged while the third, the target *t,* is XORed with the AND of the first two, which means *t* is flipped if and only if the AND of c_1 and c_2 is 1. This means that of the eight possible input states, the first six in the table leave *t* unchanged while the last two entries are switched.

Since this gate combines the function of both XOR and AND, it is a universal reversible gate in its own right, making it an alternative to the Fredkin gate. (The reason Fredkin didn't consider this gate for his conservative logic scheme was that it didn't satisfy his additional constraint of conserving the number of 0s and 1s in the inputs and outputs.)

As it happens, neither the Toffoli gate nor the Fredkin gate is universal

c_1	c_2	t	c_1'	c_2'	t'
0	0	0	0	0	0
0	0	1	0	0	1
0	1	0	0	1	0
0	1	1	0	1	1
1	0	0	1	0	0
1	0	1	1	0	1
1	1	0	1	1	1
1	1	1	1	1	0

c_1 control $c_1' = c_1$

c_2 control $c_2' = c_2$

t target $t' = t \text{ XOR } (c_1 \text{ AND } c_2)$

Figure 4.6. Truth table and circuit symbol for the Toffoli gate.

in the quantum setting, but in 1989 David Deutsch showed that a three-qubit gate that was mathematically closely related to the Toffoli gate was universal.[8] Later it was realized that, unlike in classical computation, we don't actually need *any* three-input gates such as the Toffoli gate or Fredkin gate for quantum computational universality. This was a big surprise when it was discovered—a subject we'll come back to in Chapter 7—but quantum versions of both gates still turn out to be important for reasons we shall see later on. For now, however, we will need the two-bit controlled-NOT gate and its quantum mechanical equivalent.

The quantum controlled-NOT behaves in essentially the same way as its classical progenitor when presented with 0s and 1s, but being a quantum device, its inputs and outputs can each be put into a superposition of states. This additional capability confers remarkable new powers on the quantum controlled-NOT gate.

Let's see how. Suppose the control input c is in a superposition $|0\rangle + |1\rangle$. What happens to the target t? Does it get flipped or not?

The answer is that it does both. If t starts out in the state $|0\rangle$, it's tempting to say that it finishes in the state $|0\rangle + |1\rangle$, but that's too simplistic. The state of t certainly does finish in a superposition, but the superposition is dependent on the superposition held in c. The quantum states of c and t actually become *entangled*. The way such entangled states are denoted is to write down the complete state for both c and t:

$$\left.\begin{matrix} c = |0\rangle + |1\rangle \\ \\ t = |0\rangle \end{matrix}\right\} \xrightarrow{\text{controlled-NOT}} \underbrace{|00\rangle + |11\rangle}_{\textbf{entangled state}}$$

The state $|00> + |11>$ is a superposition of states in which c and t are either *both* $|0>$ or *both* $|1>$. What this means is that if we measured the outputs and repeated the experiment many times, we would *always* see either 00 or 11 and *never* 01 or 10. This idea becomes more concrete if you imagine what is happening in parallel universes (even if you don't subscribe to that interpretation). In one universe c is 0 and the value of t is unchanged by the gate, so the outputs are $c = 0$ and $t = 0$. In another universe c is 1 and t is flipped by the gate, so that the output is $c = 1$ and $t = 1$. Note that it would be incorrect to write the outputs as:

$$c = |0> + |1>, t = |0> + |1>$$

These superpositions are *not* entangled. The whole state, in this latter case, is equivalent to:

$$(|0> + |1>)(|0> + |1>) \quad \textit{or} \quad |00> + |01> + |10> + |11>$$

This is the state we got by independently setting two qubits as superpositions (see page 126). This latter state includes $|01>$ and $|10>$, states that do not arise from the controlled-NOT gate with the inputs we have considered. One definition of entanglement is to say that when the state of two qubits cannot be factored into two separate and independent states, such as $(|0> + |1>)(|0> + |1>)$, they must be entangled.

The entangled quantum state $|00> + |11>$ is actually the very state we examined in the EPR experiment. Thus, using only two quantum gates, a quantum version of a controlled-NOT gate and an H gate (to create the initial superposition for the control input c), it is very easy to create entangled EPR states with all of their attendant properties. Entanglement, as we have noted before, is the key to the power of quantum programming, and with these gates' rotations and the controlled-NOT gate we now have sufficient functionality to do universal quantum logic.

Implementing the controlled-NOT gate in hardware is trickier than the single-qubit rotations we have examined because it depends on getting a physical interaction between two qubits. However, there are several schemes for achieving the appropriate interaction, which we will examine in Chapter 7.

Playing the Markets with a Quantum Computer

The first quantum program to demonstrate the potential of quantum methods over classical was proposed by David Deutsch in his 1985 paper on quantum parallelism. It concerned the calculation of a mathematical function $f(x)$. Suppose, Deutsch asked, you wanted to know whether this function f took the same value for two different input values, $x = 0$ and $x = 1$. Now, because this was only a toy example, not only was the input restricted to the two values 0 and 1, so also was the output. The function thus operates on a single qubit (although it requires a second qubit to perform the calculation). Despite the limited range of the values f can take, it could still be a very complicated function to evaluate. At its simplest, the question "Will it rain this afternoon?" has only one of two answers, "yes" or "no," but the program to compute the answer would probably be very complicated.

Instead of the weather, Deutsch invited us to imagine a function predicting tomorrow's stock exchange movements. Now, suppose it took 24 hours to work out f for each value of x and you needed to know for your investment strategy whether the two values of the function—$f(0)$ and $f(1)$—gave the same answer, without necessarily needing to know what the answers were. The scenario could be that you will buy some stock in a company, but only if two economic indicators $f(0)$ and $f(1)$ agree with each other.

On a classical computer you would need to do two calculations (one each for 0 and 1) and compare the answers to find out. Unhappily, these two computations would take 48 hours to complete—hardly useful for tomorrow's investment decisions. Deutsch showed how you could cut the computation time to 24 hours using quantum parallelism.

An outline of the procedure is as follows. Imagine you have written a program for calculating f. The program will take as its input x a number placed in one qubit of a memory register. In a quantum computer it is possible to load the memory register with 0 and then "rotate" the state into a quantum superposition of 0 and 1 using a Hadamard gate.

The program is now able to operate on 0 and 1 *simultaneously,* provided it processes the numbers quantum mechanically and preserves the quantum superpositions. The program will then grind its way through the calculation, finally producing 24 hours later an answer that is a quantum superposition of $f(0)$ and $f(1)$. To read the answer, we need to make a measurement of the output. However, making a measurement "collapses the wave function," so that we would only get one of the answers, $f(0)$ or $f(1)$. That would be no better than a classical computer. Deutsch showed that if you could get these numbers to *interfere* with one another, you could read off an answer to whether $f(0)$ equaled $f(1)$ immediately. The interference, it turns out, is possible to achieve with two extra Hadamard rotation operations.

In his 1985 paper Deutsch said very little about how one might actually implement any of his quantum programs. As a result they were presented in a rather abstract mathematical form. However, in 1989 he made his ideas easier to interpret physically by introducing the notion of quantum logic gates, which could be wired together to create a quantum circuit or network. The wiring was only metaphorical because, in practice, the network would be realized by performing a sequence of quantum operations on a set of qubits. The same qubits would carry the input data at the beginning and the output data at the end.

A glimpse of what form a circuit might take (see Figure 4.7) shows one way of solving the Deutsch problem. (For an explanation of how the circuit works and how it can be improved, see Appendix D.)

So here was an example of a program that could run faster by exploiting quantum parallelism. The increase in speed here was only a factor of 2, and even then the quantum program didn't guarantee to find the answer every time. In fact, the algorithm has only a 50-50 chance of success, which means that half the time you can't act on its calculation. But the program at least told you when it had failed, so you could still use it for playing the market on those days when it did succeed—something that would not be possible with a conventional computer. And as later transpired, it is possible to modify the program so that it works *every* time (see Appendix D).

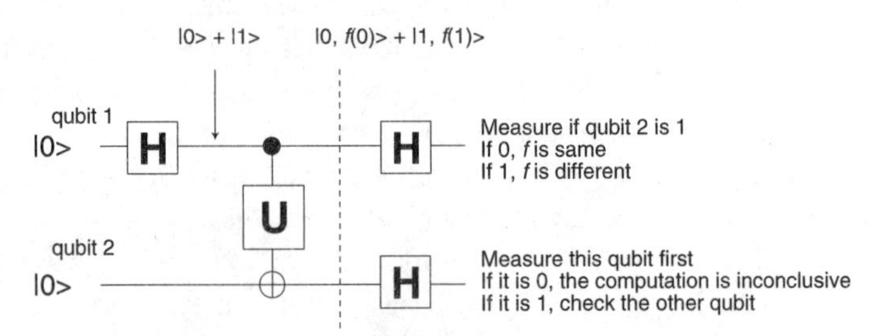

Figure 4.7. Circuit for Deutsch's problem

The circuit is read from left to right so that the inputs are both |0>. The boxes marked **H** are Hadamard transforms, and **U** is the main processing element that calculates $f(x)$, where x is the value on qubit 1 and f is the stock market function we need to calculate. The output of **U** is placed on qubit 2.

Turbocharged Algorithms

Remarkable though Deutsch's original quantum program was, it was neither particularly useful nor amazingly fast. Despite the stock market scenario, it was hard to see how such an algorithm could be applied to real-world problems in practice. Furthermore, with computers doubling in computational power nearly every 18 months, the "speedup" offered by these quantum programs looked positively puny. However, from the many-universes perspective, his algorithm exploited only two parallel universes, so even if you didn't accept that view, the program still had plenty of unused capacity. What was needed to demonstrate the real potential of quantum parallelism was an algorithm that took advantage of a much larger number of parallel universes.

Such programs were noticeable by their absence in the years following Deutsch's 1985 paper. Although he had provided evidence that quantum algorithms could be more powerful than classical algorithms, the feeling grew that perhaps the capabilities of quantum computers would remain confined to a very limited range of arcane mathematical problems. If so, quantum computing might have remained little more than an intellectual curiosity.

The ice began to break in 1992 when Deutsch and Richard Jozsa, also then at Oxford, showed how a quantum computer could solve a more general version of the problem addressed by the previous algorithm.[9] Jozsa, a highly competent mathematician, had spotted a flaw in one part of Deutsch's 1985 quantum parallelism paper, in what purported to be a proof that no quantum algorithm for computing functions could operate *on average* faster than a classical algorithm. Much to Deutsch's delight, Jozsa showed that there was no such restriction on quantum computation. Deutsch had already developed the basic idea for a more general algorithm that would be faster than any classical algorithms, but had been reluctant to publish it because it plainly violated his own earlier claim. Jozsa not only showed that the claim was incorrect but also demonstrated how the algorithm could be improved.

In this new version of the quantum algorithm, instead of being limited to the values 0 and 1 at its input, the function f was allowed to take any integer input value. The question to answer was whether f was *constant* as in the stock-market example, or *balanced.* By "balanced," we mean that for all possible inputs, f gives 0 as its answer as many times as 1. Say, for example, f was allowed up to four input wires (or qubits). It could then take any number between 0 (=0000) and 15 (=1111) as its input. Suppose the function f simply determined whether the input was greater than or equal to 8 (=0100); f would then be *balanced* because inputs 0 to 7 give the answer no and inputs 8 to 15 give the answer yes, and the yes and no answers would therefore be equally represented.

Deutsch and Jozsa showed how a quantum computer could distinguish between constant and balanced functions very rapidly. The solution in the form of a circuit is actually very similar to the circuits for the original Deutsch problem, though that only became clear much later, with the unifying work of Deutsch's colleague Artur Ekert and some of his students (see page 298). Again, this particular problem was rather contrived because it's hard to think of any real application for the solution. Nevertheless, the importance of this work by Deutsch and Josza was that it established for the first time that a quantum computer could solve a problem *exponentially* faster than a classical machine.

What does that mean? To solve the problem classically you might need to calculate f for all the possible values of the inputs and add up the number of 1s and 0s to see whether they balance. The solution on a quantum computer, on the other hand, can be done in one pass by using a superposition of all possible inputs. In the case of our previous example function, the quantum calculation would be up to 16 times faster than the classical calculation. If you consider functions with n input wires, the increase in speed will be up to 2^n, which is an exponential speedup. if $n = 100$, for example, the increase in speed will be approximately 10^{30}, which is a million trillion trillion.

Tractability vs. Intractability

The issue of exponential speedup is particularly important in the context of a branch of computational science known as *complexity theory.* Here the issue is to understand which problems are solvable in a reasonable amount of time on a computer, a question with very important practical implications. What do we mean by "reasonable"? We already briefly met this question when examining Feynman's arguments over why it was hard to simulate quantum physics on a classical computer. The problem there was that the task became exponentially harder the larger the system we wanted to simulate.

Ideally, we don't want computer programs to take too long solving larger problems. This statement can be made more precise by a classification that was devised in the 1960s. A problem of size N is said to be *tractable* if its solution takes a length of time that depends on a *polynomial* function of N—that is, an algebraic power of N such as N squared, N cubed, and so on. The computational resources required for tractable problems generally scale with the numbers in a moderate way. If, on the other hand, the time taken to solve the problem blows up exponentially with the size of the input—for example, of the order of 2^N or greater—then the problem is *intractable.* See, for example, the behavior of polynomial and exponential functions plotted in Figure 4.8.

Computer scientists have devoted considerable energy to trying to discover which problems are tractable and which intractable. This has turned out to be a very rich area of study, although getting definitive answers has been notoriously difficult. There is, for example, a famous class of problems known as *NP-complete,* or *NP-hard,* which are *thought* to be intractable; no one, however, has actually managed to prove them so.

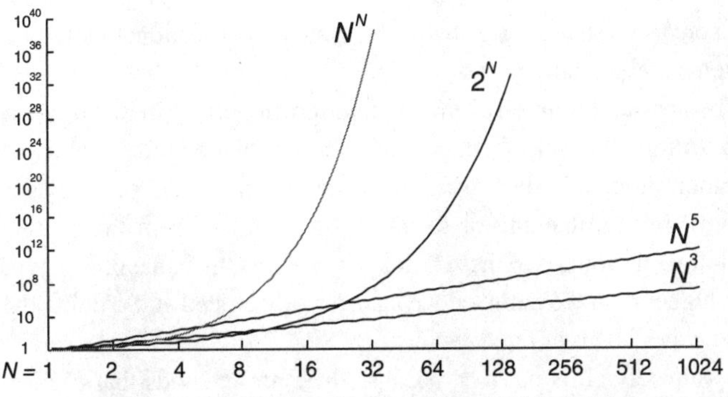

Figure 4.8. Polynomial and exponential functions

The difference between the behavior of polynomial and exponential functions as N increases is exhibited in the graph above. The polynomial functions N^3 and N^5 increase relatively gradually compared with the exponential functions 2^N and N^N. Problems that take a polynomial amount of time are defined as tractable, while problems that require an exponential amount of time are intractable.

The Traveling Salesman Problem

The classic example of an NP-complete problem is the conundrum faced by the traveling salesman. A salesman has to visit a number of cities, each only once but to save time and money he wants to minimize the amount of traveling by calculating the shortest possible route. Surprisingly, given how simple the problem is to state, no one has found a surefire method of producing the answer on a computer in a reasonable (that is, polynomial) amount of time.

The obvious method is to consider every possible route and then see which is the shortest. For a small number of cities and roads, this is not an unreasonable task. With five cities, for example, there are twelve different ways of doing the tour. However, with longer itineraries, the number of possibilities rapidly increases. With ten cities there are 181,440 possibilities. With twenty-five cities the problem becomes truly daunting; with a computer that could search through a million routes per second, the problem would take 10 *billion* years to solve, roughly the age of the universe. The size of the problem increases in proportion to $(N-1)!$, the factorial of N-1. The factorial function ($N! = N \times (N-1) \times (N-2) \ldots 3 \times 2 \times 1$) actually increases much faster than an exponential function like 2^N, so this problem is really hard.

The absence of an efficient solution has important implications not so much for traveling salesmen, perhaps, but for a wide range of practical optimization problems, such as in the design of printed circuit boards or microchips, in the planning of telephone networks, and in the routing and scheduling of aircraft. In these kinds of problems the value of N is typically much higher than the numbers we have already looked at. So doing exhaustive searches is out of the question.

Fortunately all is not lost, because there are methods that can shortcut the search if we allow the rules to be relaxed somewhat. For instance, there are so-called *approximation algorithms,* which can produce solutions that are less than perfect yet still of considerable practical value. One algorithm, for example, runs in cubic time (which is polynomial) and produces a tour that is guaranteed to be no longer than one and a half times the length of the optimal tour.[10]

Another method is to introduce *randomness* into the search. The idea is to get the computer to start by making guesses at solutions with which it then tinkers to see how far they can be improved. Once again there is no guarantee that the solution will be optimal, but on average it can be very good. In 1994 researchers at British Telecommunications Laboratories in Martlesham Heath solved the traveling salesman problem for 35,000 cities by taking advantage of these ideas using a technique based on a combination of genetic algorithms and neural networks.

Neural networks are computer simulations of groups of nerve cells, which can be trained or programmed to learn patterns of behavior. The randomness comes from the genetic algorithms, which imitate the process of Darwinian natural selection by "breeding" new solutions from old and then selecting the most promising newcomers. The reproductive process is a computer simulation that relies on solutions to the problem being coded in the form of genes. These genes are then subject to random mutation and recombination. It is the controlled use of randomness that can make genetic algorithms so powerful.

Does P Equal NP?

Algorithmic techniques that rely on randomness to help them along are called *nondeterministic*. In the 1960s, when computer scientists first began to realize that nondeterministic algorithms could be more powerful than conventional algorithms, they found it helpful to consider a special "magical" form of nondeterminism. With this resource, whenever a program needs to make a random choice, it always chooses the best option, even if there is no obvious way of deciding which really is the optimal choice until you've seen the final solution. Another way of looking at this idea is to imagine a computer with an unlimited number of parallel processors so that whenever a program meets a branch point, it can follow all possible computational paths simultaneously.

Armed with this fictitious form of algorithmic device, researchers found that certain problems, like the traveling salesman problem, could be solved in polynomial time. Of course, that didn't mean these problems were tractable after all. Rather it meant that they would be tractable if only we had access to this "intelligent" kind of randomness.

The class of problems amenable to magical nondeterminism is known as NP, where N stands for nondeterministic, and P for polynomial. They include a wide range of optimization problems, such as sorting items into order, constructing timetables for teachers and students, packing objects into bins, and working out optimal routes for salesmen, couriers, electronic circuits, telephone networks, and so on. An interesting feature of NP problems is that, though they are hard to solve efficiently, answers are easy to check. For example, solving a jigsaw puzzle, which is an NP problem, is hard, but verifying a candidate solution is easy. We can do it at a glance. A solution to the traveling salesman problem might not appear to be so easy to check, but it depends on how the original problem is phrased. The correct expression of the problem from this point of view is to ask whether there is a route less than a certain specified distance. If the answer is yes, it's such and such a route, we can check the length of the route against this criterion very easily. Of course, this route may not be the optimum route, but we can progressively lower the distance threshold until we think we're close to the limit.

Problems that are genuinely tractable (that is, they can be solved with deterministic algorithms in polynomial time) are said to belong to the class P. These are easy problems, such as looking up words in a dictionary or finding someone's number in a telephone directory. An important question arises over the precise relationship between P and NP. Clearly all problems in P also belong to NP because deterministic algorithms are simply a special case of nondeterministic algorithms. But what about the other way round? Is there any way NP problems can be solved using deterministic polynomial algorithms? On the face of it, the NP problems are all sufficiently hard for people to think the answer must be no. But so far no one has proved it, making this a tantalizingly unresolved problem. The relationship between the classes P and NP is illustrated in Figure 4.9. P must be contained within NP but, as I'll explain shortly, we can only presume that it is not within NP-complete. Beyond NP there are problems that take exponential time but whose answers, unlike those of NP, cannot be checked in polynomial time.

This may all seem rather odd. After all, why should we get worked up about something that depends on some weird kind of nonexistent nondeterminism? One reason is that the class NP contains a wide set of problems of great theoretical and practical interest. Other complexity classes have been defined to capture other problems, but NP is certainly one of the most important. Another reason for the intense interest concerns the special group of problems known as NP-complete, which, as I mentioned earlier, in-

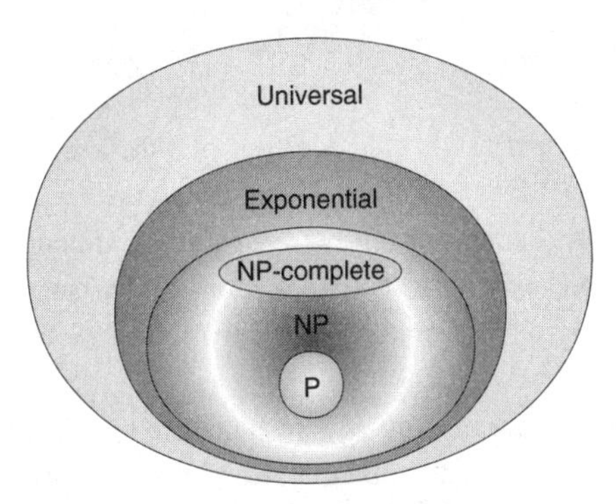

Figure 4.9. Complexity classes

All solvable computational problems can be classified according to their level of difficulty into one of the subsets or classes shown. Those in P require only a polynomial amount of time; NP problems include P but also take in problems that appear to require an exponential amount of time but whose answers can be checked in polynomial time; exponential problems include NP and problems whose answers cannot be checked in polynomial time; the universal class includes programs that have the ability to act like any program whatsoever. Examples of such programs are computer language compilers or interpreters.

cludes the traveling salesman problem. NP-complete problems are the hardest members of the NP class. They share the following remarkable property: If any one of them could be shown to be tractable, then *all* NP problems would be tractable. So if someone were able to produce an efficient deterministic algorithm for the traveling salesman problem, he or she could break open the champagne and conclude that all NP problems, ranging from jigsaw puzzles to bin-packing problems, could be solved much more efficiently than is known today.

The fact that all NP-complete problems stand or fall together (bringing down the rest of NP as well if they fall) is one of the reasons that proving P does not equal NP is so difficult. The problem is that the proof would have to envisage all possible NP algorithms and demonstrate that they were all inefficient. Thousands of NP-complete problems have been studied, but all of them appear to be pretty much equally difficult. According to the mathematician and writer Ian Stewart, the most plausible methods for proving that $P = NP$ or that $P \neq NP$ simply won't wash. His conclusion: "Nobody has the foggiest idea. It's one of the biggest unsolved mysteries in mathematics."[11]

Consulting the Oracle

Given the intense interest in the question of whether P is equal to NP, Deutsch and Josza's discovery in 1992 of an algorithm that showed how quantum computers could apparently transform an intractable problem into a tractable one caused some excitement. Could it be that a quantum computer could transform NP problems into P problems? If so, this truly would be a revolutionary discovery.

Expectations were tempered by the fact that the Deutsch-Josza algorithm, as we saw, concerned a rather strange-looking problem, not one that belonged to the elite class of NP-completeness. What class did it belong to?

Recall that what their algorithm did was to categorize an unknown function f as *either:*

(a) constant (always either 0 or 1)

or

(b) balanced (exactly half the possible output values are 0 and the rest 1).

Actually, the algorithm couldn't even do this properly. All it could do was report that one of (a) or (b) was *false.* Therefore, to conclude anything positive about the function one had to specify that *only* constant or balanced functions would be supplied to the algorithm. If you gave it a function that was *neither* constant *nor* balanced, it would report at random that either one of (a) or (b) was false. It couldn't tell you both were false, so its output in this case would be ambiguous.

The problem of deciding (a) can be done efficiently using nondeterministic methods because you need to find only one exception to prove that f isn't constant. However, deciding (b) is more difficult. You may need to test every value of f before you can rule it out. If N is number of input wires or bits, the size of the computation will increase in proportion to 2^N and is therefore exponential even with nondeterministic methods. So (a) is therefore in NP but (b) is not.

So here was an algorithm that was tougher than NP in one respect and yet on a quantum computer could be solved efficiently—that is, in polynomial time. This showed, incidentally, that the power of the quantum computation could not simply be ascribed to the indeterminism inherent in quantum theory. Deutsch and Josza proposed a new class to capture such algorithms: QP for quantum polynomial time—the quantum analog of P.

Deutsch and Jozsa also pointed out that the circuitry that computed the function *f* could usefully be regarded as an "oracle." An oracle in this context is a kind of hypothetical assistant to the computer that can answer certain types of questions to help the calculation. The oracle is assumed to respond more or less instantly and could simply be a lookup table of values stored in a ROM (read-only memory) chip, or a black box device whose contents remained a mystery.

Alan Turing introduced oracles into computing theory in the 1930s as a way of studying questions that would otherwise be unanswerable or at least very difficult. Turing was interested in certain problems that were provably *uncomputable.* The famous *halting problem* was one such example (see Chapter 3, page 114). Turing proved that it was generally impossible for a computer to be able to predict whether a computer program would halt or run forever simply by inspection. Nevertheless, as an exercise in free thinking, he wondered what a computer would be able to do if we were given access to some magical device, an oracle, that could answer the halting problem for any program with which it was presented. It turned out that such a computer could crack many of the biggest problems in mathematics.

Oracles were later extensively studied in investigations of the question of P = NP. In 1975 three American mathematicians, Theodore Baker, John Gill, and Robert Solovay,[12] showed that if a computer used a certain kind of oracle, it was possible to *prove* that P was the same as NP. In other words, such a machine could solve all of its NP problems efficiently.[13] However, they also proved that if it used another kind of oracle, then P was definitely not the same as NP. According to the authors, the fact that they could get different answers using different oracles was further evidence of the difficulty of the real P = NP question.

After Deutsch and Jozsa's 1992 discovery, other researchers pursued the idea of using oracles in quantum computers further. At that time, though, quantum computers were merely computational thought experiments, so many of the ideas remained pretty abstract. Nevertheless, three papers appeared within the next year that progressively enlarged the scope of what was possible using quantum algorithms. These were by André Berthiaume and Gilles Brassard of the University of Montreal, Ethan Bernstein and Umesh Vazirani of the University of California at Berkeley, and Dan Simon, then at University of Montreal, of Microsoft.

For the most part, these papers consisted of fairly technical developments that we need not explore in detail here. All of them concerned either the original Deutsch-Josza problem in an oracle setting or, in Simon's case, a new variant of the problem. Oracle problems are interesting as far as they go but ultimately not all that useful in providing answers to practical questions. Nevertheless, these papers—Simon's paper in particular—laid the foundations for the biggest and perhaps most revolutionary discovery yet. This was the work of Peter Shor at AT&T Labs, who in 1994 discovered a way of applying a naked quantum computer, stripped of any hint of oracles, to a real and very exciting problem.

5

Code Breaking and the Shor Algorithm

> The additive structure of the whole numbers 1, 2, 3 . . . is a bit too simple to hold any great mysteries, but the multiplicative structure poses problems that still inspire creative work after thousands of years.[1]
>
> IAN STEWART

The Problem of Factorization

No one knows who first described mathematics as the queen of the sciences, but it was the great nineteenth-century mathematician Carl Friedrich Gauss who called the theory of numbers the queen of mathematics. The theory of numbers is concerned with the behavior of the whole numbers 1, 2, 3, and so on. Its beauty stems from its underlying simplicity. Yet from humble beginnings emerges a subject of extraordinary richness and complexity.

Preeminent in this respect are the *prime* numbers, numbers that cannot be obtained by multiplying two smaller numbers. The primes form an infinite sequence, 2, 3, 5, 7, 11, 13, 17, 19, 23 . . . , that has captivated mathematicians since ancient times. The Greeks recognized and proved that every number was the product of a unique combination of primes. Every whole number therefore has its own unique signature composed of prime numbers. But this raises the question of how one can find this signature for any particular number.

To find the divisors, or *factors,* of a number, n, the obvious method is to try dividing it by all the numbers from 2 up to the square root of n. If none of them divide exactly, then n must be prime. Alternatively, if you had a list of primes, you could save some effort by dividing only by primes (up to the square root). Generating prime numbers is not so easy because they cannot be expressed in terms of an algebraic formula. As the number theorist Don Zagier put it, primes "grow like weeds among the natural numbers, seeming to obey no other law than that of chance, and nobody can predict where the next one will sprout."[2]

Nevertheless, the Greeks discovered a method. What you do is write down in sequence all the integers up to n. You then strike out all those that are multiples of 2, then all those remaining that are multiples of 3, and so on until all composite numbers have been eliminated. This method, known as the *sieve of Eratosthenes,* catches in its mesh all the primes up to n.

Simple though they are, these exhaustive techniques for testing primality and for factoring are hopelessly inefficient for large numbers because they take superpolynomial time. If you had a computer that could perform 10 billion trials per second (which is much faster than most present computers can do), a 30-digit number would take about a day to factor, a 50-digit number would take 3 million years, and a 60-digit number would take longer than the age of the universe.

One point to note, incidentally, is that the running time of these algorithms is always compared with the *size* of n in terms of the number of digits rather than its *value.*[3] The simple method of trial division described above grows in proportion to $\sqrt{n}$, which is much slower than the $N!$ we saw in the traveling salesman problem in the last chapter. But if L is the number of digits, then n roughly equals 10^L and the growth rate is $\sqrt{10^L}$ or $10^{L/2}$, which in spite of the square root is still exponential. (An exponential growth rate merely requires the parameter L to appear in the exponent.)

Primality testing and factorization are both in the class of NP problems, which means they *can* be solved in polynomial time using "magical" nondeterminism. However, it is not known for sure whether they also belong to P or, going to the other extreme, to the NP-complete problems, the hardest members of the NP class.

Until the mid-1970s, the fastest algorithm for testing primality was one that grew according to $n^{1/8}$ (the eighth root of n), which like the square root also grows exponentially with respect to the number of digits: for each three extra digits in n, the running time would more than double. But in 1976 Gary L. Miller discovered an algorithm that requires only a polynomial amount of time.[4] The reason we cannot now classify primality testing as belonging to P is that the algorithm depends on an unproven theorem in mathematics known as the *extended Riemann hypothesis.* The Riemann hypothesis poses one of the most famous problems in mathematics because many important results depend on it. But, although most mathematicians think the hypothesis is probably correct, so far no one has found a proof.[5]

Nevertheless, testing for primality appears to be a lot easier than factorization because it is possible to produce powerful tests that say, "yes, this number is prime" or "no, this isn't prime" without actually revealing the factors. In the so-called Miller-Rabin test, for example, this is achieved probabilistically. Whenever the test shows a number to be composite it is always right, but if it suggests the number is prime there is a chance it could be wrong. However, this error rate can be made as small as you like by repeating the test in different ways.

The Miller-Rabin test runs in polynomial time and offers a very high degree of statistical certainty. But a test discovered by Len Adleman and Robert Rumely in 1980 offers a *guarantee.* This test avoids any probabilistic or conjectural fudges at the expense of taking a little longer to run. For a number with L digits the test running time grows according to $L^{\log \log L}$, which is "almost polynomial" because log log L grows very slowly.

In contrast to testing for primality, factorization has proved to be a much tougher problem. Whereas a 1,000-digit number could be tested for primality quite easily in 1999, factoring a 200-digit number was still well beyond the scope of any computer.

Secret Codes

Why should we care about factoring and primality? Apart from the mathematical interest, there is a powerful practical reason. Some of the most important techniques in modern cryptography depend on the fact that it is very difficult to factor large numbers.

Codes and ciphers have played a crucial part in protecting communications throughout history. They date back at least as far as the ancient Egyptians in 2000 B.C. and were used by civilizations in India, Mesopotamia, Babylon, Greece, Rome, and so on up to modern times. In times of strife and warfare the security of codes has often been a matter of life and death. Mary Queen of Scots, for example, lost her life in the sixteenth century because an encrypted message she sent from prison was intercepted and deciphered. Prior to the Second World War, Polish mathematicians fathomed the secret to the Enigma code used by the Germans. Alan Turing and his colleagues, working for the British Foreign Office, built upon this discovery and used some of the world's first digital computers to decrypt messages conveyed by the Germans. This, perhaps the greatest success ever in code breaking, provided the Allied forces intelligence of incalculable value.

In modern times codes and ciphers continue to play a vital role not only in military intelligence but also in protecting information exchanged by governments, banks, commercial companies, and private citizens. With the growth of commercial transactions on the Internet and the spread of digital forms of cash, the security of encryption systems is likely to become increasingly important.

In recent years, the development of secure codes has been transformed by new mathematical techniques and by the use of modern computers, which can do all of the hard work involved in translating messages into and out of code. At the same time, cryptanalysis—breaking codes—has become increasingly sophisticated using the same weapons as those of the cryptographers to expose weaknesses in new encryption systems. The construction of a quantum computer could be one of the most important developments yet in this cryptographic "arms race."

The importance of factorization to cryptography emerged in the late 1970s with a new kind of code called a public-key system. Prior to this, cryptographic systems had invariably depended on a secret rule, or *key,* agreed to in advance by the sender and receiver. One of the simplest examples of such a key is found in the substitution cipher. In it, each letter of a message is substituted by a different letter. Julius Caesar, for example, concealed his communications from the Gauls simply by shifting letters of the alphabet three places to the right. So using this cipher, the message "PLVDVSB" would mean "MISASPY." Such a cipher seems laughably simple today and would be cracked by any self-respecting cryptanalyst in her sleep.

However, substitution ciphers were still in use in the 1930s, albeit with much stronger keys. Typically a key would take the form of a sequence of random numbers, each of which would be "added on" to a letter in the message. So if the letter was an A and the random number was 4, the substitute letter would be an E.

Such private-key systems depend on the sender and receiver being able to agree on a key without anyone else discovering what it is. The most secure system is the *one-time pad,* which was invented in 1917 by Major Joseph Mauborgne and AT&T's Gilbert Vernam. One-time pads have long been used in intelligence work and in high-level intergovernmental communications links, such as the hot line between the White House and the Kremlin in the days of the Soviet Union. The idea is that two parties are issued identical pads of cipher keys. Each time they want to communicate a message, each party uses the key detailed on the top sheet of the pad, which they then tear off and destroy. The next message is encoded using the next encryption key in the pad. Provided the keys are genuinely random, the pads don't fall into the wrong hands, and, as their name suggests, they are used only once, the system is totally foolproof. However, it is rather cumbersome. To be secure, the keys need to be as long as the messages sent, which means having to produce an endless supply of pads. These then need to be distributed to the appropriate parties without being compromised. The number of pads required also increases very rapidly according to the number of people who want to use the scheme.

With the development of electromechanical machines and then computers, it became possible to use much shorter keys to encode long messages without fear of their being broken. The best known of the modern private-key algorithms include the Data Encryption Standard; RC2, RC4, and RC5; IDEA; and Skipjack. The Data Encryption Standard, DES, which was adopted as a U.S. government standard in 1977, quickly became the encryption workhorse of the banking and financial world. It is still the most widely used system in the late 1990s, though its security was breached in 1997 by brute-force attacks carried out by groups of people collaborating on the Internet. Despite such attacks, though, the security of private-key systems can easily be enhanced by using longer keys.

Private-key algorithms, no matter how mathematically secure, still depend ultimately on the safe distribution of keys. In the United States, government agencies used to distribute cryptographic keys to foreign embassies by sending couriers handcuffed to locked briefcases. Such encumbrances in exchanging keys makes these kind of cryptographic techniques impractical for small companies and ordinary citizens. Furthermore, in large organizations there is a major problem in being able to supply enough keys, because you need to have a different key for each possible pair of people who want to talk to one another. The number of possible pairs increases approximately as ½ times the square of the number of people in the organization. That means an organization of 20,000 people, for example, would need around 200 million keys.

Public-Key Cryptography

In the mid-1970s the answer to the problem of distributing keys arrived with the development of a revolutionary new approach to cryptography known as *public-key* systems. The idea behind these systems was beautifully elegant. Instead of relying on a symmetric system in which the same key is used for encryption and decryption, a public-key system uses a pair of *asymmetric* keys. One is the *public* key, which can be widely distributed, and the other is the *private* key, which is kept secret.

The strength of the system relies on the fact that when a message is encrypted using the public key, it can only be decrypted using the private key. The public key therefore enables anyone to encrypt messages, but only the owner of the associated private key can decrypt them.

Let's take an example using Bob and Alice, two fictional characters who have a habit of turning up in discussions of cryptography. Suppose Alice wants to talk to Bob using a public-key encryption system. Bob must first publish his public key. He could do this by sending it directly to Alice, or he could publish it in a newspaper, on a notice board, on the World Wide Web, or anywhere he likes. Alice could then use the key to encrypt a message that she then sends to Bob. Bob then decrypts the message using his private key, which only he knows. Meanwhile, Alice publishes her own public key, which Bob can use to encrypt his messages to her. Only she possesses the private key necessary to decrypt Bob's message.

The attraction of this system is that only public keys need be exchanged. It is as if the public key unlocked the opening to your mail drop, allowing *anyone* to insert items, but the private key lets only *you* get at the contents. Provided you guard the private key, there is no danger of anyone else's reading messages intended for you. One of the surprising features of these public-key systems is that even though the procedure for encrypting messages is completely within the public domain, it is exceedingly hard for anyone to reverse it *unless* they have the private key.

The inspiration for public-key systems came from ideas about so-called trapdoor functions, mathematical functions that are easy to calculate in one direction but hard in reverse unless you have a special key that unlocks a trapdoor. The first published idea on this track was a method proposed in 1974 by Ralph Merkle while he was a graduate student at the University of California at Berkeley. His system involved one party, Alice, sending another, Bob, lots of mathematical puzzles, each containing an encryption key. Bob would then select one of these puzzles at random and solve it to reveal the key. He would then send Alice a prearranged message encrypted using the key. Alice would then try decrypting the message with every key until she found the one that worked. After that Alice and Bob could communicate, encrypting their conversations with the key. A third party, Eve, listening in to their conversations, would have to solve a large number of puzzles to find the key. So provided the puzzles were reasonably hard, Eve would need a long time and a lot of computing power to unscramble Alice and Bob's conversation. Note that the puzzles couldn't be made too hard, since otherwise Bob would have difficulty getting at his one single key.

The idea of a public system for exchanging keys was totally new, so Merkle had trouble getting anybody to take his idea seriously to begin with. But the following year Whitfield Diffie and Martin Hellman of Stanford University wrote a paper called *Multi-User Cryptographic Techniques* in which they outlined how a public-key system might work in principle, although they didn't offer any specific algorithms. After seeing their paper, Merkle sent a copy of his ideas to Diffie and Hellman. The following year Diffie and Hellman announced what was thought to be the first practical public-key system.

The history of cryptography had to be hastily revised in 1997, however, when the British Government published documents showing that the first people to discover public-key systems had actually been mathematicians working at the Government Communications Headquarters, GCHQ, in England. James Ellis, Clifford Cocks, and Malcom Williamson had developed public-key systems several years before the ones that emerged in the academic world. For reasons of national security nothing was announced to the outside world, and so these people lost their chance to bask in the glory of being considered the true discoverers. That, it seems, is part of the price of working in secrecy.

The Diffie-Hellman algorithm proceeds as follows. Bob and Alice first agree on two numbers, *M* and *N*, which can be published openly. Alice generates a random number that she keeps to herself as a private key. Using the private key and the numbers *M* and *N*, she produces, according to a mathematical formula, a public key that she sends to Bob quite openly. Meanwhile Bob does likewise, generating his own private key and sending a public key to Alice. Alice is then able to send messages to Bob using a combination of her private key and Bob's public key. Bob doesn't know Alice's *private* key, but because of the cunning way the mathematics works, he is able to decode the message using Alice's *public* key and his *private* key. A similar situation operates in reverse for Bob to send messages to Alice. Anyone monitoring their encrypted conversations would be unable to decode them because they would have access only to the public keys and the numbers *M* and *N*. The strength of the algorithm depends on the fact that you need one of the private keys to unlock the code and that the private keys cannot easily be derived from *M*, *N*, and the public keys.

The Diffie-Hellman algorithm has one drawback: *both* participants must exchange information before any communication can take place. This makes it acceptable for telephone conversations but far less suitable for e-mail encryption and other kinds of one-way communication.

How Diffie-Hellman Works

Before proceeding to the next development in public-key cryptography, let's examine the mathematics that makes the Diffie-Hellman procedure possible. Some of the concepts will prove helpful in understanding how quantum computation works.

The Diffie-Hellman procedure depends on the rather magical properties of whole numbers. An amazingly rich source of ideas in number theory is to consider what happens when you divide one number by another and look at the remainder. Families and friends may encounter this issue when they try to divide portions of food, money, land, and numerous other items without causing a squabble over who gets most. For example, a pack of four doughnuts among a family of three will leave one unspoken for, while twenty-four cans of beer among five people will leave either one person short or four cans in the fridge.

In the nineteenth century Gauss established an elaborate body of theorems based on the idea of these arithmetical reminders. He adopted a notation that is widely used by mathematicians today:

$$y \equiv x \bmod a$$

What this means is that if you divide y by a you get a remainder x. The symbol "≡" is called a congruence relation and simply means "equivalent to" (you can use "=" if you prefer), while *mod* is short for *modulus,* or *modulo,* derived from the Latin for "a measure." We actually use the idea of a congruence relation when we talk about the time of day. One o'clock in the afternoon is also 1300 hours on the 24-hour clock. In this mathematical sense, 1 and 13 are congruent modulo 12, or using Gauss's notation:

$$13 \equiv 1 \bmod 12$$

The point of these congruence relations is that they create a new arithmetical system in which numbers are limited to a finite range of 1 to 12 or whatever. If you multiply two numbers in this system you also still get a number between 1 and 12. For example,

$$3 \times 7 \bmod 12 \equiv 9 \bmod 12$$

This then is just notation.

Now for the interesting part of the story concerning the Diffie-Hellman algorithm. Alice and Bob agree publicly to use randomly generated numbers M and N. Alice then generates a further random number, S_{Alice}, which she keeps secret; Bob does likewise, generating S_{Bob}. These are the private, or *secret,* keys. Alice now calculates her *public key* using the following formula:

$$P_{\text{Alice}} = M^{S_{Alice}} \bmod N \qquad (1)$$

What this means is that the number M is raised to the power S_{Alice} (Alice's secret key). The result is divided by N and the remainder presented as the public key, P_{Alice}. Note that if M, N, and P_{Alice} are publicly known, it might be possible for an outsider to reconstruct the secret key, S_{Alice}, from them. Indeed, the reverse version of this formula is known as a *discrete logarithm.* However, it turns out that if the numbers are all sufficiently large, it is very hard to calculate the discrete logarithm in a reasonable time. The security of the Diffie-Hellman algorithm depends on this fact.

Now Bob generates his public key with:

$$P_{Bob} = M^{S_{Bob}} \bmod N \qquad (2)$$

The public keys are exchanged. Alice now computes the number K_{Alice}, which is the *session key* she uses to encrypt and decrypt all subsequent messages to and from Bob, using the following formula:

$$K_{Alice} = P_{Bob}{}^{S_{Alice}} \bmod N \qquad (3)$$

Bob likewise calculates his session key:

$$K_{Bob} = P_{Alice}{}^{S_{Bob}} \bmod N \qquad (4)$$

Now we discover an amazing fact if we do a little mathematical work. If we replace P_{Bob} in (3) with $M^{S_{Bob}} \bmod N$ from (2):

$$K_{Alice} = (M^{S_{Bob}} \bmod N)^{S_{Alice}} \bmod N = M^{S_{Bob}S_{Alice}} \bmod N$$

Likewise, if we replace P_{Alice} in (4) with $M^{S_{Alice}} \bmod N$ from (1):

$$K_{Bob} = (M^{S_{Alice}} \bmod N)^{S_{Bob}} \bmod N = M^{S_{Alice}S_{Bob}} \bmod N$$

Now we see the really clever part:

$$K_{Alice} = K_{Bob}$$

The session keys are the same! This is because the order of the numbers in the exponent of M is unimportant: $S_{Alice}S_{Bob} = S_{Bob}S_{Alice}$. Consider, for example, $(2^3)^2 = 8^2 = 64 = 4^3 = (2^2)^3$. The fact that Alice and Bob have the same session key means that they can use it to encode and decode messages between them just like a normal private-key system.

This procedure has therefore enabled Alice and Bob to share a session key without ever having to reveal it directly either to the outside world or to one another. Alice and Bob *do* reveal some information: namely, their public keys, M and N. However, as we saw, it is very hard to derive the session key from these numbers alone. Therefore, anyone monitoring their conversation will be left in the dark. A highly secure system—highly secure, that is, in a world without a quantum computer.

The RSA Algorithm

The RSA public-key algorithm is the most famous and successful public-key system yet. The algorithm is more versatile than the Diffie-Hellman algorithm and today forms part of an increasing number of cryptographic systems.

RSA was developed in 1977 by Ronald Rivest, Adi Shamir, and Len Adleman, who were all then young professors at the MIT Labratory for Computer Science. They began looking for such a system after reading Diffie and Hellman's early paper on multiuser cryptography. According to Simson Garfinkel in his book, *PGP: Pretty Good Privacy,* "Rivest came up with ideas; Adleman shot them down; Shamir sometimes traded sides, sometimes making codes, sometimes breaking them."[6] In 1977, after several false runs, Rivest eventually produced a method that the others confirmed worked.

What emerged was one of the most startling and remarkable developments in cryptography ever. Yet with RSA, as in the discovery of the Diffie-Hellman algorithm, history has had to be rewritten because GCHQ got there first, only they didn't reveal so until 1997. Its mathematician, Clifford Cocks, produced essentially the same algorithm in 1974. The great thing about RSA was that it did not require any active participation between receiver and sender. To send an encrypted message to Alice, all that was necessary was for her to publish a public key. With that alone anyone could send her a message that only she could read.

Another great feature of the system was that it offered a way of "proving" one's identity electronically using a *digital signature.* This relies on the fact that the role of the public and the private keys can be swapped in the RSA algorithm. If Alice publishes a public key, she can sign a document she sends to Bob by appending a short message that she encrypts using her private or *decryption* key. Alice then encrypts the whole document, plus the encrypted message, using Bob's public key. Her signature is now doubly encrypted. When Bob receives Alice's message, he applies the first round of decryption using his private key. He reads the document and sees the digital signature at the end. To verify the signature, Bob applies a second round of decryption using Alice's *public* key. Provided the message in the signature is specific to him and the document is suitably dated, he will know that Alice must indeed have been the author of the accompanying document because only the person who knows Alice's *private* key could have produced the signature. The message must be specific, otherwise somebody else receiving another document from Alice could copy her signature and attach it to a fake document.

Of course, the digital signature cannot actually prove that Alice is the person she claims to be. If Eve published a key in Alice's name, then she could digitally sign messages in Alice's name. Bob would have to make sure that Alice was indeed the owner of the public key before accepting the digital signature on any subsequent communications. But given that the public keys can be widely published, it shouldn't be too difficult for Alice to ensure that nobody is publishing fake keys in her name.

The concept of digital signatures is extremely powerful and is likely to play an increasingly important role in transactions in daily life, especially as commerce grows on the Internet.

In 1983 a U.S. patent on the RSA algorithm was granted to MIT and Rivest, Shamir, and Adleman, but because the algorithm had been published before the patent had been filed, its creators were unable to secure the foreign rights to their invention. Nevertheless, Rivest, Shamir, and Adleman formed the company now known as RSA Security, Inc. After a few rocky years the company began to thrive as its secured licensing deals with Lotus Notes, Digital Equipment Corporation, Novell, and others. Ironically, though, it was the controversial release of the e-mail encryption program PGP that probably did the most to publicize the importance of the RSA algorithm.

The main disadvantage of RSA is that the algorithm is computationally expensive and therefore too slow to use as a general means of encrypting information. Its main application to date has been in encrypting keys used for other algorithms, such as DES and IDEA, which are much faster to implement. Nevertheless, by offering a safe way to exchange keys, RSA has transformed modern cryptography.

How RSA Works

Factorization provides the basis for the RSA algorithm because it depends on the fact that multiplying numbers is easy but factoring them is hard. The procedure works as follows. If Alice wants to receive messages encrypted using RSA, then she must first create a pair of keys, one public, the other secret. To create these keys, she must first choose two very large prime numbers p and q at random. She must also choose two large numbers d and e such that $(de - 1)$ is divisible by the number $(p - 1)(q - 1)$. Alice then publishes the *modulus* pq (without revealing either p or q separately) and the number e as her public key. She keeps d to herself as part of her private key.

If Bob then wants to send a message to Alice, he breaks his message up into a series of numbers that he then encrypts using the following formula:

$$c = m^e \bmod pq \quad (1)$$

where m is a number representing (each chunk of) the message, and c is the encoded message, known as the *ciphertext.*

To decrypt the message, Alice uses:

$$m = c^d \bmod pq \qquad (2)$$

The reason decryption is hard for anyone other than Alice to perform is that a code breaker needs the number d, which was chosen to satisfy the requirement that $de - 1$ is divisible by $(p-1)(q-1)$. Calculating d is straightforward if $(p-1)(q-1)$. But knowing this is equivalent to knowing the secret primes p and q. So ultimately the security of the RSA algorithm depends on the fact that it is very difficult to find the prime factors of n. (For more details on how the algorithm works, see Appendix E.)

Let's take a numerical example of the RSA algorithm in action. For simplicity we'll use small numbers, although obviously for security much bigger numbers would be required.

Bob chooses $p = 7$, $q = 13$, $e = 5$. So $(p-1)(q-1) = 72$ and $d = 29$ (this satisfies $de - 1 = 144 = 2 \times 72$). He publishes his public key, which has two parts, the modulus pq, which is 91, and the exponent e, which is 5. Let's say Alice wants to send a simple message that is simply the number 4. So $m = 4$ and the ciphertext is given by:

$$4^5 \bmod 91 = 1{,}024 \bmod 91 = 23$$

Alice sends the number 23. To decrypt it, Bob uses his private key, where the modulus is 91 and $d = 29$:

$$23^{29} \bmod 91 = 4$$

Bob reads Alice's message: the number 4. Note that even though we have used very small numbers, we still have to calculate some rather hefty exponentials. You may wonder how you can check the value of 23^{29} mod 91 given that you need to know all the digits exactly. Doing it on a calculator won't work because 23^{29} has 39 digits! The trick is to use repeated squaring or multiplying in small steps, taking the modulus 91 at each stage. So:

$$23^2 \bmod 91 = 529 \bmod 91 = 74$$

Now square again:

$$23^4 \bmod 91 = 74^2 \bmod 91 = 16$$

Square yet again:

$$23^8 \bmod 91 = 16^2 \bmod 91 = 74$$

Notice we have now got a repeating cycle of 74 and 16. Expressing 23^{29} in terms of these lower powers of 23, we can write:

$$\begin{aligned} 23^{29} \bmod 91 &= (23^{16} \bmod 91)(23^8 \bmod 91)(23^4 \bmod 91)(23 \bmod 91) \\ &= 16 \times 74 \times 16 \times 23 \bmod 91 \\ &= 435{,}712 \bmod 91 = 4 \end{aligned}$$

Using this technique it is possible to do enormous modular exponentiation efficiently using only moderate-sized numbers. In practice, all of the numbers we have considered would be much larger, but the same techniques apply.

All it would take to crack the above example would be to factor the number 91. When using RSA for real, it is common to use keys of at least 512 bits, which corresponds to around 155 decimal digits.

Cryptography and the Real World

Strong cryptography has long been a politically sensitive issue in the United States because it can be used for criminal purposes and as a weapon of war. In 1954 it was brought within the jurisdiction of the Munitions Control Act, which meant any exports of cryptographic technology would be subject to tight controls. This remained in force until 1996, but even now only weak forms of encryption are generally allowed for export, a restriction that has caused immense frustration among U.S. software companies because it has made it difficult for them to incorporate strong encryption in communications software. In many cases they have had to either release specially weakened versions of their software for export or not export at all. Netscape and Microsoft, for example, were obliged to release different versions of their Internet browsers to avoid infringing the export laws. The U.S. versions have long offered 128-bit keys (for the private-key system known as the RC4 stream cipher) for handling secure transactions such as the transmitting of credit card details, while the export versions were for many years limited to 40-bit keys, which are now regarded as very weak.

In the 1990s U.S. law enforcement agencies became increasingly concerned about the way new communications technology was making it more difficult to wiretap suspect criminals. In part this was because of complications introduced by digital telephony. The spread of sophisticated new encryption technology only made things worse. If gangsters, terrorists, and even wayward warmongering nations were to have access to strong encryption, the government agencies responsible for gathering domestic and international intelligence—the FBI and the NSA—argued that they would be put at a severe disadvantage in fighting crime and defending the nation.

In 1991 Congress attempted to pass a clause in an anticrime bill that would have obliged telecommunications companies and equipment manufacturers to ensure the government could always break encrypted messages when authorized by law. This move caused a storm of protests from civil liberty activists and from organizations such as the Electronic Frontier Foundation.

It was the threat of such legislation that led to the release onto the Internet of the freeware encryption program PGP, an e-mail software package that offered "military-strength encryption for everyone." PGP was conceived and written by Phil Zimmermann, a cryptography consultant and freelance programmer living in Boulder, Colorado. Something of a political activist, Zimmermann was concerned that unless someone made such software widely available, the authorities would be able to strike a devastating blow against individual privacy. Because PGP exploited a combination of the RSA algorithm for key management and IDEA for data encryption, it represented powerful technology. Its global spread via the Internet amounted to a prima facie violation of the export laws, but Zimmermann said he had not been responsible. He had given a copy to a friend who had posted the program on the newsgroups of the Internet.

Zimmermann's predicament became a cause célèbre when for several years he was the subject of a federal crime investigation by the U.S. Customs Department: If found guilty of breaking the export laws, he could have faced a four-year prison sentence. He was also in some trouble over the program's infringement of the U.S. patent for RSA, which belonged to RSA Security, Inc. Fortunately for him, both of these matters were later resolved. The investigation was dropped in 1996, and RSA Security released a version of its algorithm that could be freely used in noncommercial software. Either way, though, there was little the U.S. government or RSA Security could do to stop PGP, since it was already in global circulation.

Useful though PGP is for protecting e-mail against snoopers, it cannot be the whole answer to ensuring privacy. After all, far more people communicate via the telephone than via e-mail. Although Congress eventually dropped the controversial antiprivacy clause in its crime bill, the idea resurfaced in a different guise in 1994, when the FBI succeeded in persuading the Clinton administration to back a bill on digital wiretapping. The bill required the nation's telecommunications companies to comply with the technological demands of the Justice Department. What that meant was that the authorities would have to be given easy access to the new digital highways to carry out any necessary surveillance activities.

At the same time, the NSA had developed a new encryption system, known as the Clipper chip. This chip was intended for manufacturers who wanted to include strong encryption in voice-communications equipment such as digital telephones. The Clipper used an 80-bit encryption algorithm known as Skipjack, whose details were classified. The Clipper chip proved highly controversial because its algorithm included a mathematical trapdoor—a way of breaking in that could be made available to government agencies. The concept is known as *key escrow.* The way it works is that each chip is manufactured with its own secret key, a copy of which is split into two pieces and stored by two trusted third-party agencies. Then, if the law enforcement agencies need to listen in on a suspect's telephone conversations, they would apply for the two halves of the key after appropriate authorization through the judicial system.

The Clipper chip was only voluntary, but it was made a government standard in the hope that it would be widely taken up by the industry. Manufacturers, however, hated the idea. One reason was that the chip presented a hardware-only solution, which meant it wasn't at all attractive to software companies. Another was that without international agreement to adopt key escrow, U.S. manufacturers would be stuck with a system that effectively had, as one industry executive put it, "Uncle Sam inside." If foreign software companies were free to develop unadulterated encryption software, then who would buy these compromised products?

In October 1996, after intense lobbying by industry, Vice President Al Gore announced a plan to ease export controls. In particular, the plan allowed for the export of public-key systems but with the limitation that key lengths for the public modulus be no longer than 512 bits. For private-key systems the rules stipulated a maximum key length of 56 bits, an improvement over the old limit of 40 bits.[7] Such an increase was made imperative because in 1996 two French students managed to crack a 40-bit key used by the export version of Netscape Navigator.

The plan also indicated that the administration would lift all key size restrictions on cryptographic exports provided that manufacturers offered a viable means of key recovery for legitimate government access. The goal was to promote "a worldwide key management infrastructure with the use of key escrow and key recovery encryption items to promote electronic commerce and secure communications while protecting national security and public safety."

U.S. companies still complained that the rules worked against them, given that competitors in many foreign countries were able to produce strong encryption products without the need for key recovery. In 1997 a bill introduced in Congress, known as the Security and Freedom through Encryption (SAFE) Act, looked as though it would eliminate the Clinton administration's requirement to build key recovery systems into encryption products. This was followed in 1999 by the Cyberspace Electronic Security Act, which could give law-enforcement agencies the right to enter homes and offices secretly to obtain passwords and install covert software.

In the long run, however, U.S. policy is likely to be shaped less by Washington politics than by international pressures. Few countries have shown any interest in the idea of key escrow. The OECD and European Union have, in particular, argued for the liberalization of controls on cryptography in order to promote the development of market-based cryptography products and services. Nevertheless, in December 1998 thirty-three nations agreed to participate in the so-called Wassenaar Arrangement, which calls for restrictions on the export of weapons, including cryptographic hardware and software using key lengths greater than 56 bits. The restrictions are primarily aimed at preventing the export of so-called dual-use technologies to pariah nations. Nevertheless, countries such as Switzerland and Germany have declared that they will not abide by any such controls.

While the export of technologies that could ultimately risk the lives of many people is a very serious issue, advocates of personal freedom claim that governmental regulation of cryptographic security techniques seriously endangers our personal privacy. In the increasingly networked environment in which we live, it is going to become all the more difficult to strike a balance between these two conflicting concerns. Either way, though, cryptography is likely to play an increasingly important role in our lives.

The Challenge of RSA-129

One of the complicating factors in deciding on cryptographic policy is that the technology on which it is based refuses to stand still. Decisions about maximum acceptable key lengths have to take into account rapid improvements in code-breaking technology. In the last twenty years advances in cryptanalysis have been breathtaking. Few examples serve better to illustrate than the extraordinary tale of RSA-129.

Soon after Rivest, Shamir, and Adleman published the details of their RSA algorithm in 1977, the science writer Martin Gardner contacted the authors so he could write about the idea in his "Mathematical Games" column for *Scientific American.* At Gardner's suggestion, the authors produced a challenge for his readers: a secret message encoded using a 129-digit public key, which came to be known as RSA-129. Readers were offered $100 if they could crack the message.[8] Gardner warned that Rivest had estimated the running time required to factor the key would be about 40 quadrillion years! Rivest was therefore confident that he would never see the message decrypted in his lifetime.

The encryption exponent (e) was 9,007 and the 129-digit public modulus (pq) was:

114381625757888867669235779976146612010218296721242362562
561842935706935245733897830597123563958705058989075147599
290026879543541.

Anyone who wanted to decrypt the message needed to factor the 129-digit public modulus into its prime factors p and q. With these they could then calculate $(p - 1)(q - 1)$, from which they could then calculate the decryption exponent d.

Rivest's estimate of 40 quadrillion years[9] to do the factorization had allowed for a substantial increase in computational power with improving technology by assuming a hypothetical machine capable of performing 1 billion modular multiplications per second, which was still well beyond anything available on a single machine in the late 1990s. So Rivest had good reason to believe the task to be impossible in his lifetime.

What Rivest and his colleagues hadn't anticipated was the development of some radical advances in mathematical techniques for factoring numbers. Several emerged in the 1980s. The first, known as the *elliptic curve method,* was enormously faster than previous methods but still not quite fast enough to crack RSA-129 in a reasonable time. The next method was the *quadratic sieve,* which looked more promising. Then, in 1990, an even faster method, called the *number field sieve,* was developed by John Pollard, a mathematician in England who had already invented two other factoring methods.

Someone who was well aware of these developments was Arjen Lenstra, a Dutch researcher then working at Bellcore in New Jersey, the research arm of the U.S. Baby Bell telephone companies. In 1990 he and several colleagues were the first to factor the ninth Fermat number, a number with 155 decimal digits.[10] This number, which can be expressed as either $2^{2^9} + 1$ or $2^{512} + 1$, was the first great success with the number field sieve. Furthermore, Lenstra had enlisted the support of numerous enthusiasts on the Internet to participate in the computation. With this 155-digit monster cracked, you might think a 129-digit number such as RSA-129 would have been easy to factor, but at that time the number field sieve was suitable only for special numbers like the Fermat numbers. Nevertheless, in 1993 Pollard found a generalization of the number field sieve that would work for any number. Lenstra was keen to apply the new version of the sieve to RSA-129. His plan was to get a summer student to write the software and run the program on a new supercomputer that was being built for Sandia Industries. The machine was a 1,800-node massively parallel Intel Paragon machine, the biggest and most powerful computer to be built at that time. Lenstra was hoping to run the program during its testing phase.

Unfortunately, the software wasn't ready in time, and he had to abandon the plan. However, at around the same time, Lenstra received an inquiry from Paul Leyland, a computer systems expert at Oxford. He had been talking to a group of "cypherpunks" on the Internet who wanted to demonstrate that the use of 384 bits on PGP (one of the optional settings on early versions of this program) was insecure simply by using ordinary desktop computers and workstations to crack the code. Leyland was interested in the idea and wanted to use the quadratic sieve software Lenstra had written in the 1980s, which would enable the task of factoring to be divided among lots of computers on the Internet.

Lenstra approved of the plan but suggested that the team devote its efforts to cracking RSA-129 because 116-digit numbers (which corresponds to 384 bits) had already been shown to be breakable the year before, albeit using powerful computers.

So began a collaborative tour de force that eventually recruited some

600 volunteers from 25 countries on the Internet to participate in what became the world's largest computation. Michael Graff of Iowa State coordinated the volunteers, and Derek Atkins of MIT handled queries and collected results. Lenstra and Leyland supervised the whole process and dealt with the more difficult technical problems.

Altogether the team marshaled the computational horsepower of 1,600 workstations, desktop computers, and even the odd supercomputer. The organizational effort was all the more impressive given that the team had to ensure that the software worked on many different platforms. This point was amusingly illustrated in one part of their paper: "Our code was run successfully on machines as disparate as 16 MHz 80386sx PCs and Cray C90s. An attempted port to a Thinking Machines CM-5 failed, but one U.S. corporation managed to get the sieving code running on a couple of fax machines!"[11]

After eight months of number crunching, the results were collected in the form of a database, dumped onto a data tape by Leyland in Oxford, and sent by Express Mail to Lenstra at Bellcore. Ironically, it was faster to send the tape that way than transmitting many gigabytes of data over the Internet. On receiving the tape, Lenstra processed the data for two days on a MasPar supercomputer to finish the calculation. Eventually, on April 2, 1994, the answer appeared on Lenstra's screen. The computer had finished the calculation and had factored RSA-129. And there was Rivest, Shamir, and Adleman's decrypted message: THE MAGIC WORDS ARE SQUEAMISH OSSIFRAGE.

The team won the $100 prize and donated it to the Free Software Association. The announcement of their breakthrough caused a press sensation, with headline coverage in newspapers around the world.

Although Lenstra was disappointed not to have been able to do the factorization using the number field sieve, he subsequently developed the necessary software for the next challenge: RSA-130. This was completed on April 10, 1996, and was the first number to be factored using support recruited through the World Wide Web (as opposed to e-mail). Although the event drew far less attention, the prize, at least, was a little more exciting—$11,000. This was one of a number of prizes now offered by RSA Security, Inc. Similar prizes await anyone who manages to crack RSA-140, RSA-150, RSA-160, and so on. Using the number field sieve meant that factoring RSA-130 took only about one fifth of the computational effort required to crack RSA-129. On larger numbers still, the relative performance of the number field sieve gets even better.

Given these amazing improvements in factoring technology, Lenstra thinks RSA codes depending on 512-bit numbers, corresponding to 155 digits, are within reach of organizations with powerful large machines. This is somewhat alarming because some financial encryption systems already depend on keys of this length. One example is the system called CREST, introduced in 1997 by the Bank of England. This is a paperless trading system for handling shares on the London Stock Exchange. CREST uses RSA technology to protect the digital signatures attached to the electronic share certificates but uses key lengths of only 512 bits.

Lenstra is not impressed by the generally slack attitude toward data security found in many organizations. Keys of 512 bits might even be within the reach of cypherpunks, he told me. "In principle," he said, "they could crack such numbers overnight. Using the new method, these numbers are only eight times harder to solve than RSA-129 with the old method. If you wrote a virus or worm that spread itself over the Internet and started computing there, you could get hundreds of thousands of machines. If the program destroyed itself after one night, you would get a lot of computing power without anyone ever knowing. These considerations are things that people should take into account when they are putting their systems into place. But they don't."

Factoring by E-Mail

One of the clever aspects of the way RSA-129 was factored was that so many people were able to contribute to the solution no matter how modest their machines. Quite apart from the organizational headaches this imposed, the algorithm had to be structured in such a way that many machines could work on different parts of the problem without duplicating one another's efforts.

The quadratic sieve is almost ideally suited for such division of labor (as is also true for the number field sieve). It involves two main steps:

1. Sieve and collect a large database of numerical "relations."
2. Process the database in a giant mathematical matrix and get the factors.

Step 1 requires much more computational work than step 2, but it's the step in which anyone with a computer can contribute. The process is also fault tolerant because it is much easier to check sieved data than to produce them. So if someone tried to sabotage the calculation or screwed up the way his program ran, it would be easy to eliminate any rogue data. For RSA-129, step 1 took around eight months of work and terminated when the participants had collected around 8 million relations, a heuristic figure based partly on theory and partly on previous experience of factoring smaller numbers.

Step 2 needed to be carried out on a powerful machine with plenty of memory but took only two days.

Lenstra later offered a much-simplified account of how the factoring algorithm worked for a report that appeared in the journal *Science.*[12] The following is based on the example he offered there.

The quadratic sieve depends on finding two numbers x and y such that $x^2 - y^2$ is a multiple of n, the number to be factored. In other words:

$$x^2 \equiv y^2 \bmod n \tag{1}$$

Rearranging this expression, we get:

$$(x - y)(x + y) \equiv 0 \bmod n \tag{2}$$

which means that either $x - y$ or $x + y$ is divisible by n, or each of $x - y$ and $x + y$ are multiples of the two prime factors of n. Therefore, by random sampling there is a chance that $x - y$ will contain one of the prime factors.

Suppose we wanted the factors of $n = 143$. It is difficult to guess numbers that satisfy the equation for x and y straight off, so what we do is build up a database of numbers that satisfy a weaker relationship.

If we guess $x = 17$, we get $x^2 = 289 = 3 + 2 \times 143$. So $x^2 \equiv 3 \bmod 143$. Since 3 is not a square, it doesn't satisfy equation (1), but it is a small prime, which, according to the rules, makes it useful for the database.

Our next guess is $x = 18$: $x^2 = 2 \times 19 \bmod 143$. Since 19 happens to be a rather large prime in this situation, it's thrown out.

Third guess: $x = 19$; $x^2 = 3 \times 5^2 \bmod 143$. Since 3 and 5 are small primes, this relation goes into the database.

The two database entries $17^2 \equiv 3 \bmod 143$ and $19^2 \equiv 3 \times 5^2 \bmod 143$ turn out to be enough to factor 143. If we multiply the two entries together, we get:

$$17^2 \times 19^2 \equiv 3 \times 3 \times 5^2 \bmod 143$$

This can be rewritten as:

$$(17 \times 19)^2 \equiv (3 \times 5)^2 \bmod 143$$

which satisfies equation (1) with $x = 17 \times 19 = 323$ and $y = 3 \times 5 = 15$.

The final step involves finding the greatest common divisor of n and $x - y$, where $n = 143$ and $x - y = 308$. Over 2000 years ago, the Greek mathematician Euclid devised a very efficient method for finding greatest common divisors. It works in this case as follows: Divide 308 by 143 and take the remainder, which is 22. Now divide 143 by 22 and take the remainder, which is 11. Now divide 22 by 11 and there is no remainder. According to the Euclidean algorithm, 11 is therefore the greatest common divisor and is therefore also a prime factor of 143. Dividing 143 by 11 reveals the other factor, 13. QED.

The number field sieve is difficult to explain without resorting to more advanced mathematics. However, it is interesting to note that it depends on the manipulation of *irrational* numbers, numbers involving terms like $2 + \sqrt{7}$. In the case of the ninth Fermat number, the sieve involved numbers including the fifth root of 2. These form an exotic arithmetical system in which every number can be expressed as $a + bx + cx^2 + dx^3 + ex^4$, where $x = \sqrt[5]{2}$, and a, b, c, d, and e are rational numbers (which means they must all be integers or fractions).

Such numbers can be uniquely factored, rather like the integers, and some behave like prime numbers too. It turns out that by establishing parallels between the world of ordinary numbers and these exotic numbers, it's possible to sieve for factors faster than is possible by ordinary means alone.

Despite the impressive strides made with these new mathematical techniques—an improvement of 10^{19} in speed over what Rivest had estimated—these algorithms still do not run in polynomial time. The number field sieve and the quadratic sieve are thought to be, strictly speaking, "subexponential" in running time, which means the increase in running time is faster than any polynomial but not quite the full-blown exponential.[13] In fact, the running time of the number field sieve very approximately doubles for each three extra digits in the number we want to factor.

The absence of a polynomial time solution means that RSA encryption codes are safe for the time being, provided people use sufficiently long keys. In 1997 RSA Security said that 512-bit keys no longer provided sufficient security because of the advent of new factoring algorithms and distributed computing. Instead, it suggested that 768-bit keys be used for personal security, 1,024 bits for corporate security, and 2,048 bits for extremely valuable keys, such as those used by a certifying authority. It reckoned 768-bit keys would probably remain safe until at least 2004. However, these days few experts are willing to give any guarantees about the long-term strength of cryptographic systems.

One overriding concern is that no one has found a proof that factoring has to be intractable. It is possible, though most mathematicians think it is unlikely, that someone could find a polynomial-time algorithm. If so, then the security of all data currently protected by the RSA system would disappear overnight.

Factorization Takes a Quantum Leap

In onc of those coincidences of fate, the very month in which Lenstra and his colleagues factored RSA-129, another event of perhaps even greater long-term significance to the world of cryptography was unveiled to an unsuspecting audience of mathematicians, physicists, and computer scientists. This was the discovery in April 1994 by Peter Shor at Bell Labs in New Jersey of a way to factor numbers efficiently using a quantum computer. His work was entirely theoretical. Then as now, no one had built a quantum computer. But Shor had an algorithm: a program to run on a machine that might be built in the future.

What was a truly astonishing about his discovery was that the algorithm could solve the factorization problem in polynomial time. What that meant was that factoring numbers could be done in a reasonable amount of time *no matter how large the number.* As we saw with current methods for cracking RSA codes, the time increases by a factor of roughly 2 for each three extra decimal digits in the RSA key. With a quantum computer the computation time would hardly increase at all.

One consequence was that if somebody were able to build a quantum computer, all RSA public-key systems would immediately become vulnerable. So here, at last, was the killer application for a quantum computer, a program that went far beyond the contrived mathematical problems we saw in the last chapter. Shor's result also confirmed that Deutsch's early intuitions had been right: Quantum parallelism offers genuinely interesting new computational capabilities.

Nevertheless, the discovery came as a surprise even to Deutsch. A year before Shor obtained his result, Deutsch had joined other researchers at a conference in Turin to discuss the latest developments in quantum computing. At that time most researchers, it seemed, regarded the idea of solving the factorization problem as an impossible dream. "It was one of the best meetings I'd ever been to," Deutsch recalls. "There were a lot of major advances announced there, and a great deal of discussion about how to push them even further. But at various moments we said, 'This can't possibly work, because if it did, you'd be able to factor numbers.' "

How then did Shor conjure up his result? Unlike many of his fellow researchers in this discipline, Shor was a relative newcomer to quantum mechanics and had barely made contact with the subject of quantum computing. He was a computer scientist and number theorist rather than a physicist and was employed by Bell Labs to conduct research into faster algorithms. He wasn't even aware of Feynman's contributions to the subject of quantum computing. He had, however, read Deutsch's papers and some of the papers prompted by them. What set him thinking about the factoring problem was a paper published by Dan Simon in 1993.

As we noted at the end of the previous chapter, Simon and others developed variants of Deutsch and Josza's balanced versus constant problem. Simon's particular variant was to show how a quantum computer could efficiently decide whether a function was 1-to-1, meaning every input produced a distinct output, or 2-to-1, meaning two different input values could produce the same output. Think, for example, of the function $f(x) = x^2$. This is 2-to-1 because for every positive number, $x = 3$, for example, there is a negative number, $x = -3$, that produces the same answer—that is, 9. However, Simon's algorithm was a little more exacting in that only certain types of 2-to-1 functions were allowed. Specifically, for each pair of values of x that produced the same answer, the values had always to be separated by the same amount. The function $f(x) = x^2$ does not obey that rule, but a function like a sine wave, $f(x) = \sin x$, when looked at over two complete oscillations, does. For any value of x, the output f is the same for the input $x + 2\pi$ (or $x + 360°$) because a sine wave repeats itself after one complete oscillation.

Simon showed how a quantum computer could not only distinguish between 1-to-1 and 2-to-1 functions but, in the latter case, could also reveal the gap between repeating values of x, which in our sine wave example is 2π. In other words, if the function was 2-to-1, Simon's algorithm was able to measure its *periodicity.* Such a result was in many ways more interesting than the original Deutsch-Josza problem. Measuring periodicities is akin to measuring frequencies, a common requirement in a wide range of practical problems, especially in electronic signal processing. The usual method for deriving frequencies from a signal such as an audio waveform is to apply a mathematical transformation known as a *Fourier transform.* The Fourier transform is an immensely powerful tool that crops up in a huge variety of scientific disciplines: mathematics, physics, crystallography, acoustics, optics, and electronic engineering, to name but a few. Sure enough, Simon's algorithm exploited a simple kind of Fourier transform to pluck out the periodicities in 2-to-1 functions.

The use of Fourier transforms, as it turned out, was the key to the algorithm Shor produced for factoring numbers. Although Shor was inspired by Simon's paper, the idea that Fourier transforms could be implemented on a quantum computer was first explicitly stated in an earlier paper than Simon's, one by Bernstein and Vazirani (mentioned in the previous chapter). Furthermore, the essence of the Fourier transform idea, it was later realized, was actually already hidden within the Deutsch-Josza quantum algorithm solving the constant versus balanced problem.

Heat, Sound, and Fourier Series

If you drew a squiggly line on a sheet of paper and were asked what mathematical function described the shape of the line perfectly, you might be tempted to say that there wasn't one. After all, it's exceedingly unlikely that you would happen to draw a perfect parabola corresponding to the function $y = x^2$, or any other curve corresponding to an algebraic function. There would surely be some error at some point along the curve. In the eighteenth century, mathematicians as distinguished as Leonhard Euler held precisely that view. Yet they turned out to be wrong. The man who showed that there was a way of describing *any* shape of curve precisely using an algebraic formula was the French mathematician Jean-Baptiste Joseph Fourier (1768–1830).

Fourier's achievement, for which his name will be eternally famous, was to demonstrate that any shape of mathematical function could be expressed as an infinite sum of sine (and cosine) waves. On the face of it, this was an extraordinary discovery because sine and cosine are trigonometric functions that relate to the properties of triangles. Why should they turn up in the description of squiggly lines? It's hard to give a concise answer other than to say these functions turn up virtually everywhere in mathematics.

In his treatise *The Analytical Theory of Heat,* Fourier showed how it was possible to describe discontinuous functions in terms of these Fourier series. This was useful because heat flow is sometimes subject to almost instantaneous changes (such as the sudden removal of a source of heat). The problem was that heat flow was described by differential equations—equations that relate the way variables like temperature and energy change with respect to time and distance. Think for a moment, for example, of the heat that builds up inside a typical car engine. Automotive designers may use differential equations to model the heat flows to ensure that there won't be any hot spots, but differential equations generally require smoothly continuous variables. Discontinuities involve an infinite rate of change, and whenever infinities arise in mathematical equations, there is likely to be trouble. So the question arose in Fourier's day of how abrupt changes in conditions could be modeled satisfactorily.

Consider, for example, heat flowing through a metal wire. If one end is kept at room temperature while the other end is kept at 0 degrees Celsius, eventually the wire will settle down to a steady temperature profile in which the temperature gradually decreases from one end to the other. However, if the cold end is suddenly plunged into boiling water, the temperature along the wire now has a discontinuity at its hot end. The wire in the boiling water is at 100 degrees C, but immediately beyond, the temperature is still at 0 degrees C because it takes a little time for any heat to pass down the wire. How, then, does the system return to a steady temperature profile? By expressing the initial discontinuous temperature profile as a Fourier series—that is, a sum of sine waves—it's possible to plug the series into the heat equation and produce an answer very easily. The remarkable aspect of Fourier's discovery was that you could apply Fourier series to any initial temperature distribution to analyze how it would change as heat flowed.

Today we are more likely to come across the idea of Fourier series in the study of sound waves rather than in the analysis of heat flow. And it's useful to consider a few examples before we see how Fourier's insight applies to the quantum computer. One reason musical instruments such as a violin, a piano, and a trombone sound different from each other is that each instrument generates its own characteristic range of harmonics, or overtones. If you listen carefully to a piano play a single note such as middle C, you can hear some of the harmonics, including the C one octave above and the G above that. These harmonics are, in fact, the higher terms in the Fourier series describing the original waveform. The structures within the cochlea of the inner ear process incoming sounds by separating them into their frequency components. So it's perhaps not surprising that if we concentrate hard, we are able to pick out some of these Fourier components. Normally, though, we are not aware of doing so because the brain has a way of regrouping frequency components so that a piano, guitar, singer, or whatever sounds like an integrated whole rather than a collection of notes and their harmonics.

Electronically generated sounds often exploit everything from pure tones to signals whose waveforms have sharp edges. Square waves and sawtooth waves, which were common on analog synthesizers, sound harsh and abrasive compared with the clean and mellow quality of a sine wave. The reason for the difference is that square and sawtooth waves are heavily endowed with harmonics (see Figure 5.1).

Some forms of electronic synthesizer rely on what is in effect the opposite approach to Fourier analysis to generate musical sounds. Starting with

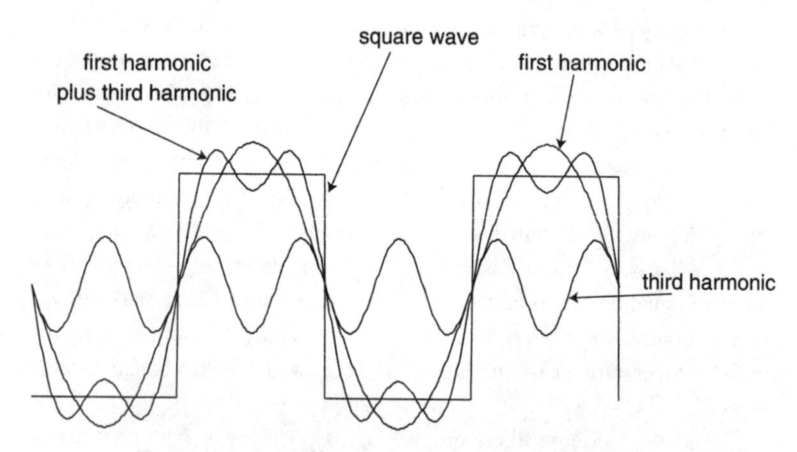

Figure 5.1. A square wave and its first two harmonics

When a square wave is decomposed into its Fourier components, it is found to consist of every odd-numbered harmonic: the first harmonic, or fundamental, the third, the fifth, and so on. The amplitude of the higher harmonics gradually diminishes with frequency. By adding the first two harmonics together, we see a waveform that roughly approximates the original. The more harmonics we include, the better the approximation.

pure musical tones, sine waves, it is possible to add them together in the appropriate proportions to evoke reasonably authentic sounding instruments such as pianos, harpsichords, organs, and winds. These techniques don't work too well for string instruments like the violin and cello because synthesizer manufacturers pay little or no attention to these instruments' dynamic qualities.

When a violin bow bites on the string in order to play a note, it often generates a slight gritty noise initially. This is something a good player will take advantage of to add emphasis. In a musical passage marked fortissimo the player may attack the strings with gusto, producing quite a lot of this percussive effect. In a pianissimo passage, on the other hand, the approach will typically be much more delicate, and the player will normally try to avoid any extraneous noise. Most synthesizers, however, completely ignore this effect and instead produce the same homogenized string sound regardless of loudness. Simply adjusting the amplitude is not enough to simulate loud and soft playing (which explains why sampled digital synthesizers are not much better). Similar comments apply to vibrato. Most synthesizers offer only limited means for adjusting vibrato. More expensive models, which allow for some adjustment on the fly, still tend to be very limited compared with the infinitely more subtle color and shading a violinist or cellist adds to performance by continuously modulating the speed and depth of vibrato according to mood and taste.

These shortcomings of electronic synthesizers do not undermine the analytical powers of Fourier transforms. If synthesizer manufacturers went to the trouble of offering more dynamic methods of altering the shading of their sounds, it would be possible to offer a much more realistic string sound. In their book *The Mathematical Experience,* Philip Davis and Reuben Hersh take this idea to its limit as they delightfully (but fancifully) point out, "To give a performance of Verdi's opera *Aïda,* one could do without brass and woodwind, strings and percussion, baritones and sopranos; all that is needed is a complete collection of tuning forks, and an accurate method for controlling their loudness."

Light, Music, and Fourier Transforms

Fourier series show how periodic functions such as square waves can be broken down into individual sine waves. The Fourier transform extends this idea so that any function, regardless of whether it includes any period-

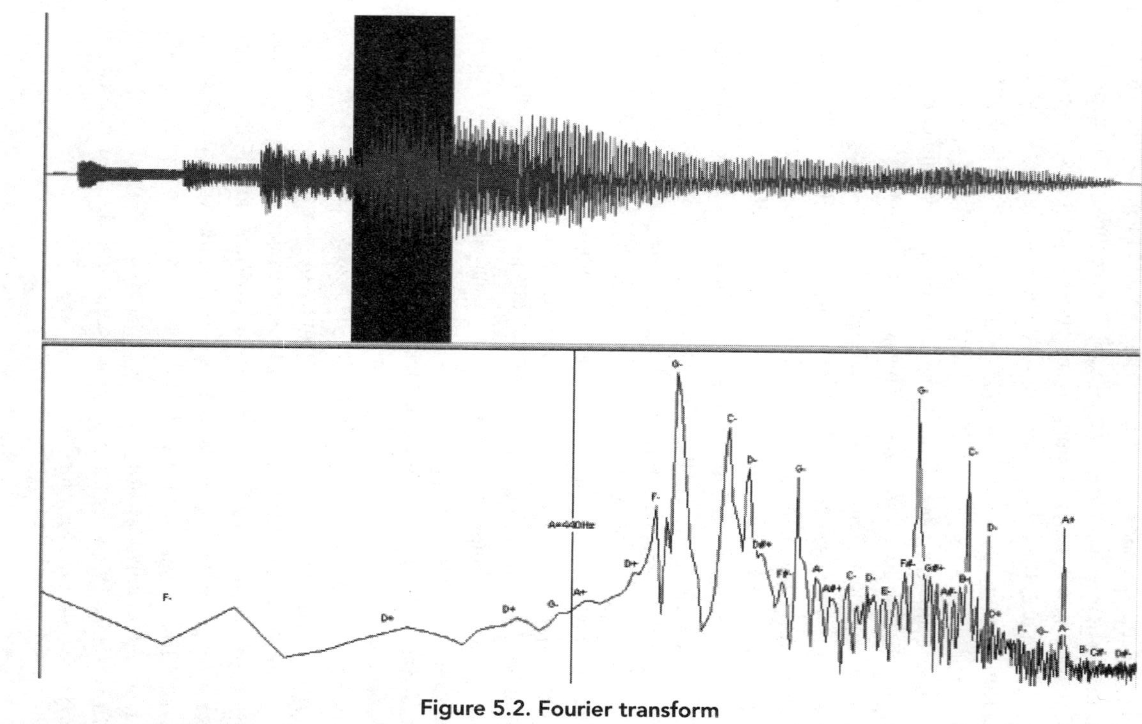

Figure 5.2. Fourier transform

The spectrum of frequencies *(shown in the lower pane)* in a sample of music *(highlighted in the upper pane)* was produced by computing the Fourier transform of the sample.

icity, can be expressed in terms of an infinite sum of sine waves. Unlike the Fourier series, in which every term or harmonic is an exact multiple of the fundamental frequency, the Fourier transform consists of all possible frequencies. Whenever you see a spectrum analysis of a sound sample, you are seeing its Fourier transform into the frequency domain (see Figure 5.2).

Fourier transforms, as we noted before, have numerous applications in many different branches of science and mathematics. They are widely used in X-ray crystallography, quantum mechanics, and electronic engineering. They also play a big role in understanding various optical phenomena such as diffraction and interference. The fringes seen in Young's two-slit experiment, for example, are, in fact, a Fourier transform of the two slits. The way this works is as follows. If we plotted a graph of the light intensity immediately next to the back side of the mask, we would see two rectangular pulses corresponding to the bright light emerging from the two slits (see Figure 5.3). (These are pulses in space rather than in time.)

To calculate the Fourier transform, it turns out that we can regard the waveform consisting of two square pulses as a combination, technically known as a *convolution,* of a single rectangular pulse with two infinitely thin spikes known as *delta* pulses. We can now calculate the Fourier transforms of these latter two waveforms separately. The transform of two delta pulses is a simple sine wave, and the transform of a square pulse is the shaped wave packet shown in Figure 5.4.

The Fourier transform of the original waveform, which represents the light emerging from the two slits, is given mathematically by multiplying the two transforms shown below. The final result is the familiar pattern of Young's fringes (shown in Figure 5.5).

The important message of these examples is that Fourier transforms are crucial to interference experiments. In Young's two-slit experiment, for example, the Fourier transform occurs in the physics of the apparatus. But, as we shall see, in a quantum computer, where it is necessary to perform inter-

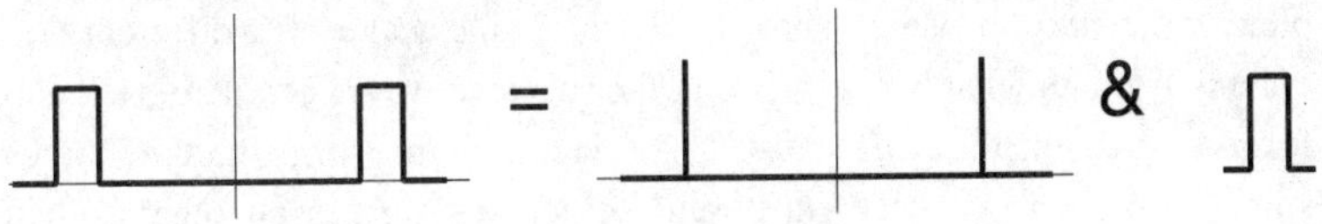

= &

Figure 5.3. Two slits

A convolution of two delta pulses and one rectangular pulse.

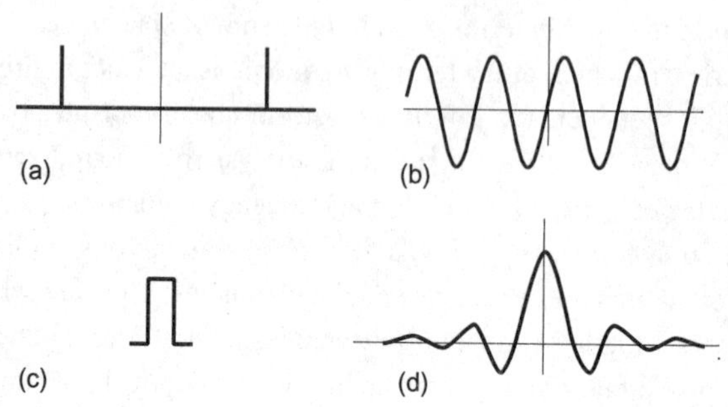

Figure 5.4. Waveforms and their Fourier transforms

Two delta pulses (a) transform to a pure sine wave (b) [14], and a rectangular pulse (c) transforms to shaped sinusoidal wave packets (d).

ference measurements on computer data, we will need to implement the Fourier transform explicitly using quantum logic.

The mathematics for calculating Fourier transforms requires knowledge of integral calculus, but I can offer a loose analogy. Think of the principle of a decoy duck, a wild duck tamed and trained to entice other ducks into a trap. With the Fourier transform, the strategy is very similar because we detect the presence of each frequency component in a waveform by mathematically introducing a sine wave of the very frequency we are looking for. If $f(t)$ is the function describing the original waveform, where t is time, then to measure the strength of any particular frequency ω, what we do is multiply the waveform by a sine wave, $\sin \omega t$, giving the product $f(t) \times \sin \omega t$. This product is then averaged over many cycles to give the answer. The process has to be repeated for many different frequencies to plot a full spectrum or transform. The reason this procedure works at all is that when $f(t) \times \sin \omega t$ is averaged over many cycles, the result is zero unless $f(t)$ contains an oscillation of the *same frequency* as ω.

Why is that? If $f(t)$ is a constant, the answer is obvious. Because $\sin \omega t$ oscillates between 1 and -1, when $f(t) \times \sin \omega t$ is averaged over a complete cycle, the answer is zero. If $f(t)$ is a sine wave of a different frequency from ω, what happens is that the two sine waves oscillate back and forth at different times. Because of the lack of synchrony, the two waves sometimes reinforce each other and sometimes fight each other. When averaged over many cycles, the result is again zero. But if $f(t)$ carries a sine wave component of exactly the *same* frequency as ω, then the average over many cycles is definitely greater than zero. This is because the square of

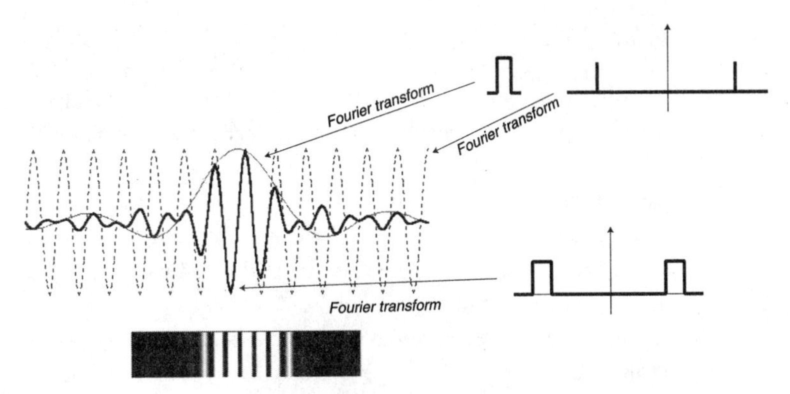

Figure 5.5. Young's fringes, produced by the Fourier transform of two slits

Mathematically, the fringes produced by Young's two-slit experiment can be explained in terms of the Fourier transform of the waveform representing the amplitude of the light emerging from the two slits. The original waveform can be decomposed into two waveforms that can be transformed separately and recombined by multiplying the two transforms. The sinusoidal wave packet shown in bold represents the result, which is the amplitude of the light reaching different points on the final screen. The corresponding image of the fringes is shown directly beneath. Large swings of the final waveform correspond to bright regions, while crossing zones (that is, points where the waveform is close to 0) correspond to dark regions.

sin ωt is always positive and therefore when averaged over one or more cycles, we get a positive answer. Hence multiplying by sin ωt flushes out the presence of the same frequency in the original waveform.

It is possible to carry out Fourier transforms using something akin to the above procedure on a conventional computer, but the technique is not very efficient when transforming large sets of data. Instead there is a more efficient technique, known as the Fast Fourier Transform, or FFT. This is a rather clever mathematical algorithm usually attributed to James Cooley and John Tukey in 1965 although it appears the idea was first discovered in the early 1900s, long before computers were ever built.[15] The technique involves multiplying data points from the original waveform by complex numbers representing the "decoy" signals. With additional data shuffling, the Fast Fourier Transform is able to cut some corners by avoiding unnecessary recalculation of similar terms.

The Quantum FFT

Shor's great insight[16] in 1994 was to see how it might be possible to implement a Fourier transform on a quantum computer and apply it to the problem of factoring large numbers. Shor knew that there was a method for factoring numbers that involved finding things known as *orders,* or *periods.* These are periodicities that arise when you take a sequence of powers, $x, x^2, x^3, x^4 \ldots$, and look at the remainders when the powers are divided by the number you are trying to factor.

Suppose, for example, we wanted to factor the number 15. If we take $x = 2$, we get the sequence 2, 4, 8, 16, 32, 64, 128, 256. . . . Now divide by 15 and if the number won't go, keep the remainder. This produces the repeating sequence 2, 4, 8, 1, 2, 4, 8, 1 . . . etc. The *order* in this case is 4 because the sequence repeats itself every four terms. Likewise with $x = 7$, we get 7, 49, 343, 2,401 . . . , which (mod 15) gives 7, 4, 13, 1, 7, 4, 13, 1. . . . Again the order is 4. Whatever number we choose for x, we always get a repeating pattern (although not necessarily of the same length). This is closely related to the fact that fractions when expressed in decimal either come out exactly or produce recurring sequences of numbers.[17]

These orders turn out to have very interesting properties. Although the two we just looked at turned out to have the same value, the pattern of orders for different values of x is usually somewhat haphazard. See, for example, the table illustrating sequences of powers mod 35. Here the orders are 1, 2, 3, 6, and 12—all divisors of 12. The pattern varies for different values of the modulus n, and although the orders do conform to some rules, in general evaluating them for larger values of n becomes increasingly time consuming. Indeed, in 1976 Gary Miller showed (in the same paper that offered a polynomial-time algorithm for primality testing) that finding such orders when the modulus n gets very large was as hard as factoring n. The problem of finding orders was, in his words, "computationally equivalent" to the problem of factoring integers. What that meant was that if we could devise an efficient algorithm for calculating orders, we ought to be able to factor numbers efficiently too.

The connection between finding orders and factoring is as follows. For a given number n we want to factor, choose a random value of x less than n and calculate its order f. If f is odd, try again with another value of x until you get an even number for f. Then calculate $x^{f/2} - 1$ and compute the greatest common divisor of this number with n. The answer will have a

Table 5.1. Calculating the orders of numbers modulo 35.

x	x^2	x^3	x^4	x^5	x^6	x^7	x^8	x^9	x^{10}	x^{11}	x^{12}
2	4	8	16	32	29	23	11	22	9	18	1
3	9	27	11	33	29	17	16	13	4	12	1
4	16	29	11	9	1						
5	25	20	30	10	15						
6	1										
7	14	28	21								
8	29	22	1								
9	11	29	16	4	1						
10	30	20	25	5	15						
11	16	1									
12	4	13	16	17	29	33	11	27	9	3	1
13	29	27	1								
14	21										
15											
16	11	1									
17	9	13	11	12	29	3	16	27	4	33	1
18	9	22	11	23	29	32	16	8	4	2	1
19	11	34	16	24	1						
20	15										
21											
22	29	8	1								
23	4	22	16	18	29	2	11	8	9	32	1
24	16	34	11	19	1						
25	30	15									
26	11	6	16	31	1						
27	29	13	1								
28	14	7	21								
29	1										
30	25	15									
31	16	6	11	26	1						
32	9	8	11	2	29	18	16	22	4	23	1
33	4	27	16	3	29	12	11	13	9	17	1
34	1										

good chance of being a divisor of n. If not, repeat the procedure until you obtain a divisor. In practice you will need to repeat the procedure only a few times before you find an answer. The expression $x^{f/2}+1$ is also a factor. (See Appendix F for an explanation of why this works.)

Let's try to factor the number 15 with the procedure. If we start with $x = 2$, we saw that the order f was 4. This is even, so we can proceed to the next stage:

$$x^{f/2} - 1 = 2^{4/2} - 1 = 3$$

which is a factor of 15. In this example with small numbers, we didn't need to compute the greatest common divisor (gcd), but again we can use the rapid algorithm discovered by the Euclid (see page 169).

Let's see how it works with another example: $n = 35$. If we try $x = 2$, the order f comes out at 12 (see Table 5.1). So plugging the numbers into the formula, we get:

$$x^{f/2} - 1 = 2^{12/2} - 1 = 63$$

We now compute the gcd of 63 and 35 using Euclid's method. First divide the smaller number into the larger and keep the remainder, which is 28.

$$\frac{63}{35} = 1 + \frac{28}{35}$$

Now divide the remainder 28 into the previous divisor 35 and keep the new remainder, which is 7.

$$\frac{35}{28} = 1 + \frac{7}{28}$$

Now repeat the process until we get no remainder. In this case we get the answer straight off because dividing 28 by 7 leaves no remainder. According to Euclid's algorithm, the gcd is 7, which is therefore one of the factors of 35.

To recap, what we have just seen is a method for factoring numbers given information about orders or periods. But classically this method would not offer any speed advantages over other methods of factoring because calculating orders becomes increasingly time consuming for large n. Peter Shor was aware of this, and he realized that a quantum algorithm might do the job more efficiently only after tackling a different but related problem: how to find discrete logarithms. We met the idea of the discrete logarithm before, in connection with the Diffie-Hellman encryption scheme (see page 154). When Shor saw Dan Simon's paper, which exploited a quantum algorithm to pick out a periodicity in a mathematical function, he thought there might be a way of adapting it to find discrete logarithms efficiently.

"I guess there was a flash of inspiration saying that you should be able to find discrete logs this way, but after this flash of inspiration it took a lot of work to figure out how to do it," Shor told me. "There is a strange relation between the discrete log and factoring numbers in number theory because any method that has worked for one has worked for the other. So once I had got the answer for the discrete log, I naturally started thinking about applying it to factoring." Oddly enough, solving the factoring problem is in some senses easier than the discrete logarithm problem because the main thing you need to do is find the period of a number. Nevertheless, Shor's insight was one that only someone familiar with these aspects of number theory was likely to have.

Shor's quantum program uses two quantum registers for storing and manipulating numbers. Each of these consists of a bank of qubits, just as an ordinary computer register consists of a bank of memory bits. The program then proceeds in three main steps. First, use H gates (or square root of NOT operations) to set the first quantum register into a superposition of states representing all possible numbers. Second, compute x^a mod n in the second register, where x is a number chosen at random and a is the (superposition of) number(s) in the first register. Third, carry out a Fourier transform on the first register and then observe its content. From the result, calculate the order of x. This number has a good chance of predicting one of the factors of n.

Let's examine this sequence of steps in a little more detail. The first step we have already seen in Chapter 4. We start with the quantum register set to all 0s and then apply a Hadamard transform to each qubit in the register. This puts each qubit into a superposition of 0 and 1. The register then as a whole represents all values from 0 to $2^q - 1$, where q is the number of qubits. In the many-universes interpretation, what we have are 2^q different universes each containing a different number in the register.

In the second stage of the algorithm, the quantum computer reads off the content of the first register, computes x^a mod n, and places the result in the second register; x is chosen randomly but with the proviso that it shares no common factors with n (easily checked using Euclid's algorithm). In the many-universes interpretation, the computer performs a different calculation in each universe. What it does is calculate all of the powers of x, i.e., $x^0, x^1, x^2 \ldots x^{2^q-1}$ in different universes. This has to be done using quantum logic to avoid disturbing the superposition in the first register.

We now have in the second register a superposition of all of the powers of x. As we saw before, these form a repeating sequence of length f. The third stage of the algorithm is to subject the first register to a Fourier transform. This, as we saw earlier, is a mathematical way of bringing about interference between the different universes or quantum states within the superposition. The transform involves processing the information in each qubit of the register via a series of operations designed to make the periodicity f—the harmonies within the superposition of powers, if you like—pop out of the calculation.

Now, it might seem odd that the Fourier transform is applied to the first register rather than the second. After all, the first register is left unchanged by the second step in the program, which places its result in the second register. An essential aspect of the whole procedure is that at some stage we need to carry out a measurement not only to read off the answer but also to see the effect of interference between the different universes or different quantum states. In Shor's original paper, he suggested making the measurement at the end of the process by looking at the result of the Fourier transform on the first register. But in a later paper,[18] Artur Ekert and Adriano Barenco pointed out that a measurement can also be made on the contents of the second register before carrying out the Fourier transform.

So what is going on here? Remember that at the end of the second stage the second register contains all of the powers of x. Because these form a repeating sequence with a period or order f, there are, in fact, only f different values represented. If we were trying to factor the number 15 and had chosen 2 as our value of x, then the powers (mod 15) form the sequence 2, 4, 8, 1, 2, 4, 8 1 . . . If we then measure the contents of the second register, we would see one of four values: 2, 4, 8, or 1. Suppose we saw the number 8. This observation collapses the quantum state onto that value, but there are many values of the exponent a that could have produced it: 3, 7, 11, and 15 . . . This is because $2^3, 2^7, 2^{11}, 2^{15} \ldots$ (mod 15) are all congruent to 8.

So an interesting thing has happened here. Although we have performed a measurement, the quantum state has only partially collapsed. The first register is still in a superposition of states representing the values 3, 7,

11, 15 . . . We now perform an interference experiment by implementing a Fourier transform on this superposition to extract the periodicity lurking within those numbers. Strictly speaking, though, the measurement on the second register isn't necessary. The Fourier transform still works regardless of whether the superposition has been partially collapsed.

This is an intriguing subtlety because if we don't make a measurement on the second register, the calculation of the powers stored in it looks rather pointless. Indeed, a conventional computer programmer searching for ways to make a program more efficient would regard any calculations that were never referred to again as redundant and would probably excise them from the program. Quantum programs are different in that though the results of a calculation may never be referred to again, they can still influence the rest of the calculation. The reason is interference. Those universes in which the powers stored in the second register are the same are able to interfere with one another in the first register. Where the powers differ, no interference can take place. So the calculation of the powers in the second register is indeed necessary. It's here we see how universes cooperate.

How is the Fourier transform implemented? Shor's paper set out the mathematical principles of his algorithm rather than any practical implementation. Nevertheless, other researchers quickly spotted ways of turning his ideas into quantum circuits. Don Coppersmith at IBM's Research Center in Yorktown Heights, in particular, offered a version of the Fourier transform based on the standard classical Fast Fourier Transform. He also suggested how it could be executed using simple kinds of quantum logic gates.

We have already seen some quantum circuits for carrying out Deutsch's "stock market" program. The Fourier transform requires a more complicated arrangement, but one example of an implementation for handling numbers using up to four qubits (that is, between 0 and 15) is shown in Figure 5.6. Two kinds of logic gate are used. The H gates, as we saw before, do the same sort of thing as a square root of NOT function. The boxes marked ϕ are two-qubit gates that each perform a "conditional phase shift." This means they change the phase of the incoming signals by a predefined amount—but only if both qubits are in their $|1>$ states. Such gates can be constructed from a combination of controlled-NOT gates and single-bit rotations, although they could possibly be implemented much more directly in some hardware settings.

It is difficult to explain how this circuit works without resorting to

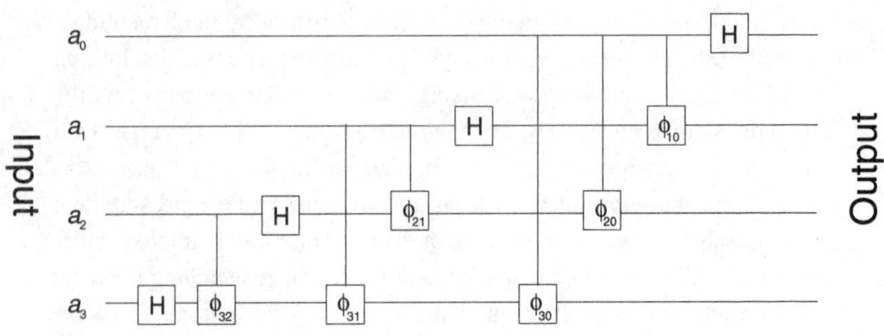

Figure 5.6. Circuit for a 4-bit quantum Fourier transform

The quantum Fourier transform is closely based on the classical Fourier transform. The incoming qubits (from the first register in Shor's algorithm) are subject to a network of H gates and conditional-phase-shift gates, marked ϕ. The subscripts of ϕ determine the size of the phase shift according to the formula $\phi_{ij} = \pi/2^{i-j}$, where ϕ is measured in radians. So, for example, ϕ_{32} is $\pi/2$, which is the same as 90 degrees; ϕ_{31} is $\pi/4$, or 45 degrees; and ϕ_{30} is $\pi/8$, or 22.5 degrees.

more advanced mathematics. However, the principle is based on the classical Fast Fourier Transform approach. This involves shuffling of signals and multiplication by complex numbers that happen to be the roots of unity. The *n*th roots of unity are those numbers that when raised to the *n*th power equal 1. For example 1 and -1 are the square roots of unity; 1, *i*, -1, $-i$ are the fourth roots of unity; and 1, $(1 + i)/\sqrt{2}$, *i*, $(-1 + i)/\sqrt{2}$, -1, $(-1 - i)/\sqrt{2}$, $-i$, $(1 - i)/\sqrt{2}$ are the eighth roots of unity; and so on. All of these roots of unity can be represented by points on a circle in the complex plane (see Figure 5.7).

Multiplying a signal by the fourth root *i* is equivalent to rotating the signal on this plane by 90 degrees, or $\pi/2$ radians. The phase shifts in the quantum circuit above correspond to rotations of this kind. With larger number of qubits, the phase shifts become correspondingly smaller as we go for higher roots of unity.

Subsequent independent work by Don Coppersmith at IBM, Adriano Barenco and colleagues[19] at Oxford, David Deutsch, and others showed that the circuits required for computing the quantum Fourier transform can be simplified by making certain approximations. Indeed, it turns out that it's possible to dispense with many of the smaller-valued phase shifts in the circuits without disrupting the calculation. This is important because to factor large numbers, some of the phase shifts would be extremely small. If

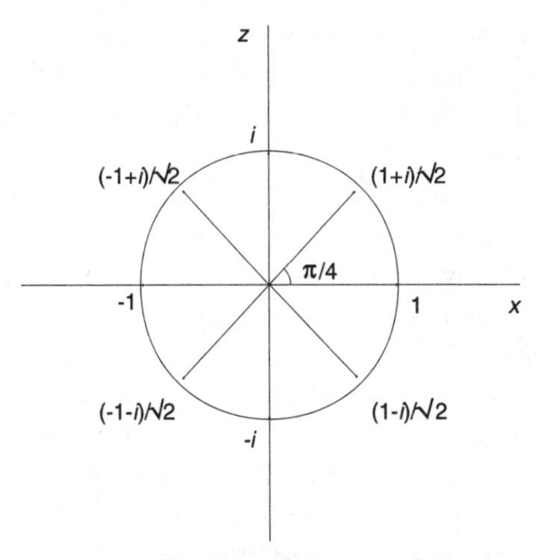

Figure 5.7. The eighth roots of unity

To compute the Fourier transform of a waveform represented by a discrete number of points, it is necessary to multiply the data by the nth roots of unity, where 2^n is the number of points. For 256 ($=2^8$) points, we need the eighth roots of unity, which are shown on the unit circle above. For the quantum Fourier transform circuit shown in Figure 5.6, there are 4 input wires, which allows for $2^4 = 16$ data points and requires the fourth roots of unity. The roots of unity are combined with quantum data by applying phase shifts corresponding to the angles of rotation each root makes on the complex plane. The reason the nth roots arise is that they represent different points in the sine and cosine waves required to reveal frequencies within the input data.

the Fourier transform were sensitive to tiny errors in these phase shifts, we would quickly run into severe practical problems in implementing Shor's algorithm. This turns out not to be the case. In fact, the procedure appears to be amazingly resistant to phase errors. I'll have more to say about this point in Chapter 8 when we look at the problem of quantum error correction.

Although the quantum Fourier transform is the most interesting aspect of the Shor algorithm, thanks to the various shortcuts we've discussed it is not the most computationally demanding. The real bottleneck, it turns out, is the procedure required for calculating the powers stored in the second register. Although we saw how modular exponentiation (that is, the calculation of powers modulo n) can be performed efficiently using repeated squaring, the whole calculation still requires more processing time than any other aspect of the Shor algorithm. Fortunately, though, the running time is still only polynomial in the number of digits, with the best implementations requiring a running time proportional to the cube of the number of digits, which is much less than any known procedure for factoring numbers on classical machines.

As we saw, Shor only had the factoring problem partly in mind when he began work on adapting Simon's quantum program. He finished by solving two problems, both with major implications for cryptography. His solutions to the problem of finding discrete logarithms and to the problem of factoring numbers threatened to blow a hole through both the Diffie-Hellman encryption scheme and the RSA system. Nevertheless, cryptographers appear not to have been particularly rattled when they heard of Shor's discovery. Many, after all, felt they could continue to sleep easily in the knowledge that no one had yet built a quantum computer or looked likely to do so in the immediate future. But some also knew that in the unlikely event of such a machine's being constructed, they could turn to another aspect of the quantum information revolution for a new weapon to protect their secret codes.

6

Privacy Lost, Privacy Regained

The quantum taketh away and the quantum giveth back![1]

ASHER PERES

Messages from Across the Quantum Channel

While quantum mechanics could spell disaster for public-key cryptography, it may also offer salvation. This is because the resources of the quantum world appear to offer the ultimate form of secret code, one that is guaranteed by the laws of physics to be unbreakable. That is the remarkable promise of *quantum cryptography,* a discipline that has flourished since the 1980s, initially independently of quantum computing but now as an integral part of it. Indeed, quantum cryptography may turn out to be the first real application of quantum computing because both depend on the same technology and the same conceptual foundations.

Quantum cryptography actually originated in the late 1960s when Stephen Wiesner, who was then at Columbia University, dreamt up a way to use quantum mechanics to produce a form of money that would be impossible to counterfeit. He also suggested a way to combine two classical messages into one quantum message from which the receiver could extract either one message or the other but not both. This latter idea might sound like a rather unproductive thing to do, but it turns out to have surprisingly interesting ramifications. Wiesner wrote up his ideas in 1970 and tried to get them published, but his scientific paper was rejected. It wasn't until 1983 that it saw the light of day.

That it was finally published was due in part to an event that took place in October 1979. Gilles Brassard, a cryptography researcher at the University of Montreal, was swimming in the ocean around Puerto Rico when, as he recalls it, a complete stranger swam up and started telling him about a scheme for producing unforgeable banknotes based on quantum mechanics.

The stranger was IBM's Charles Bennett, and the encounter was the start of a collaboration that finally got the subject of quantum cryptography off the ground. "This was not a completely chance meeting," Brassard recalls. "We were attending a conference in Puerto Rico, and I was scheduled to give a talk on classical cryptography. So Charlie thought it would be a good opportunity to tell me about these crazy ideas." Bennett had learned of Wiesner's ideas when, as undergraduates, they had shared rooms.

The unforgeable banknote scheme, though strikingly novel, was hardly going to take the banking world by storm, because, as Brassard says, "it required storage of quantum information, which is hard enough today but was unthinkable in 1979, let alone 1970. It was estimated that the average banknote would have to be the size of a small truck and would last for a tenth of a second before it decayed! So it wasn't very practical, but still it was a neat idea. When Bennett told me about it, the following day I had an idea to improve it. In Wiesner's original idea, the banknote could only be spent at the bank that created it, and I thought of a way in which it could be spent at the corner store."

Bennett and Brassard continued to meet regularly and during their brainstorming sessions concluded that instead of using quantum mechanics for information storage, it was going to be much easier to apply it to information transmission. That, after all, is what photons, which by definition travel at the speed of light, are good at. Among the ideas they considered was the application of quantum mechanics to public-key cryptography. After a few years they developed a protocol for implementing a system of *quantum key distribution.*

In 1984 they published a version of this idea known in the trade as BB84,[2] which offered a practical method of using quantum mechanics to exchange secret keys securely over public channels. The scheme, like

Wiesner's idea for protecting banknotes, depended on the uncertainties intrinsic within quantum information to keep information secret.

So how does it work? The scheme involves sending photons prepared in different states of polarization. Using a polarizing filter, it is possible to select any polarization angle with respect to the horizontal. In this scheme four particular angles are chosen: 0, 45, 90, and 135 degrees, which we can illustrate as:

To exchange a secret key, the sender, Alice, first generates a sequence of photons whose polarizations she chooses at random from each of these four states. When her friend, Bob, receives the photons, he decides randomly whether to measure the polarizations along the rectilinear (horizontal/vertical) directions or along the diagonals. He can do this using a crystal of calcite, which separates incoming photons along two separate paths according to their state of polarization.

Note that it is possible for Bob to discriminate with certainty only between perpendicular (that is, orthogonal) states and only if he has his measuring apparatus correctly oriented. If Alice sent a stream of only diagonally polarized photons, and Bob set his receiver to distinguish rectilinear polarizations, he would see 50 percent of the photons taking the horizontal route while the other 50 percent took the vertical route. But the path for any particular photon would be completely random and would thus not convey any useful information. Even if Alice switched between the two diagonal

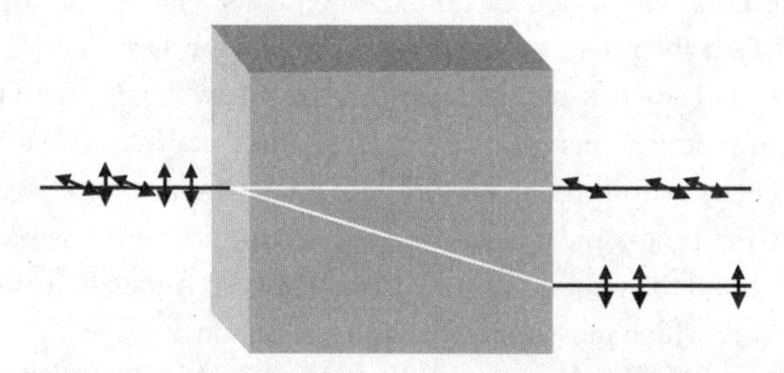

Figure 6.1. A calcite crystal will separate photons of different polarization

If Alice sends a message in a series of photons using a sequence of horizontal and vertical polarizations, Bob can read the message with near 100 percent accuracy by separating the photons using a calcite crystal. If she uses diagonal polarizations, however, Bob will get random results unless he reorients the crystal.

states, Bob would see no impression of her activity in the signals he measured. They would remain completely random. But if he reoriented the calcite crystal for diagonal states, he would then be able to distinguish the two polarizations with 100 percent accuracy, assuming he had a perfect detector and crystal. Bob would by this means be able to extract one bit of information from each photon.

The rectilinear states and diagonal states are examples of the conjugate quantum states we saw in Chapter 3. Conjugate properties are subject to Heisenberg's uncertainty principle in that the more we know about one, the less precise will be our knowledge of its conjugate. This means that if someone makes the wrong kind of measurement on a particle, its quantum state will be disrupted, making it impossible to recover the information.

So if Alice sent a vertically polarized photon followed by a horizontal one, then, provided Bob had his crystal oriented for detecting rectilinear polarizations, he would see one photon go along one path and the other go along the second. If Bob was looking for diagonally polarized photons, though, there would be no way he could make further measurements to find out what state the photon had been in originally. The uncertainty principle intervenes so that the very act of measurement disturbs the state of the system Bob is trying to measure.

In BB84, Alice and Bob both choose between rectilinear and diagonal polarizations at random, so it's inevitable that some of Bob's measurements will be of no use. But once Alice has sent her initial sequence of photons, Bob announces which type of measurement he made for each of the photons. He only reveals the *type* of measurement, not the measurement *result.* Alice then tells him which of his measurements were of the right kind. Alice and Bob then discard all cases in which Bob has made the wrong measurement. They can now be confident that they should share the same sequence of polarizations for the remaining photons. Unless, that is, somebody has been eavesdropping on their communications.

If Eve, the prototypical eavesdropper, surreptitiously measures the polarizations of all the photons as they travel between Alice and Bob, she too will be unsure which measurements to make. So, on average, 50 percent of the time she will make the wrong ones, but not necessarily on the same occasions that Bob gets it wrong. In fact, on average she would make the wrong measurement 50 percent of the time Bob is making the right measurements. So 50 percent of Bob's final data will have been measured incorrectly by Eve. The disturbing effect of Eve's measurements will lead to random results in half of Bob's data. Half the time these random results will produce the *correct* answer simply by chance. Therefore the net effect of Eve's measurements will be to cause errors in one quarter of Bob's final data.

Nevertheless, this is enough to betray Eve's intrusion. To check whether anybody is listening in, Alice and Bob can randomly select a subset of their polarization data and compare them. If the comparisons reveal differences, Alice and Bob can assume that somebody has been attempting to eavesdrop. If not, then they can use the remaining data, which they have not revealed to one another, to construct a secret key that can be used to encode messages using a one-time pad cryptosystem. Since Alice produced her data randomly, they can generate the secret key from the data directly without further ado. Alice and Bob can then communicate using the secret key to encrypt their messages, confident that no one else will have been able to deduce their key.

In their description of the scheme, Bennett and Brassard discussed two different types of communication channel that would be required between Alice and Bob. One is the quantum channel, which carries the polarized photons (that is, the qubits), and the other is a conventional classical channel, such as a telephone wire, which could carry all of the conversations in which Alice and Bob discuss the measurements they've made and the data they want to compare. Notice that these conversations can be public because none of that information will reveal the contents of the qubits exchanged along the quantum channel, apart from the sacrificial qubits used for checking the integrity of the quantum channel.

Eve could be a little smarter in bugging the quantum channel by making measurements of only a small proportion of the photons passing between Alice and Bob. She could calculate that provided she looks at only a subset of them, Bob might not notice the occasional error, especially as her measurements affect only a quarter of Bob's final data anyway. Of course, by doing this she would sacrifice some information about the photons, but she might figure that some information about the secret key is better than none. To avoid such circumstances, Alice and Bob need to compare enough data to ensure that they can detect low-level eavesdropping.

Although direct comparison would work, it isn't very efficient because Alice and Bob would have to waste a lot of data even if they thought somebody might listen in only occasionally. To overcome this problem, Bennett and Brassard suggested a better method for detecting eavesdroppers. They pointed out that Alice and Bob could compare the *parity* of certain randomly selected bits of their data. Parity represents whether a number is even or odd. So, for example, Alice could tell Bob that she had looked at the 1st, 4th, 5th, 9th, 12th . . . of her 100 bits of data and found that they included an odd number of 1s. Bob would then count the number of 1s he had for the same bits, and if he got an even number, he could conclude that his data had been corrupted in some way. With this technique, the chances of any corruption being undetected are halved for each bit sacrificed, so it is very efficient. Parity checking of this kind is a standard feature of conventional communications systems for eliminating errors.

All About Eve

When Bennett and Brassard were thinking about quantum cryptography in the early 1980s, most physicists (including Bennett and Brassard!) regarded their ideas as science fiction or, at best, more akin to thought experiments than practical laboratory concepts. However, by 1989 Bennett and Brassard had built an experimental prototype at IBM's Thomas J. Watson Research Center that led to the first experimental demonstration of quantum cryptography. They were assisted by François Bessette and Louis Salvail, who helped write the software for running the experiment, and John Smolin, who worked on the electronics and optics. This was a truly historic moment: it represented the first time that a computational task beyond the repertoire of the universal Turing machine—that is, any classical computer—had ever been performed.

The system demonstrated the secure exchange of data via a beam of light traveling over a grand distance of 32 centimeters. At Alice's end was a green light–emitting diode, a lens, a pinhole, and some filters that produced a collimated beam of horizontally polarized light. The beam passed through a pair of Pockels cells, electro-optical devices that change the direction of light polarization according to electrical signals sent to them. At the other end of a lightproof box that housed everything, Bob's receiving apparatus used a Pockels cell to allow him to choose the type of polarization he would measure—that is, his measurement basis. After traveling through the cell, the light was split by a crystal of calcite into two perpendicularly polarized beams that were directed toward two photon multiplier tubes, used to detect individual photons. Control of the whole apparatus was supervised by a single computer, but Alice's software and Bob's software were designed to work independently, even though they ran on the same machine.

The protocol was closely based on the BB84 scheme, but modifications were necessary to overcome some practical problems. No one knew how to build a controllable source of individual photons,[3] so the experimenters used dim flashes of light instead. This was done by pulsing the LED and using enough filtration to reduce the light intensity to one tenth of a photon per flash. (In other words, only one in ten flashes on average yielded a photon.) The receivers were not totally efficient either: they didn't respond to every photon that hit them, and the detectors sometimes triggered spontaneously in the dark.

If one were not careful, these practical deficiencies could open up new possibilities for eavesdropping. If, for example, Alice's light flashes were too bright, Eve might be able to split the light beam and divert some photons for her own measurements, leaving the rest to make their way to Bob. She could then make measurements without any danger of disturbing Bob's. By reducing the intensity to one tenth of a photon per flash, the researchers reduced to 1 in 20 the chance of Eve's being able to split a non-empty pulse.

The "dark count" problem of the detectors, on the other hand, leads to errors even when there has been no eavesdropping. Because this happens quite often, it would slow the information flow to a crawl if Alice and Bob simply rejected their data whenever they discovered an error, as the ideal BB84 protocol would require. Instead of throwing the qubits away, they can use some form of error correction. This entails public discussion between Alice and Bob to narrow down the location of an error so that it can be remedied. The only trouble is, such discussion inevitably leaks information to Eve about the key (assuming Eve has access to their conversations). Furthermore, Alice and Bob are unable to distinguish between noise on their quantum channel (such as the dark count) and Eve's attempts at eavesdropping.

Overall, then, Alice and Bob have to live with the fact that some leakage of information about their key will occur. But Bennett and Brassard had another trick up their sleeves. In 1988, along with Jean-Marc Robert, then a student of Brassard's, they developed a mathematical technique called *privacy amplification.* With it, Alice and Bob could work out by public discussion how to purify their secret key, eliminating the part that was most likely to have been made vulnerable to any potential eavesdropper.

The procedure is based on computing techniques known as *hashing* functions. These are widely used in computer programs for compressing big numbers into smaller ones. To see how they work, imagine a hotel in a coastal resort notorious for petty thieves. The manager decides to engrave door keys with random-looking numbers instead of the room numbers so that if one gets lost or stolen, an outsider won't automatically know which door it will open. The front desk can remind forgetful guests by checking on a large lookup table. Or the hotel staff could use a hashing algorithm.

Consider, for example, the function $f(n) = 137 \times n \bmod 239$. Now, if you were given hotel keys with numbers in the sequence 1450, 1451, 1452 . . . 1459, you'd expect all the rooms to be on the same floor and most likely strung along the same corridor. However, in this strange hotel where they use the hashing function above to convert the key numbers to room numbers, you find that the corresponding rooms are scattered all over the place: 41, 178, 76, 213, 111, 9, 146, 44, 181, and 79. Now suppose a potential thief, Steve, discovers a room key on the beach. Unfortunately for him the last digit has been worn away, so he can make out only the first three digits: 145. Even if he knows the hotel's hashing algorithm, after he's done the conversion he won't be sure of any of the digits. His uncertainty in one digit gets blown up into an uncertainty in all the digits. This is the general idea behind privacy amplification.

In the IBM/Montreal experiment, Alice and Bob use this principle, although the hashing function is different. They select random subsets of bits, replacing them with just one bit representing the parity of the removed bits. Alice and Bob tell each other which random bits to measure and replace, but without revealing the parities. Provided Alice and Bob start off with exactly the same data, they will finish with the same answers. On the other hand, Eve, who has only partial information initially, finds that she has virtually no information about the key by the end of the hashing procedure. Although Eve may know what hashing function Alice and Bob are using, her uncertainty over the final form of the key is magnified as more of the bits are thrown away and replaced by parities. In fact, Bennett, Brassard, and Robert showed that in conventional circumstances such techniques of privacy amplification could virtually eliminate any danger of Eve's being able to exploit information leakage via the public and quantum channels.

However, as Bennett freely admitted when I talked to him, that still doesn't make the system totally foolproof. "It's hard for a cryptosystem to be totally secure," he said. "You have to be aware of all sorts of possible attacks. For example, from an eavesdropper's point of view, our quantum cryptosystem wasn't secure at all, because the Pockel cells were running from a power supply that hummed. And it hummed at a different loudness depending on the voltage that was applied to the cells. So although you wouldn't be able to eavesdrop on the photons, you could just listen to the hum to find out what data were going through the system!"

Dial Q for Qubits

Surprisingly, the IBM/Montreal work did not attract much attention from other physicists, at least initially. The situation changed when Artur Ekert at Oxford produced a different approach to quantum cryptography as part of his 1990 Ph.D. dissertation. His proposal, which we'll examine in more detail soon, was closely related to the EPR experiment (see Chapter 3, page 94) in which entangled particles were seen to exhibit strong correlations that go beyond anything that is possible by classical means. A crucial feature of Ekert's idea[4] was that he proposed exploiting the same kind of hardware that had already proved its worth in demonstrations of the EPR effect—the technology of parametric down-conversion crystals, optical fibers, and interferometers. Indeed, in 1992, it was while skiing in the Italian Dolomites that Ekert devised a plan with John Rarity of the United Kingdom's Defence Research Agency (DRA) for implementing a practical method[5] of entanglement based cryptography.

The introduction of these techniques led to rapid progress in quantum cryptography with new schemes that greatly increased the range over which it could be made to work. The technologies were also applicable for schemes based on Bennett and Brassard's approach. In 1994 Paul Townsend and Christophe Marand at British Telecom's research labs in England used a version of the Bennett and Brassard protocol to demonstrate the first truly long-range distribution of quantum keys. The BT systems operated over 30 kilometers. In spite of the impressive distance, though, the experiment remained a bench-top affair because the cable lay wound on a spool in the same room as the send and receive apparatus.

Following the work at BT, a team from the University of Geneva led by Nicholas Gisin and funded by Swiss Telecom was the first to demonstrate a system in which the sender and receiver were actually separated by a significant distance. Their signals ran along a cable running beneath Lake Geneva between Nyon and Geneva, 23 kilometers apart. Since then the British team successfully tried out its system on spare fiber-optic cables on BT's network in Ipswich, a large town in eastern England. They even managed to convey their cryptographic signals alongside simulated telephone traffic carried on the same fiber-optic cables.

In 1999 Richard Hughes and colleagues pushed the limit further by distributing quantum keys through the Los Alamos National Laboratory Fiber Network across a 48 kilometer range.

In the meantime, Townsend took things one step closer to reality by demonstrating quantum cryptography over a passive fiber-optic network in which the sender could distribute keys to many different receivers. The network is described as "passive" because there is no active switching to route the signals. Instead the fiber is used a little like a broadcast medium. The broadcaster at one end sends signals down the fiber over a wide range of wavelengths. At the other end, the fiber spreads out like a tree with receivers connected to each of the branches. In principle, everyone gets the same signals, but each receiver tunes into the data or the telephone call that's meant for that one person, just as a radio receiver filters out one radio station from the rest. Passive optical networks are already in use and are of great interest to telecommunications companies because of their simplicity and potential low cost. Their only flaw is an intrinsic lack of privacy. Anyone can listen in to other people's conversations.

However, there are ways of protecting digital data using standard encryption techniques. It is these that now protect digital mobile phones from "scanners," receivers specially designed for picking up mobile phone calls. The older analog cellular phones in Europe didn't use any encryption, a fact that led to some rather high-profile embarrassments. The "Squidgygate" press sensation of 1990, for example, broke because somebody with a scanner caught Princess Diana talking on an analog cellular phone with a boyfriend who called her "Squidgy."

Passive networks could be made reasonably safe provided, of course, the encryption systems are sufficiently strong. How then can quantum cryptography be implemented on such a network? The BT team took advantage of the fact that when the signals are in the form of individual photons, they get randomly distributed by the splitters at nodes on the network tree. The result is that only one person gets any one photon. Alice, in effect, broadcasts her cryptographic keys onto the fiber-optic network while splitters along the way randomly apportion different qubits to different recipients. The splitters therefore ensure that each person gets a unique assortment of qubits. There is little danger, therefore, of eavesdroppers finding out what somebody else got.

But how does Alice know who got which qubit? The answer is in the timing. Each receiver must carefully time the arrival of each photon. (This also applies, of course, to the original Bennett-Brassard system because not all of Alice's photons make it to Bob's receiving apparatus or are detected.) When each person on the other end of the network reconciles his measurements with Alice, he will tell her the arrival time of each of his photons. Alice can then determine which qubit is which in the sequence she sent. By this process Alice can distribute a unique secret key to everyone on the network. After that she can communicate with anyone privately using the appropriate secret key to encrypt conversations in the normal way.

Such developments have raised hopes for the possibility of using quantum cryptography commercially, but there are still many issues to be resolved. A big limitation at the moment is the distance over which these systems will work. BT's Paul Townsend reports a typical error rate in his quantum transmissions of around 3 percent over 30 kilometers of fiber-optic cable. Such errors are bad news for security because it's hard to distinguish them from the intrusion of an eavesdropper. A possible solution is to use privacy amplification to clean up the signals, but this technique has its limits.

Inevitably, as the length of fiber increases, the number of photons making their way through the pipe drops. This in itself would not be such a bad thing if it were not for the problem of the "dark counts": this refers to the unfortunate tendency the detectors have for spontaneously firing at random intervals even in the absence of photons. The combined effect of the fiber loss and the dark counts is that over longer distances the error rate on photons that survive the journey goes up. Once the signals become too noisy, even privacy amplification will not be able to put things right. So an important objective is to improve the efficiency of the transmitters and receivers, which so far have mostly been built using off-the-shelf components.

In 1997 Gisin and his colleagues in Geneva demonstrated an elegant noise cancellation technique that dramatically cut the error rate over their 23-kilometer fiber-optic link. The principle works as follows. Bob sends

Alice weak pulses of light that carry no information. Alice manipulates the pulses to encode information into their phases, rotates the polarization of the pulses by 90 degrees using a device known as a Faraday mirror, and sends them back to Bob. Bob measures the phase information in the pulses and obtains Alice's information. The reason this procedure reduces errors is that getting the pulses to retrace their journey with their polarizations flipped causes many of the errors arising on the forward journey to be canceled out by equal and opposite errors on the return journey. This method not only reduced errors to 1.3 percent, it also greatly simplified the setting up and running of the equipment because it eliminated the need for continual realignment of the equipment, a problem that normally arises as a result of thermal drift. The error cancellation technique's practical advantages make it very attractive.

Even with such improvements, Paul Townsend doesn't think it will be possible to distribute quantum keys via fiber-optic systems over intercontinental distances. "That's inconceivable," he says. "Or at least it's inconceivable with a single link. You could probably get up to distances of, say, 60 to 100 kilometers—this would be an ultimate limit because the fiber loss will then be so high that you just don't get enough photons."

Ordinarily, on long fiber links such as transatlantic cables, for example, the answer is to use repeaters every 30 kilometers or so to boost the signals. The problem is that to amplify a quantum signal, you first need to measure it and that inevitably disturbs its state. So conventional repeaters would be of no help.

Nevertheless, there are a number of possible solutions to the range limitation. One idea is that you could arrange for different keys to be set up between each of the repeater stations. Alice would establish a quantum key with the first station, while Bob agreed on some other quantum key with the last station. Different keys would be established along each leg of the journey in between. So once this chain of keys was in place, Alice and Bob's conversations would pass back and forth using different encryption keys at different points along the communications path.

Such a system could work, but a drawback is that the signals would be vulnerable to eavesdroppers at the repeater stations because it is at these points that the messages would be translated classically from one encryption code to another. Anyone who had access to these repeater stations could eavesdrop.

Another possibility is to transmit photons through the air or through space rather than through fiber-optic cable. This approach is being explored by Richard Hughes and co-workers at the Los Alamos National Laboratory in New Mexico. The idea is that by using lasers and satellites it might be possible to bounce signals around the Earth and thereby establish a global system of cryptographic links. In principle, this could be done simply using mirrors on the satellites because, again, it wouldn't be possible to amplify the quantum messages using photoelectronics.

The difficulty with this method is that the photons have to travel through the atmosphere for at least part of their journey. As any amateur astronomer knows, the Earth's atmosphere is not very kind to faint sources of light. Even on a clear night stars twinkle because of atmospheric disturbances to the photons as they travel toward us. The molecules of air cause photons to be scattered off course, while variations in pressure and density cause timing and hence phase jitter. Even if a photon makes it through the atmosphere, it may well have interacted with a molecule and lost its initial quantum state. Once the light beams have escaped the atmosphere, though, they are much less likely to be disturbed as they travel through space. So quantum links between satellites would represent less of a problem.

At the moment, free-space experiments have got further to go before they catch up with the fiber-optic demonstrations. Nevertheless, Hughes is confident that his team will be able to overcome the problems. In 1999 they demonstrated a quantum key distribution system working over a distance of ½ kilometer in full daylight. The main problem Hughes faced in extending the range was simply to find a suitable site to locate the experiment. So the idea does hold out some hope that it might one day be possible to communicate globally using quantum cryptography.

Another distinct possibility that could eventually emerge from attempts to build quantum logic circuits would be the production of genuine quantum repeater stations. They would exploit the nonlocal effects of entangled particles to boost the quality of quantum channels. If this could be made to work, the distance limitation of cryptography could be lifted completely, which would probably totally transform its potential as a serious player in the cryptography market.

Of course, whether any of these technologies make it into the real world depends ultimately on economics as well as practicality. Satellite-based methods of quantum cryptography will obviously be very expensive, but they could come into their own for ultrasecure military and government communications. For business applications at the moment, classical cryptography is cheap and easy to implement. With the enormous growth in business transactions expected on the Internet, encryption is bound to become an increasingly important feature for people using this medium. At the moment there's no obvious way of implementing quantum cryptography on the Internet, and it's doubtful whether many people would be willing to pay the high costs required for its use on dedicated links. However, if public-key cryptography were suddenly brought down by the development of quantum computers, then clearly quantum cryptography would suddenly look very attractive.

Quantum Clones and Counterfeit Coins

Cryptography has many more applications than just sending secret messages. It can help solve a wide range of problems that involve secrecy, authentication, integrity, negotiation, and making agreements. Above all, it is an essential weapon in fending off dishonest people. Researchers have therefore wondered whether quantum cryptography could be applied to problems other than quantum key distribution.

Wiesner's original ideas in 1970 are a case in point. They offered novel applications that went beyond anything that was known in classical cryptography. His idea for unforgeable money, impractical though it was, is a good example of the so-called no-cloning theorem, which shows that it is impossible to clone or copy quantum states. Though the idea was known long before, the proof of this theorem by William Wooters and Wojciech Zurek, which appeared in a paper published in the journal *Nature*[6] in 1982, was another defining step in the development of quantum information theory.

Of course, if cloning were possible, quantum cryptography would be a total nonstarter because Eve would be able to replicate any photons she saw passing between Alice and Bob. She would then be free to make whatever measurements she liked on the cloned photons without disturbing the original ones, which could be sent on their way to Alice and Bob. Fortunately, Heisenberg's uncertainty principle stands as an impregnable fortress against such artifice. If cloning were possible, it follows that we could make conjugate measurements on a photon and its clone. We could, for example, measure both the vertical and horizontal polarizations of one particle and its clone and thereby gain knowledge about the two in defiance of the uncertainty principle. This is ruled out by the laws of physics.

Actually, to say it's ruled out is a bit of an oversimplification because as we saw in the EPR experiment (see Chapter 3, page 94), it *is* possible to produce entangled particles that are, in a sense, clones of one another. With an entangled pair of particles it is, of course, possible to perform a pair of conjugate measurements—one on the first particle, and the other on the second particle. With this arrangement we could say that the second particle, in effect, acts as a proxy for the first. The reason this procedure doesn't conflict with the uncertainty principle is that as soon as the first particle is subjected to its measurement, quantum mechanics demands that the conjugate measurement on the second particle will yield a completely random answer, just as you would expect from the uncertainty principle. The mysterious aspect about this phenomenon is how the second particle "knows" what has happened to the first. We'll come back to this point a little later in this chapter. The main message for now is that although it is possible to generate particles in the form of identical twins, it isn't possible to clone a particle given to you in an unknown quantum state.

The inability to clone quantum states explains why it isn't possible to build conventional repeater stations for fiber-optic methods of distributing quantum keys. Nevertheless, this inability clearly opens the way to some interesting applications.

Wiesner's idea, for example, was that a coin or note of quantum money would contain a number of isolated two-state particles, such as atomic nuclei of spin ½. When such nuclei are subjected to measurements, they would have either a spin up or a spin down. However, just as photons can have intermediate polarizations, these nuclei could have intermediate spin directions.

Wiesner imagined a banknote (or coin) harboring twenty of these atomic spins, each of which had been carefully oriented at the mint. The people at the mint would do this by first generating two random binary sequences of twenty digits each. Each bit in the first sequence, M_i, is used to determine which measurement basis is used to orient each atomic spin: 0 = down/up, and 1 = left/right. The second sequence, N_i, is then used to determine which of the two opposite directions is selected: 0 = down or left, and 1 = up or right. So if M_i was 001011 . . . and N_i was 100110. . . , the sequence of spins would be:

$$\uparrow \downarrow \leftarrow \uparrow \rightarrow \leftarrow \ldots$$

Each banknote is also engraved with a serial number in the usual way, while the two binary sequences describing the state of the atomic spins are kept on record at the mint. If a bank needs to check the authenticity of a note, it checks the state of the spins against the records held at the mint.

Now, imagine that you wanted to duplicate this note illegally. What would you do? If you attempted to measure the orientations of the spins in the note, you would need to decide which measurement basis to use for each atomic nucleus. Your dilemma would be whether to orient your apparatus left/right or down/up. For each incorrect choice of measurement, you would get the wrong answer 50 percent of the time, so overall there is a 1 in 4 chance of each digit you read being wrong. Overall, the probability that the counterfeit note will pass inspection is $(3/4)^{20}$, which is around 1 in 300. Obviously the odds could be made even higher by adding more atomic spins. The forger's lot is made all the more troublesome because his attempt at measuring the spins is very likely to corrupt his genuine banknote.

How to Send a Quantum Valentine

Imagine a man sitting in a singles bar. His eyes alight on a particularly alluring woman sitting by herself. He wonders whether she is with anybody, but after quite a while she is still quite alone. He believes he caught her eye for a moment, but he's not sure. She certainly looks like she might like some company, but you never can tell—perhaps she just wants to be alone. The trouble is, the more he looks at her, the more attractive she becomes, so that by now he's in turmoil. Has he the courage to go over and talk to her, he wonders. Judging from previous experience he'll blow it and she'll tell him to get lost. But if he doesn't ask, he'll never know what he missed. If only he could contrive some seemingly innocent reason for talking to her, he could then assess his chances without making his intentions so obvious. But all he can think of are really improbable openings such as "Hi! Weren't you at the quantum gravity conference in Vegas last month?" How can he get to talk to her without making a complete fool of himself?

Strangely enough, quantum cryptography could offer a solution to this dilemma. It helps, though, if the man's predicament is presented in the more formalized guise of the "dating problem," in which two singles want to find a way to make a date with each other but only if they both like each other. If, for example, Bob likes Alice but she doesn't like him, then no date should be arranged; moreover, Alice mustn't find out that Bob liked her. Of course, in this circumstance it is inevitable that Bob will discover that Alice isn't keen on him, because otherwise the date would be on. But at least he doesn't suffer any embarrassment by revealing his feelings.

How can this arrangement be brought about? This scenario and a wide variety of other situations belong to what are known in cryptographic circles as *discreet decision* problems. Another example is the millionaire's problem. Two millionaires want to know who is the richest without revealing how much money each has got. A more serious variant of that would be two companies that wanted to embark on a joint venture if and only if they had sufficient combined capital. They would like to know whether this condition is fulfilled but without revealing how much money each has. Similarly, two military powers might want to know who had the most missiles without divulging how many each has exactly. All these problems can be solved very simply by introducing a referee to arbitrate. But can the referee be trusted to keep his or her mouth shut afterward? Alice or Bob may not be willing to take the risk.

These and other problems have aptly been described as "post–cold war" applications of cryptography, purposes that go well beyond the cloak-and-dagger world of sending secret messages between spies. These applications are amenable to classical cryptographic solutions involving public-key systems, but like the key distribution protocols, they are based on unproven assumptions about the difficulty of factoring large numbers and other related problems. What hard-core cryptographers would like is a totally bulletproof system that's reinforced by the best guarantees you can get—like the laws of physics. That's where quantum physics comes in.

In the 1980s cryptography enjoyed a golden period that saw the arrival of many new ideas about how to construct solutions to such problems from basic cryptographic building blocks. One of these building blocks was the notion of *oblivious transfer,* proposed in 1981 by Michael Rabin. In this, Alice sends Bob a one-bit message in such a way that it has exactly a 50 percent chance of arriving and only he will know whether it did. At first this might seem like a rather strange thing to do. However, it turns out to be potentially very useful, as we'll see in a moment.

Rabin's oblivious transfer can be brought about by exchanging pairs of public keys,[7] but the idea had already been foreseen in a quantum guise by Stephen Wiesner in his unpublished paper of 1970. Recall that Wiesner's idea was that Alice could send Bob a pair of messages encoded in a quantum state such that Bob could read either one but not both messages and

Alice wouldn't know which one.[8] That idea later resurfaced under the name *1-out-of-2 oblivious transfer.* The two versions of oblivious transfer turn out to be equivalent in the sense that each can be constructed from the other. More important, Claude Crépeau, one of Brassard's collaborators, and Joe Kilian showed that oblivious transfer could be used as a building block for solving *any* two-party problem requiring discretion.

An analogy for 1-out-of-2 oblivious transfer is as follows. Imagine you are offered two jars of cookies. You are told that you can choose a cookie from either jar, but in order to stop you getting fat, as soon as one is opened you are no longer allowed to have any cookies from the other. If the owner of the jars of cookies isn't allowed to look at which jar you've opened, then you've got 1-out-of-2 oblivious transfer.

Wiesner's quantum method of doing this was to encode a series of photons using different types of orthogonal polarizations (for example, ↔ ↕ or ⤢ ⤡, as in the banknote scheme). So Alice encodes each photon with a bit from the first message using rectilinear states, say, or a bit from the second message using diagonal polarizations. Note that it isn't possible to encode each photon with both bits *simultaneously,* so she chooses randomly between each message for each bit. Bob can choose to read bits from the first message by setting his detector always for rectilinear polarizations, or he can read bits from the second message by setting his detector always for diagonal polarizations. Of course, Bob will only be able to read a random selection of bits from one message or the other because the rest were measured in the wrong basis. But if Alice tells Bob afterward which basis she used for each bit, he can, at least, work out which are the good bits. However, he mustn't reveal the dud bits to Alice because she will then know which message Bob was trying to read, which would go against the rules of the protocol.

Well, the scheme isn't really a flier because Bob can get quite good information about *both* messages by setting his detector at an intermediate angle of 22½ degrees. Nevertheless, there are much more secure but necessarily more complicated quantum schemes[9] that also take account of noise in transmission and detector inefficiencies.

Having established that we can, at least, see a way of doing oblivious transfer using quantum mechanics, let's see how we apply it to a real problem like our singles searching for a date. I'll warn you that the explanation gets a little more complicated than you might expect, but the solution is really quite cute.

Suppose Alice starts off by sending a pair of random binary numbers A_1 and A_2 to Bob using 1-out-of-2 oblivious transfer. Bob is told that if he wants to go on a date with Alice he must choose to read Alice's first number; otherwise he should read the second. (He is told what measurement basis to use for each number.) He then has to calculate two new numbers by combining Alice's number with one of two random numbers he generates: B_1 and B_2. Now, the method of combination he must use is the logical function *bit-wise XOR*. What this means is that each bit of one number is XORed with the corresponding bit of the other number. The rule for XOR is very simple: if both input numbers are the same (that is, 00 or 11), the output is 0; if they're different (01 or 10), the output is 1. So, for example, the bit-wise XOR of 0010 and 1011 is 1001. So Bob calculates $A_?$ XOR B_1 and $A_?$ XOR B_2, where $A_?$ is either A_1 or A_2, depending on Bob's intentions about the date.

Bob now returns both these two numbers to Alice using 1-out-of-2 oblivious transfer. When Alice receives this message from Bob, she has to make her choice. If she wants the date, she must read the first number, $A_?$ XOR B_1, and if not, she must read the second, $A_?$ XOR B_2. Notice that when she reads either message she cannot tell which of her numbers Bob chose, because they have been disguised with his random numbers using the XOR function.

Now comes the exciting part. Whichever message she chooses, Alice is told to combine it with her first number, A_1, using the XOR procedure again. She now declares this final number while Bob declares his first random number, B_1. If the numbers are the same, they are in luck and the date is on! If they are different, the date is off and whoever voted against it will not know which way the other person voted, so embarrassment is saved.

What is going on here? If Alice and Bob like each other, then they will both select the first messages in each pair sent to them. Bob will receive A_1 and Alice will receive A_1 XOR B_1. Alice then calculates A_1 XOR (A_1 XOR B_1), which actually simplifies to B_1 because whenever you apply the XOR function twice with the same number the effect is canceled out. (Try the result of our previous example: 1001 XORed with 1011 gives 0010.) So Alice should obtain B_1, which is Bob's first random number.

Suppose Bob cheats by voting yes to the date but then changes his mind once he sees Alice has voted yes too because she's declared B_1. He could try declaring a different number from B_1. However, Alice will be able to detect his fraud because Bob's only legitimate alternative arises if he chose her second message. In this case she would receive from Bob A_2 XOR B_1. With this she can calculate B_1 by XORing it with her second number, A_2. If Bob's declared number, B_1, fails both tests, he must be cheating.

What happens if Alice tries to cheat Bob by declaring a different result from her calculation? Bob will discover her fraud because the only legitimate alternative she can have is B_2, which she could only know if she chose to read his second message—a choice that means she doesn't want the date.

Either way, whoever votes no to the date will be none the wiser about what the other party did.

Of course, there is one major drawback to this whole scenario. Both parties must be willing to participate in this elaborate process, which rather implies there must be a willingness on both sides to consider going on a date in the first place. That would be okay for people joining a dating agency but not much help for our poor guy sitting in the bar. Not unless it was a high-tech singles bar with quantum cryptographic equipment at each table—a kind of newfangled variant of an idea some restaurants have tried: having telephones at each table so patrons can talk to each other "anonymously." Who knows, it could happen.

Although the application to the dating problem may not be entirely serious, there are many situations in which discretion is vital to making agreements. These include joint decisions between corporate or governmental organizations where the negotiating parties don't want to give away secrets. Negotiating arms treaties, forming business partnerships, and organizing mergers are good examples. For these kinds of applications, in which the negotiating parties could be in the same room, it would be possible to use a quantum cryptographic apparatus that worked over very short distances. In fact, you could almost perform the process with Bennett and Brassard's original experiment (except you would have to do something about the telltale hum from the power supply).

The Rise and Fall of Quantum Bit Commitment

In the early 1990s research into quantum cryptography began to take off in a big way. But just as researchers began to see opportunities for using quantum cryptography all over the place, as Gilles Brassard and Claude Crépeau expressed it in a journal review[10] looking back on twenty-five years of quantum cryptography, "the sky fell in."

The debacle concerned the downfall of a procedure known as *quantum bit commitment.* Like Rabin's oblivious transfer, bit commitment was originally conceived as a classical cryptographic tool, but one that was potentially as important as a cryptographic building block as oblivious transfer. In fact, it became evident that to provide totally secure quantum oblivious transfer, you would need a secure form of bit commitment.

So what is bit commitment? It's pretty much what its name suggests. Alice chooses the value of a bit—that is, 0 or 1—and "commits" it by putting it into a safe that she locks with a key. She keeps the key but hands the safe over to Bob. At some later time when Bob asks to see the bit, she gives him the key to unlock the safe. He can then examine the bit. The point of the exercise is that once Alice commits the bit she cannot change it. Why is this useful? In his book *Applied Cryptography,* Bruce Schneier offers two examples: one entertaining, the other more serious.

Alice is a magician who performs a card trick in which she guesses the card Bob will choose before he chooses it. She writes her prediction on a piece of paper that she inserts into an envelope and seals. Bob picks a card, and shows it to the audience. It's the seven of diamonds. Alice unseals the envelope and reveals her prediction, which, of course, is "seven of diamonds." To make this work, Alice has to switch envelopes at the end of the trick. If Bob had asked her to use bit commitment, though, such sleight of hand would not have been possible and she would have been obliged to stick with her original prediction.

In the second example Alice is a stockbroker who wants to convince an investor, Bob, that she can pick winning stocks. So Bob asks her to pick five stocks and says that if they are all winners after a month, he will give her his business. She complains that if she tells him the stocks, he could go ahead and invest in them without having to pay her anything. She suggests that he look at her previous month's predictions. He says he cannot be sure she won't have changed them to make them look more impressive. With bit commitment, though, she could pick her winners, commit them for safekeeping, and then reveal them a month later to Bob. He could then be confident that these were her genuine predictions.

As a very simple example of bit commitment's potential as a cryptographic building block, Manuel Blum in 1982 introduced a method for being able to flip a coin over the telephone: Alice commits to a random bit. Bob tries to guess the bit. Alice reveals the bit to Bob. Bob wins the flip if he correctly guessed the bit.

Using classical techniques of cryptography, it is straightforward to implement bit commitment. One example, using symmetric key cryptography, is the following: Bob generates a random number, *R,* and sends it to Alice. Alice creates a message consisting of the bit she wishes to commit to and Bob's random number. She encrypts the message with some random key and sends the result to Bob. When Bob wants to see the committed bit, Alice sends Bob the key. He then decrypts the message to reveal the bit and his random number. Provided the random number is correct, he can assume that Alice has not cheated and changed the bit.

However, the problem with such classical techniques is that they depend on unproven assumptions about the computational difficulty of reversing the encryption process. With the advent of quantum cryptography in the 1980s, it was natural to look for a quantum method of implementing bit commitment in the hope of providing a cast-iron protocol.

Sure enough, in the same paper where they introduced BB84 quantum key distribution scheme, Bennett and Brassard came up with a quantum method of coin flipping that implicitly relied on quantum bit commitment. It wasn't a real solution though, because they also showed how it could be broken. But it was a starting point. Work by Brassard and Crépeau presented at a cryptography conference in 1990[11] offered more elaborate protocols for coin flipping and quantum bit commitment. In 1993 the same authors, along with Richard Jozsa and Denis Langlois, presented what looked like the crowning achievement everybody had been hoping for: a protocol for quantum bit commitment that was absolutely and *provably* secure. Their paper, known after the authors' names as BCJL for short, boldly proclaimed:

> We present a new quantum bit commitment scheme. The major contribution of this work is to provide the first complete proof that, according to the laws of quantum physics, neither participant in the protocol can cheat, except with arbitrarily small probability. In addition, the new protocol can be implemented with current technology.[12]

So you get the picture. Quantum bit commitment had become the foundation for numerous applications in quantum cryptography. Oblivious transfer, and with it discreet decision making, ultimately rested on the security of quantum bit commitment. And here was this splendid protocol backed by mathematical proof. Quantum cryptographers had reason to be confident.

Then, catastrophe. In October 1995 Dominic Mayers, a student of Brassard's at the University of Montreal, discovered a subtle flaw in the "proof" of BCJL. The flaw was also discovered independently by Hoi-Kwong Lo and H. F. Chau, working at the Institute for Advanced Study in Princeton. The more people looked at this flaw, the more it became evident that the entire edifice of quantum cryptography built upon quantum bit commitment was about to come crashing down. But perhaps there was a way of patching up the proof? It is certainly not unknown for mistakes found in mathematical proofs to be subsequently corrected, leaving the main arguments intact.

No such luck. The following year Mayers showed that there was no way to fix BCJL. At the PhysComp '96 workshop on physics and computation held in Boston he delivered a paper with the in-your-face title "Unconditionally Secure Quantum Bit Commitment Is Impossible." In it he proved that it would be possible for Alice to cheat in *any* protocol for quantum bit commitment.

The way Alice can cheat is by using EPR particles—that is, pairs of entangled particles such as photons with matching polarizations. Let's see how this works.

After choosing her bit, Alice would normally send a quantum message to Bob that, in effect, described the "safe" in which her committed bit was locked. She could, for example, encode her bit in a series of photons as follows: Alice is going to send Bob a randomly chosen message using polarized photons. Before sending the message, she commits to a bit by determining whether she will send her message using rectilinear polarizations or diagonal polarizations. If she commits to 0, she sends Bob the message using rectilinear polarizations, and if she commits to 1, she sends it using diagonal polarizations. Bob is not told which kind of polarizations Alice intends to use, so he randomly orients his detector to measure rectilinear polarizations or diagonal polarizations for each photon. At this point, the first phase of bit commitment has finished. Alice has committed to her bit and has given a coded message to Bob as evidence of her commitment. Bob, however, has no way of determining Alice's bit from the information he has so far received.

At some later point, when Alice and Bob agree to "decommit," Alice unveils her bit along with the random message she sent. Bob reexamines the measurements he made to see which ones were performed in the correct measurement basis. He then checks the results of these measurements to see whether they tally with Alice's random message. If they do, all is well, but if they don't, Alice must be cheating.

That's how quantum bit commitment was supposed to work. But Alice can cheat without being caught by preparing her photons as entangled EPR pairs instead of producing them individually. She could, for example, send one from each pair to Bob while storing the other without measuring either of them. Until one or the other of the photons in a pair are measured, they have random polarizations. Only when Bob asks for Alice's bit to be unveiled does Alice proceed to measure her stored particles according to the bit commitment she now decides to make. Because her particles were entangled with Bob's, she can commit either way and still produce measurements that tally with Bob's. Bob, poor sucker, checks his readings and finds nothing amiss!

One reason it is possible for Alice to cheat is that when she sends the information describing the "safe" for her bit, she has to make sure that the information gives nothing away about the bit at that stage. Mayers showed that for any possible bit commitment scheme, the quantum states of the safe containing 0 or 1 must be either identical or very similar because otherwise Bob would easily be able to discern the difference and identify the committed bit prematurely. However, the very fact that the two states are virtually the same makes it possible for Alice to keep her options open and produce the appropriate key for either "committed" bit later on.

Given this failing of quantum bit commitment, what are the consequences? First of all, to everybody's relief, the flaw does not affect quantum key distribution because the security of the latter relies only on the no-cloning theorem. So at least one of the big applications of quantum cryptography survives intact.

But what about discreet decision making? In their 1996 review paper of quantum cryptography, Brassard and Crépeau argued that though the weakness was undoubtedly a setback, the practical implications were minimal at the present time. This was because to cheat the system Alice needs a machine with virtually the power of a quantum computer. For a start, she needs to be able to store the entangled photons for minutes, hours, or days, depending on the application. Then she needs to subject them to transformations of the kind that might only be possible on a quantum computer. Both steps are beyond anything that can be performed on current hardware. In contrast, the BCJL protocol for quantum bit commitment is well within the capabilities of current technology.

So we could envisage a period of time when discreet decision making would be carried out using quantum means in the knowledge that it was secure until quantum computers became available. Furthermore, up to that time the quantum methods would be superior to classical methods because there is no way of cracking them retrospectively. Classical methods carry the danger that any communications can be stored indefinitely and therefore cracked as and when the technology or improved algorithms become available.

The overall message, though, is probably best summed up by Brassard and Crépeau:

> The big lesson to learn from all this is that quantum information is always more elusive than its classical counterpart: extra care must be taken when reasoning about quantum cryptographic protocols and analyzing them.[13]

Cryptography by Entanglement

The idea of attacking cryptographic protocols using entangled particles was first hinted at by Bennett and Brassard in 1984. But as mentioned earlier, entanglement can also be used to *enhance* methods of key distribution. In 1991 Artur Ekert introduced a new scheme for distributing cryptographic keys by quantum means, a method based on the EPR experiment.

In Ekert's scheme Alice and Bob receive entangled particles from a central source, whereupon they both perform independent measurements on the particles along three different orientations. For each measurement, Alice chooses one of the three orientations at random. Bob does likewise, although his random choice of orientation is independent of Alice's. Afterward, they compare notes publicly on the orientations they used for each measurement. They then announce publicly the results they got for the measurements in which they adopted *different* orientations. The point of that is that these results can then be analyzed for the strength of their correlations. In the version of the EPR experiment using photon polarizations, we saw that these correlations exceed what is possible classically only when the detectors are oriented obliquely to each other and even then only for a certain range of angles. (The strongest divergence between quantum and classical correlations occurs at 22½ degrees, exactly midway between perfect correlation, at 0 degrees, and no correlation, at 45 degrees.)

Now, if someone attempts to make measurements on the particles before Alice or Bob receive them, the correlations will be disrupted. Charles Bennett has an amusing way of expressing this point that draws its references from real life. "If a third person attempts to intervene," Bennett muses, "the quality of the entanglement between two parties is inevitably degraded." The quantum version of this effect provides a crucial test for the integrity of the communications channel by which Alice and Bob receive their particles. Provided the results exhibit the superstrong correlations demanded by quantum mechanics, Alice and Bob can assume they are not being bugged. If so, they can use the results of the remaining measurements—which were the ones carried out using *identical* orientations—to construct a secret key. Alice and Bob do not discuss these results, of course, because they know that they should have exactly the same data. The result of any one measurement is random, so the end product is that Alice and Bob each have the same random key, which can then be used to encrypt subsequent communications via ordinary classical channels.

One of the curious attractions of this scheme was summed up by Ekert in his article in *Physical Review Letters* thus:

> The eavesdropper cannot elicit any information from the particles while in transit from the source to the legitimate users, simply because there is no information encoded there. The information "comes into being" only after the legitimate users perform measurements and communicate in public afterwards.[14]

What's going on here? In the practical version of the EPR experiment we examined, pairs of entangled photons are subjected to various polarization measurements. Whenever one photon is shown to be polarized vertically, say, we know that its partner must also be vertically polarized. However, what can we say about the state of the particles before they were measured? Can we assume they were vertically polarized before they were measured? The answer is no because a photon polarized at any other angle (apart from the horizontal) has a chance of passing through a vertical polarizer. Even photons without any linear polarization have a 50 percent chance of making it through, so how are we to say what polarization they originally had?

Indeed, the whole point of the EPR experiment was to test the idea that entangled photons might carry hidden information about how they will behave in subsequent measurements. That experiments rule out such hidden-variable theories means that an entangled pair of photons cannot be in any definite state of polarization *until* one or the other of the photons is measured. So that is why in Ekert's cryptographic scheme the eavesdropper cannot extract any information from the entangled particles sent to Alice and Bob: because the information has yet to "come into being."

Of course, Eve, the eavesdropper, can still make her own measurements anyway and thereby cause the information to come into being before Alice and Bob get their hands on the photons. But doing so will eliminate the superstrong quantum correlations that Alice and Bob check for in the first phase of their procedure.

Ekert's entanglement-based scheme has some potential advantages over the single photon methods invented by Bennett and Brassard. The first is that it offers a way of storing cryptographic keys securely. With the BB84 protocol, once Alice has sent her photons to Bob and they have agreed upon their secret key, they may decide to store it before use. But the longer the key is kept, the more chance there is of Eve going into Alice's or Bob's office and examining it. However, according to Ekert's scheme, if Alice and Bob were able to store their entangled particles, Eve would not gain anything by breaking in and examining them. This is because just before they needed to create a key, Alice and Bob could measure and compare some of their entangled particles to check that they were still perfectly correlated. If there were discrepancies, they would know that someone had tampered with them.

As yet, storing entangled particles is not possible for longer than a fraction of a second or so. The prospects for quantum storage depend crucially on technological advances in quantum computing, which we will come to in the next chapter. But it has proved possible to distribute entangled states over reasonably long distances. After a 4-kilometer feasibility study in 1994 by Rarity and Tapster at the DRA in England, came a demonstration in 1997 of entanglement by Gisin and colleagues in Geneva spread over a distance of more than 10 kilometers. They used a suitcase-sized generator in central Geneva to create entangled photons that were then sent through fiber-optic cables to two small villages 10.9 kilometers apart. When they compared measurements carried out at each receiving station, the correlations were just as expected from quantum mechanics, confirming that entanglement does not fall off with distance.

Another, potentially more important, advantage of Ekert's scheme involves the issue of privacy amplification. In Bennett and Brassard's experimental demonstration of quantum cryptography, account was taken of deficiencies in the detectors and noise on the line by the last stage of Alice and Bob's data reconciliation. They used privacy amplification, a mathematical procedure that has the effect of greatly reducing leaked information about their shared key. However, Bennett and Brassard were aware that these classical techniques might prove inadequate if Eve adopted a more sophisticated approach to her eavesdropping. Suppose she diverted the photons in the quantum channel to interact with her apparatus in such a way that the polarization of each became entangled with the state of a quantum system such as a string of atoms. Such entanglement would still cause some interference in the communication between Alice and Bob but Eve could gain extra information by postponing any measurements of her system until *after* she had heard Alice and Bob discuss which measurements they were going to keep. This strategy would help weaken footprints left by Eve on Bob's data while improving her knowledge of their content. So if the quantum channel was noisy, Alice and Bob would have a harder time distinguishing Eve's intrusion from the background error count.

An answer to this threat came in 1996 from David Deutsch, Artur Ekert, Richard Jozsa, Chiara Macchiavello, Sandu Popescu, and Anna Sanpera. In their paper[15] they proposed a method of *quantum privacy amplification.* Instead of processing their data using classical hashing algorithms, Alice and Bob could subject their photons to quantum processing to cleanse them of any signs of tampering by Eve. The procedure, which is applicable only to entanglement-based quantum cryptography, was actually an extension of work published previously[16] by Bennett, Brassard, Sandu Popescu, Ben Schumacher, John Smolin, and William Wootters. This latter group had already produced a method to purify entangled pairs, although they had another application in mind[17] rather than entanglement-based cryptography. With the two versions of this process of *entanglement purification,* Alice and Bob both need some simple quantum logic: a controlled-NOT gate and some single-qubit rotations. With these and some open discussion via a classical channel, they can purify their entangled particles to their heart's content.

Entanglement purification works as follows. Alice and Bob are sent two pairs of entangled particles via a noisy quantum channel. Alice receives two particles, each of which is entangled (imperfectly) with one or the other of the two particles Bob receives. Alice and Bob now carry out some simple

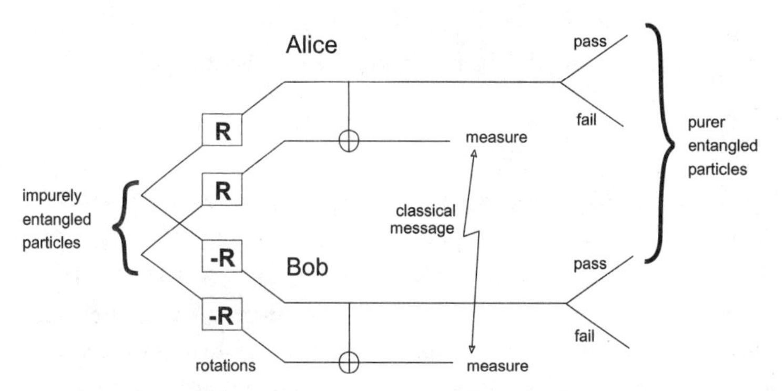

Figure 6.2. Entanglement purification

Entanglement purification requires two pairs of impurely entangled particles. Alice and Bob both perform rotations and a controlled-NOT operation. They then measure the target qubits and compare their results. If these agree, they throw away the target qubits and keep the control qubits for further rounds of purification. If the results disagree, they throw away both control and target qubits and try again with fresh qubits

quantum operations on their particles. Alice rotates the state of both her particles using a square root of NOT type of rotation. She then subjects the two particles to a controlled-NOT operation and measures the state of the target qubit. Bob carries out a similar procedure on his particles, except that he uses an opposite rotation to Alice's. He also makes sure that his target particle is the one that is supposed to be entangled with Alice's target particle. They compare measurements and if the results are different, they discard all of the particles and try again with some new ones. If the results are the same, they discard the target particles and keep the control particles. By repeating the procedure several times using the same particles as control qubits, the state of entanglement between them becomes purer and purer.

Even if Eve managed to replace the source of entangled photons with her own specially prepared photons, the hope is that Alice and Bob could still end up with purified photons that would leave Eve out in the cold.

Of course, the advantages of entanglement-based cryptography might seem academic at the moment because quantum storage and entanglement purification both depend on quantum computing technology that has yet to arrive. Furthermore, experimental demonstrations of quantum key distribution using the original BB84 protocol have progressed further. Nevertheless, the simplicity of the logic required for quantum privacy amplification means that it could become one of the first significant goals for researchers attempting to build useful quantum computational hardware.

Quantum Compression

Einstein, as we saw in Chapter 3, was greatly puzzled by the notion of entangled particles, an idea he brought to the fore with the original EPR thought experiment. He once referred to the particles behaving "telepathically" because it appears that making a measurement on one particle seems to affect the outcome of measurements made on its twin *instantaneously,* no matter how far away it has traveled. This nonlocal feature raises the question of whether such particles could be used to transmit information faster than the speed of light. Such a possibility would fly in the face of Einstein's special theory of relatively and would mean we would be able to send signals backward in time, opening up a Pandora's box of paradoxes concerning causality.

Careful analysis shows that such quantum trickery isn't possible. If Alice receives one EPR particle while Bob receives the other, what can she do with her entangled particle to send information to Bob? Alice is free to choose the *type* of measurement she makes, but she cannot decide the outcome. If she measures the polarization of one of a pair of polarization-entangled photons, for example, the measurement outcome will be entirely random no matter what orientation she chooses. What's more, her measurements won't have any statistical effect on Bob's measurements when his data are examined *on their own.* So Bob won't be able to tell what Alice is doing.

It's only when Alice and Bob examine their data *jointly* that they can see a pattern because their outcomes are strongly correlated for certain orientations of their measurements. The reason this feature seems strange is that we know from the theory associated with the EPR experiment that the polarization of these particular photons cannot have a definite value *before* the measurements. The prohibition against hidden-variable theories rules that out. So when Alice makes a measurement on her particle before Bob, she causes Bob's particle immediately to take on a definite state. But the controlled randomness of quantum mechanics prevents Alice from using this influence to send information to Bob instantly.

Despite this impediment, though, it turns out that entangled particles can be used to *increase* the amount of information carried by a single qubit.

In 1992 Charles Bennett and Stephen Wiesner invented a scheme, which they called *superdense coding,* that exploits this idea. What they showed was that if Alice and Bob have an entangled particle each, Bob can manipulate his particle and send it to Alice in such a way that she receives *two bits of classical information.* Normally, each particle could contain only one qubit of information. Now, although a single qubit is undoubtedly more versatile than a classical bit, there isn't any way of squeezing two bits out of it. Remember, we can only ever make one measurement of a photon and get a yes/no answer. With an entangled photon, however, it *is* possible to convey two classical bits of information. The way this is achieved is as follows.

Bob and Alice are each sent a particle from an entangled pair. On receiving his entangled particle, Bob performs one of four quantum operations on it. He then sends it to Alice, who "combines" it with her entangled particle and performs a *joint* measurement on the two. It turns out that certain types of joint measurement can reveal which of the four operations Bob performed. Four possibilities is equivalent to two bits of information, so Bob has, in effect, sent Alice two bits of information using only one qubit. The whole process still requires two particles, but only one of them needs to be manipulated by Bob.

The biggest problem in implementing these ideas practically is to find a way for Alice to perform her joint measurement on her entangled particle and Bob's particle. How is this achieved? This question becomes a little easier to understand when seen in the light of quantum computation because almost any two-qubit gate requires the interaction of two qubits. A controlled-NOT operation is, in effect, a joint measurement on two qubits.

The first practical demonstration of the idea of dense coding was made in 1996 by Klaus Mattle, Harald Weinfurter, Paul Kwiat, and Anton Zeilinger then at the University of Innsbruck in Austria. Their approach exploited the techniques of *two-particle interferometry,* which involve interference between pairs of particles. In Chapter 3 we saw how Young's two-slit experiment involved interference between different trajectories of single photons, so that experiment was an example of *single-particle interferometry.* With two-particle interferometry experiments, entangled particles are sent along different routes and then brought back together to reveal interference patterns in their behavior. Let's see how this worked in the Innsbruck experiment (see Figure 6.3).

Entangled photons, generated by a down conversion crystal, travel

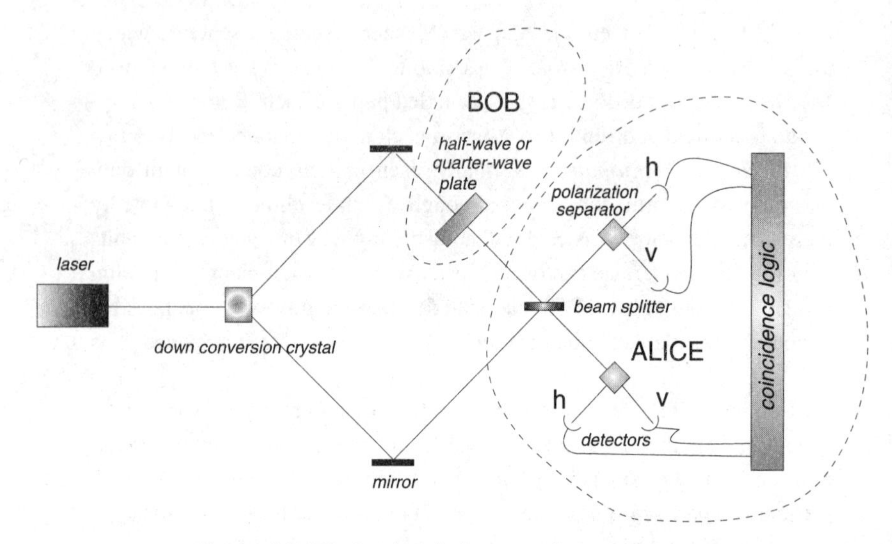

Figure 6.3. The Innsbruck dense coding experiment.

along two paths. The upper route goes via Bob, who manipulates his photons by inserting optical gadgets known as wave plates in the path of the light beam. His four choices are to use either no wave plate, a quarter-wave plate, a half-wave plate, or both quarter-wave and half-wave plates. A half-wave plate flips the polarization state of each photon, so that a horizontally polarized photon is turned into a vertically polarized photon and vice versa. A quarter-wave plate flips the phase of one polarization with respect to the other. (It does this by delaying the phase of one polarization with respect to the other by a quarter of a cycle.) Using both plates introduces both changes.

The entangled photons continue their journey to Alice, where they meet their twins at a device known as a beam splitter. As its name implies, the beam splitter, which can take the form of a semisilvered mirror, is normally used to divide a beam into two. Typically, 50 percent of photons striking such a splitter will pass through, while the other 50 percent are reflected; the decision for any one photon is random. In this experiment, however, the splitter is used to observe interference between the entangled photons traveling via the two different routes.

Let's consider what happens when Bob doesn't use any wave plates. The two paths are arranged to be equidistant so that when the two photons arrive at Alice's beam splitter they are in phase with one another. In this circumstance, it turns out that both photons always emerge *together* from one side or the other of the beam splitter. Because the two photons were generated in a down conversion crystal, they have opposite polarizations, so it's possible then for Alice to measure them on arrival using a pair of two-channel polarizers (such as calcite crystals). Both photons in a pair travel towards the same arm of the apparatus but in opposite detectors of a polarizer. All four detectors of the polarizers are wired to coincidence logic circuits that decide which detectors have fired simultaneously.

Now, if Bob inserts a half-wave plate, the result is that the photons take opposite exits from the beam splitter and arrive at detectors in different arms of Alice's apparatus. So that is easy enough to spot. Finally, if Bob chooses either the quarter-wave plate or the combination of the half-wave plate and quarter-wave plate, it turns out that the photons both finish in the *same* detector. That's a bit of a problem because it means these two possibilities cannot be distinguished from one another, although they can, as a pair, be distinguished from the other two signals Bob is trying to send.

So all in all, the results of Alice's measurements mean that she can read three out of four possible states for each photon Bob sends her. Bob is therefore restricted to transmitting information in what are sometimes called trits, units corresponding to a new kind of three-level logic—0, 1, and 2. The researchers were able to demonstrate this idea by transmitting the digital information for some ASCII characters (the standard computer code for letters and other symbols) using five trits for each character instead of the normal requirement of eight bits. So, although this experiment wasn't able to achieve the two classical bits per qubit Bennett and Wiesner's original scheme had promised, it nevertheless still managed more than one bit.

As other forms of quantum logic are developed, it's quite possible that somebody may find a way of achieving the full two bits per qubit. Given that entangled qubits are able to carry more information than ordinary qubits, theorists now sometimes refer to them as *e-bits.* In a comparison of the different types of information carriers, while qubits are superior to classical bits, e-bits are the undoubted leaders of the pack.

Quantum superdense coding may seem a rather elaborate way of compressing information. After all, why go to the trouble of squeezing two bits of information into one photon, when we can send two photons with a lot less fuss and bother. It's not as if photons were in short supply! Superdense coding is certainly not as exciting as the idea we first broached: sending signals faster than light. Nevertheless, it will help us understand how entangled particles can assist in another form of science-fiction magic: teleportation.

Beam Me Up, Atom by Atom

Talk of teleportation is likely to evoke the image from *Star Trek* of Captain Kirk and his crew (and their successors) beaming themselves from the Starship *Enterprise* onto the surface of a planet in the blink of an eye. Alas, recent work has not yet progressed that far, but it has proved possible to transport quantum states from one end of a lab bench to the other.

The idea of quantum teleportation is closely related to quantum superdense coding, and arose at a similar time. The story goes back to 1992 when Gilles Brassard invited William Wootters of Williams College in Massachusetts to Montreal to give a talk about a paper he had written with Asher Peres. Wootters raised a problem they had had about making optimal measurements. It concerned the situation in which two people, Alice and Bob, are given identical copies of a particle in an unknown quantum state. It turns out that when they measure their copies separately and pool their information classically, they cannot extract as much information about the quantum state as they can if they perform a joint measurement on the two particles together.

During the discussion after Wootters's talk, Charles Bennett asked what would happen if Alice and Bob, when working separately, were given a pair of EPR entangled particles in addition to their copies of the unknown quantum state. "After two hours of brainstorming the answer turned out to be teleportation. It came out completely unexpectedly," Brassard told me.

It became apparent in their discussions that with this extra resource of entangled particles, Alice and Bob could make an optimal measurement on the unknown quantum state even though they were working separately and communicating via classical means. But how does this idea lead to teleportation?

Imagine that Alice, who lives in Japan, is given a particle in an unknown quantum state by a stranger called Carol. Alice decides she wants to give the particle to Bob, who lives in the U.S. Owing to difficulties in sending quantum information over fiber optics or other channels without disturbance, Alice doesn't want to risk sending him the quantum state directly. What else can she do? If she could somehow measure the quantum state of the particle exactly, she could e-mail or telephone that information to Bob, who could then reconstruct the quantum state by manipulating a similar particle at his end. The trouble is, there is no way she can be sure of getting anything more than partial information about the quantum state by measurement. If Alice was given a photon polarized at 45 degrees to the horizontal, for example, she might, if she wasn't told what orientation to use, measure its rectilinear polarization and thereby get the wrong answer. And, of course, once Alice has made that measurement, the original quantum state of the photon will have been irretrievably lost.

Following Bennett's question, a team of quantum theorists—Bennett, Brassard, and Wootters, working with Claude Crépeau, Richard Jozsa, and Asher Peres—discovered the solution. What Alice has to do is perform a *joint* measurement on Carol's particle and her entangled particle. She then telephones Bob and gives him the results of her measurement. According to the information Alice gives him, Bob performs one of four different types of rotation on his entangled particle. After this rotation his entangled particle is miraculously in the same state as Carol's original particle. In other words, his particle is now a perfect replica of the original quantum state. Now, because Alice's measurement disturbs the state of her particles, her copy of Carol's quantum state is inevitably destroyed. She has therefore, in effect, *teleported* Carol's particle to Bob.

Note that it's only the quantum state of the original particle that is transported—not the actual particle. The quantum state is effectively "projected" onto Bob's entangled particle, so it makes sense to use the same kind of particles throughout if Bob wants an exact replica of the original. However, this is not strictly necessary if all Alice is worried about is sending quantum information. The particles, after all, can be viewed simply as the medium for expressing quantum information.

A strange aspect of this procedure is that, on the face of it, Alice is able to transport the quantum state by sending only classical information to Bob. However (as in the superdense coding scheme), it is clear that the entangled particles are also playing a crucial role by somehow conveying additional quantum information hidden in the original particle. This transfer of hidden quantum information appears to happen instantly, although in 1999 David Deustch and Patrick Hayden published a provocative preprint that purported to show how the nonlocal effects of quantum mechanics could all be explained locally. If correct, the paper could make much of the huge amount of debate and speculation that's focused on this particular issue look rather vacuous. I'll have more to say on this in Chapter 9, but suffice it to say for now that there is, as we said before, no way of using the allegedly nonlocal features of quantum teleportation to send signals faster than light. The teleportation process also requires the transfer of classical information, which, of course, can only be sent at speeds less than or equal to the speed of light.

The world's first demonstrations of quantum teleportation were carried out independently by two European teams in 1997. The first experiment was by Francesco De Martini and colleagues at the University of Rome; the second, following up their experiment on dense coding, was by Zeilinger and colleagues at Innsbruck. Both groups teleported quantum states a few meters from one end of their apparatus to the other using entangled photons, but by slightly different means.

De Martini's experiment exploited a suggestion by Sandu Popescu: Alice makes a joint measurement not on two photons but instead on *two different aspects of one photon.* One aspect is the polarization and the other is in the choice between traveling along one of two different paths. Bob, in the meantime, receives his entangled photon also along one of two different routes. The entanglement is in the choice of paths: if Alice receives her photon on route A, then Bob receives his on route C; otherwise, if Alice receives her photon on route B, then Bob receives his on route D (see Figure 6.4). Carol imposes her unknown quantum state by manipulating the polarization of the photon sent to Alice. Alice then performs a measurement on the (in effect, doubly encoded) photon she receives and sends the (classical) result to Bob. This determines the kind of measurement he makes, completing the circuit necessary for Carol's quantum state to be teleported to Bob.

Zeilinger's experiment used a variation of his team's dense coding scheme to achieve something a little closer to the original teleportation idea by keeping Carol's quantum state and Alice's entangled state as separate photons rather than as two aspects of one photon. Alice and Bob are sent a polarization-entangled pair of photons, while Carol sends Alice an additional photon in one of a number of different polarization states unknown to Alice. Alice's joint measurement was implemented by relying on Carol's photon and an entangled photon arriving at a measuring device at the same time. Because there was no guarantee that they would arrive at the same time, only a small proportion of Carol's photons could be successfully teleported.

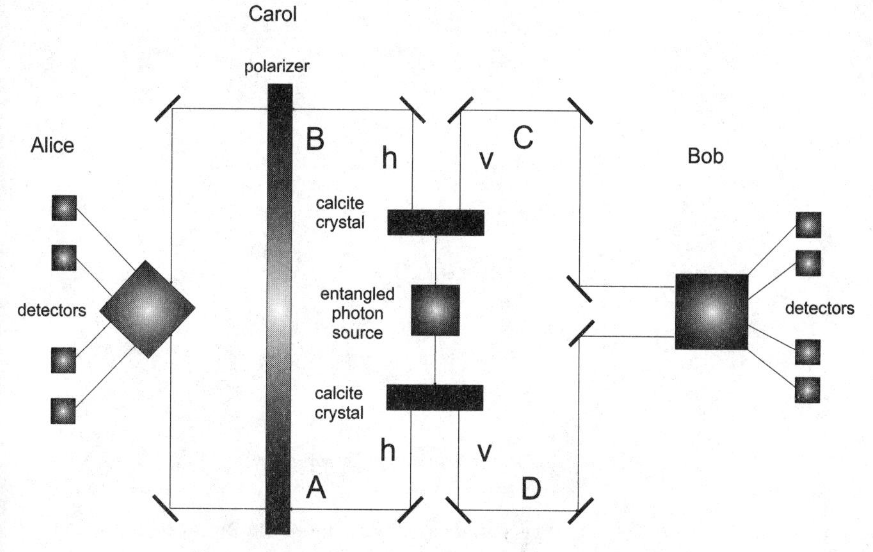

Figure 6.4. De Martini's teleportation experiment

Polarization-entangled photons are generated by a central source. They emerge with opposite polarizations and are sent along two different paths by calcite crystals. If the lower photon travels along **A** to Alice, the other travels along **C** toward Bob; otherwise, the lower photon travels along **D** toward Bob, and the other along **B** toward Alice. The calcite crystals transform the polarization entanglement into a path-dependent entanglement. Carol imposes her mystery quantum state on Alice by manipulating the polarization of the photons sent from the central source along either route. Alice makes a measurement on each photon and conveys the result by classical means to Bob. By performing a measurement on his corresponding photon, he is able to reconstruct and examine Carol's mystery quantum state.

A useful way of understanding the principle of teleportation is provided by a quantum circuit described by Gilles Brassard, Samuel Braunstein, and Richard Cleve[18] (see Figure 6.5).

In this circuit an unknown quantum state presented at input a is transferred to the output c'. The reason this circuit is analogous to quantum teleportation becomes a little clearer when we consider the state of the qubits at the point corresponding to the dotted line. The top two qubits are only involved as control bits for two controlled-NOT operations after this point. In this circumstance, it turns out that it's possible to measure the state of the qubits *before* the controlled-NOT gates.[19] The qubits a' and b' can therefore be replaced by their classical values. These values are the numbers

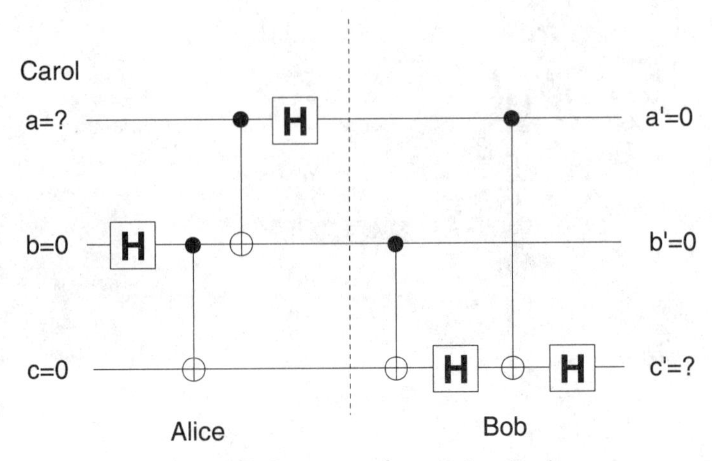

Figure 6.5. Quantum teleportation circuit.

Alice reports to Bob classically in the quantum teleportation scheme, while the qubit *c* at the dotted line corresponds to the entangled photon sent to Bob. It is entangled with the qubit *b* by the first controlled-NOT gate. The second controlled-NOT gate in this circuit corresponds to Alice's joint measurement on Carol's qubit (qubit *a*) and the entangled qubit *b*. By severing the *a* and *b* links at the dotted line and replacing them with classical links, the above circuit could be used for teleporting quantum states between different circuits.

Quantum teleportation is therefore likely to be of great importance in quantum computation because it offers a convenient and safe way to pass quantum information from one part of the computer to another or, indeed, *between* quantum computers. It's also likely to be useful in quantum cryptography because it could overcome the range limitations of fiber optics. If Alice and Bob are able to share entangled photons over long distances, then quantum teleportation would enable them to transmit quantum information to one another confidentially. They still have the problem of exchanging entangled photons, but they can afford to do that over a noisy long-distance connection, after which they can clean up their photons with entanglement purification. This was, in fact, the original application for which entanglement purification was devised: Using such techniques it would be possible to overcome the range problem of fiber optics because over very long distances you could arrange to teleport your quantum information in short hops. At each teleportation station the information is kept in the quantum domain so there is no danger if Eve listening in.

So what about the possibility of teleporting larger objects, such as human beings? This is really a totally different issue from quantum teleportation and almost irrelevant, though there is one, albeit slender, connection that makes it interesting to consider for a moment. There's no denying human teleportation would make an incredibly convenient form of transport, but the technical problems are horrendous. Teleporting a human being would entail scanning the entire body in every atomic detail, transmitting the information through space, and then rebuilding the body atom by atom at the other end. Each stage of this process is going to call for some way-out engineering. Even if we ignore the almost unfathomable technical difficulties involved in scanning the body and reassembling it at the other end, there's still a colossal information storage and transmission problem. According to Lawrence Krauss, who's examined the problems of teleporting people in his book *The Physics of Star Trek,*[20] the amount of information required to specify a human being would be about 10^{28} kilobytes, allowing 1 kilobyte to specify each atom's position and internal state. Transmitting that amount of information using the fastest current digital information transfer method of 100 megabytes per second would take about 2,000 times the present age of the universe. Still, if we are optimistic and assume that improvements in information storage and transmission will grow at a factor of 100 each decade, Krauss reckons that we may have computer technology equal to the task by around the dawn of the twenty-third century.

Krauss's estimate was actually based on sending purely classical information about each particle. Would classical information about the position and state of each atom suffice in specifying a human being? That may not be true. Indeed, in Chapter 9 we will see that some scientists believe that *consciousness* may depend on quantum processes in the human brain. Such ideas are very controversial, but suppose we wanted to guarantee that we could faithfully teleport the quantum states of people's constituent atoms, what would we do? Well, of course, the answer is quantum teleportation. So, amazingly, thanks to the work of quantum cryptographers in the 1990s, it seems the answer to a twenty-third-century problem is already in the bank.

Of course, if Alice wanted to teleport herself to Bob, she and Bob would have to have as many EPR pairs as there are atoms in her body, because each time she teleports a particle, one of the entangled pairs gets used up. So Alice and Bob would need to share a phenomenally intense source of entangled particles, although perhaps they could economize on entangled particles by restricting quantum teleportation to the brain. The rest of the body could be transmitted by classical means.

All in all, although human teleportation seems a very distant and exceedingly unlikely prospect, we can take a little satisfaction in the fact that quantum teleportation not only is already here but also offers its own exciting and immediate applications in quantum computing. And before one salutes the creators of *Star Trek* for their undoubted wisdom and foresight, it's interesting to note that according to Krauss, the main reason they adopted teleportation at all was that the meager budget of the original television show wouldn't stretch to landing a huge starship every week.

7

How to Build a Quantum Computer

I am in no doubt that the future of high technology is quantum technology.[1]

GERARD MILBURN

Coherent quantum mechanical computation may be unachievable in practice and even may be undesirable.[2]

ROLF LANDAUER

Going Universal

Since the early days of quantum computing in the 1980s, ideas about what form a quantum computer might take have changed considerably. The first models were, understandably, very abstract and in many ways totally impractical. Paul Benioff's 1981 model was based on the classical notion of the Turing machine, a type of idealized computer people assemble in their minds rather than in their garages.

Richard Feynman's 1984 model called for a machine that needed to be completely redesigned just to change the program—in some ways reminiscent of the way the very first digital computers had to be rewired each time their programmers wanted to run a new program. Actually, though, the Feynman machine would have been much worse than that, because the rewiring of the first digital computers followed a fixed algorithm, whereas Feynman's simulator would have required real ingenuity to figure out each time how its "designer Hamiltonian," the mathematical set of rules governing the computer, should be installed. Also, Feynman's conceptual model for a quantum computer operated "ballistically," which meant that any tiny errors or defects in its computational pathway would send the machine way off course.

Then there was David Deutsch's model for a *universal* quantum computer—a huge step forward—but it too required an abstract designer Hamiltonian, one that Deutsch made no attempt to interpret physically.

IBM's Rolf Landauer was one of the leading critics of those early ideas, pointing out that they took virtually no account of practicalities. In a skeptical paper entitled "Is Quantum Mechanically Coherent Computation Useful?"[3] he complained, "Why the specification of Hamiltonians has become accepted in this field as equivalent to a computer design is a remarkable mystery."

In 1989 Deutsch, heeding Landauer's advice, cleared the way toward a more realistic approach when he introduced the idea of quantum gates and networks. This brought the whole notion of a quantum computer back into the realm of conventional logic circuits. But Deutsch's idea of quantum logic wasn't quite like the sort of logic circuits you'll find inside conventional computers. In these, the logic gates are spread out in space on microchips and circuit boards. In particular, the inputs and outputs are physically distinct, taking the form either of separate pins on a microchip or of different tracks on a printed circuit board. However, because quantum mechanics is reversible, and reversible gates necessarily have the same number of inputs as outputs, Deutsch recognized that the flow of data through each quantum gate could be brought about by a quantum operation on a group of qubits. As a result, the output qubits would be the same physical objects as the input qubits.

A complete circuit or network could then be realized as a sequence of quantum operations on a register of qubits. The sequence of operations could be organized and controlled by a perfectly ordinary computer that was external to the rest of the apparatus. There was no need for the program steps themselves to take on the form of ghostly superpositions, as had been possible with Feynman's model of quantum computation.

A question remained over what were the simplest kind of logic gates or operations you would need to implement any arbitrary circuit. That was the same as asking what gates were needed to build a universal quantum computer—the quantum upgrade of Turing's universal computer. The answer determines the kind of building blocks we need to construct the quantum processor at the heart of a practical machine.

This issue of universality was well understood for classical computers: for irreversible machines, two-bit NAND gates were sufficient, but for reversible machines, three-bit gates such as the Fredkin gate were necessary. In 1981 Tom Toffoli added to the repertoire of reversible gates a reversible version of the exclusive-OR gate, known as the controlled-NOT gate, and a reversible version of the AND gate, now known as the Toffoli gate. The reversible XOR, or controlled-NOT gate was *not* universal, but the Toffoli gate was. The big difference between them was that the Toffoli gate had three inputs and three outputs while the controlled-NOT gate had only two of each.

In his 1989 paper Deutsch proved that a special quantum variant of the Toffoli gate was universal.[4] But he conjectured that almost *any* three-bit (or, more strictly, three-qubit) gate would be universal. Why the difference between the quantum and classical cases? It turns out that the extra freedom generated by performing quantum rotations makes it possible to construct any logical operation by repeated use of the same arbitrary three-bit quantum gate. But the notion of quantum universality held a further surprise that Deutsch didn't anticipate.

Given that for reversible classical logic, universality demanded a three-bit gate, it became the received wisdom that the same would be true of quantum logic. But this turned out to be incorrect. In fact, you only need *two*-bit quantum gates to make universal quantum logic. This result was discovered by David DiVincenzo of IBM Research who showed how Deutsch's three-qubit gate could be constructed from four different types of two-bit quantum gates. Similar results[5] were also discovered independently by Adriano Barenco of Oxford and also by Tycho Sleator and Harald Weinfurter of the University of Innsbruck. The surprising result was then extended by Deutsch, Barenco, and Ekert in Oxford and independently by Seth Lloyd of MIT to show that almost *any* two-bit quantum gate was good enough to build a universal quantum computer.[6] In fact, irony of ironies, the only quantum gates that *aren't* universal are the very ones that are universal classically—such as the Fredkin and Toffoli gates.

The discovery that two-bit gates were sufficient for universality was good news for people thinking about practical implementations of quantum computers because it is potentially much easier to devise two-bit gates than three-bit gates. This is because a three-bit gate requires the simultaneous interaction of three quantum states without any disturbances from other qubits in the environment. If the states are carried by individual photons, for example, you can imagine that getting two of these fleeting particles to interact at the same point in space and time is already hard enough. The problem is even trickier with three photons.

Deutsch, Barenco, and Ekert also argued that the fact that almost any two-bit gate was universal pointed to a deeper feature of the relationship between physics and computation. Specifically, it seems that the universe is "computation friendly." To quote from their paper:

> The fact that the laws of physics support computational universality is a profound property of Nature. Since any computational task that is repeatable or checkable may be regarded as the simulation of one physical process by another, all computer programs may be regarded as symbolic representations of some of the laws of physics. . . . Therefore the limits of computability coincide with the limits of science itself. If the laws of physics did not support computational universality, they would be decreeing their own unknowability. Since they do support it, it would have been strangely anthropocentric if universality had turned out to be a property of a very narrowly defined class of interactions. . . . Almost every class of physical processes must instantiate the same, standard set of mathematical relationships, namely those that are quantum computable.[7]

In other words, it would be surprising if a fundamental property of nature, namely universality, were confined to a very special class of man-made physical processes. Instead, it appears to have been built into virtually every kind of physical interaction.

Two-Bit Processors

It is all very well knowing that almost any gate can be used as a building block for universal quantum logic, but for practical purposes it is more instructive to know which kind of gate does the job efficiently. After all, if it takes dozens of gates just to realize a Fredkin or Toffoli gate, our quantum computer is going to carry an awful lot of overhead to perform even very simple calculations.

One important result[8] that emerged at a quantum computing workshop held in Turin in 1994 was that a combination of one-bit quantum gates (such as rotations) and two-bit exclusive-OR gates (that is, controlled-NOT gates) would be sufficient to build a universal quantum computer. This finding was in contrast to the classical reversible case, where the combination of the classical restrictions of these gates was *not* universal.

As an example of how these quantum gates can be used to construct a three-bit reversible logic operation, consider the network in Figure 7.1. Using three controlled-NOT, or XOR gates and four one-bit gates each of which rotates[9] the incoming quantum state by 45 degrees, the network performs a reversible AND operation on two of the input rails, *a* and *c*. The middle rail, *b*, is initially set to 0, and its final state is the result of *a* AND *c*. Note that there is no way this result could be computed using classical two-bit reversible gates. The quantum network exploits properties of superpositions to get the answer.

Another kind of two-bit quantum gate that is of practical interest is the *conditional phase shift* gate. This introduces a phase shift on the second qubit[10] when both inputs are 1. Phase, as we've noted before, is the property of waves that determines the arrival time of a wave front at a particular place. It is the key parameter when two waves interfere with one another because the relative phase determines whether the waves cancel or rein-

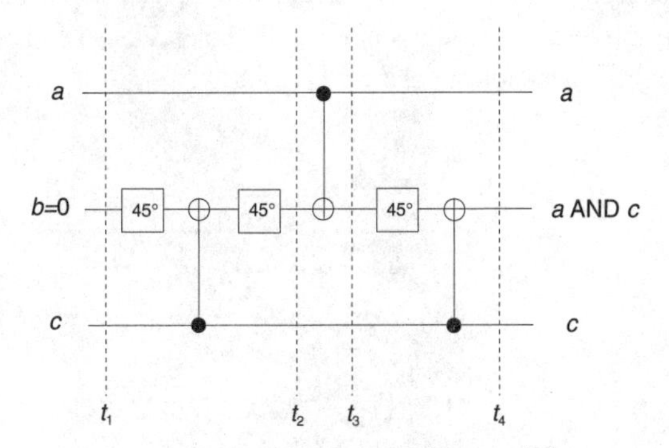

Figure 7.1. An AND gate constructed from rotations and controlled-NOT gates

See Appendix G for an explanation of how this circuit works.

force one another. The conditional phase shift gate is a very convenient building block for solving certain problems, such as the discrete Fourier transform required for factoring numbers. Although it is possible to construct conditional phase shift gates from exclusive-OR and rotation gates, some types of hardware we'll be examining make it possible to produce them directly. A noteworthy kind of conditional phase gate is one that multiplies the state by minus 1. Classically such a gate would have absolutely no effect, but the quantum version is actually very useful.

Yet another possible way of implementing quantum logic depends on a remarkable finding published in 1995 by Robert Griffiths and Chi-Sheng Niu of Carnegie-Mellon. They showed[11] how it was possible to do some powerful quantum calculations without any two-bit quantum gates at all! In particular, they showed how to perform the quantum Fourier transform necessary for Shor's algorithm using only *single*-bit quantum gates. At first this sounds well-nigh impossible because with only single-bit gates, there can be no interaction between the qubits. Without such interaction, the range of possible mathematical calculations is extremely limited. The trick Griffiths and Niu found was to make the interactions *semiclassical:* measurements are performed on certain qubits in the calculation and then *according to the results,* the phases of other qubits are adjusted using single-bit gates.

Their network for performing the quantum Fourier transform is shown in Figure 7.2. The square boxes represent one-bit gates that are controlled by classical signals and followed by measurements. Each box performs a rotation and a phase shift. The magnitude of the phase shift is determined by the outcome of the measurement made in the previous box (linked diagonally). Double lines indicate classical signals.

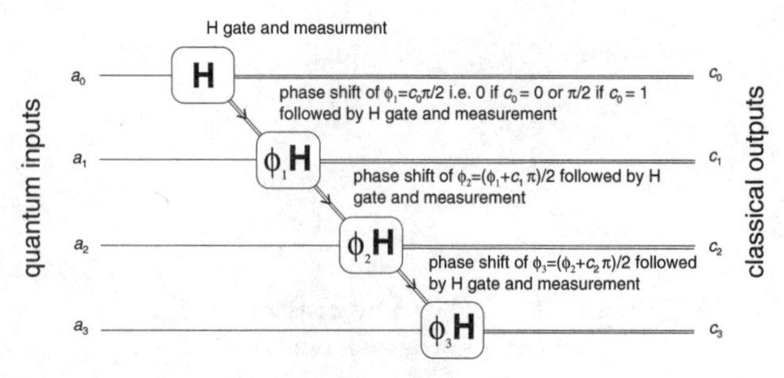

Figure 7.2. "Semiclassical" Fourier transform using only single-bit gates.

The beauty of this network is that it might be much easier to implement than a network requiring two-bit gates. Unfortunately, it's not possible to do the rest of the calculation required for Shor's algorithm, such as the modular exponentiation, using such a network. So two-bit quantum gates are still required at some point.

The Polymer Machine

"Almost anything becomes a quantum computer if you shine the right kind of light on it." So said Seth Lloyd of MIT, referring to the way that any materials can potentially be made to quantum compute if they are subject to the right kind of excitations. Yet this probably comes as a bit of shock because so far, apart from discussing the types of gates we might need, we have said very little about the actual physical hardware one might use to build a quantum computer, and when we have, it's been suggested that it's going to be very difficult. How is it that "almost anything" will do?

For many years quantum cryptography led the way in showing how to put quantum information processing into practice. Valuable though the ideas were, however, they were too limited to do universal quantum computation. In 1993 progress toward a practical model took an encouraging turn when Lloyd[12] produced a scheme for "a potentially realizable quantum computer."[13] His proposal, which built upon some ideas advanced several years earlier by Gunter Mahler and colleagues at the University of Stuttgart,[14] was to use an array of weakly interacting quantum states as the basis for a quantum memory register.

As an example, he suggested using a polymer, the sort of long-chain molecules of which plastics are composed. The way this might work would be to place a specially constructed polymer molecule inside a cavity and expose different atoms in the polymer chain to laser light for a set period of time according to some sequence of instructions. The whole thing would be like a miniature laser light show. Each molecular unit within Lloyd's polymer would have an electron that could be switched from the ground state to a long-lasting excited state by a photon of the appropriate energy. This energy is directly related to the frequency, or wavelength, of the photon, which means that the electron flips are associated with a natural resonance of the molecules.

When a unit in the ground state (which we can label as 0) is exposed to a pulse of light of a particular period of time and a particular intensity, its electron is propelled into the excited state (which we can label as 1). The same pulse will drive an excited unit back into its ground state. This is called a π-pulse because it rotates the state by 180 degrees, which is equivalent in radians to the number π.

Now, in a conventional computer there are electronic methods for accessing particular memory cells, or addresses. To achieve similar addressing capability with the polymer, the light pulses could be beamed onto specific regions of the polymer, but to make the system even more selective, the polymer's molecular units could be chosen to vary according to some specified pattern, such as ABCABCABC . . . In most plastics the repeated units consist of the same group of atoms. However, using a homogeneous polymer would mean that the units would all respond pretty much to the same frequency of light. By varying the atoms in each unit, their resonance frequencies can be separated. Units of type B, say, could be selectively excited by shining a short burst of laser light tuned to B's natural resonance frequency. It's therefore possible to tune the different molccules to respond to different frequencies rather like different notes on a keyboard.

Lloyd showed how further refinements would be possible. Because of the interatomic forces between neighboring molecular units, the resonance frequency of unit B will depend somewhat on the state of its neighboring units, A and C. If it is possible to get the quantum states of A and C to detune the resonance frequency of B sufficiently, then you might have the basis for a three-bit quantum gate. Indeed, Lloyd showed how a series of π-pulses, each of a carefully chosen frequency, could execute the operation of a Fredkin gate. He also showed that it was possible to move bits around at will, simulating the effect of wires in a network. Furthermore, by using $\pi/2$-pulses—bursts of light that last only half the time of π-pulses—he showed how the molecular units could be put into superpositions of ground and excited states.

So how would you read out the data after a computation? Here you need to resort to a little more molecular physics. If, in addition to the long-lived excited state, each molecular unit has a short-lived excited state (which we can label 2), this extra energy level can be used to measure the state of the unit. If the unit is currently in state 1, then a photon of energy $E_2 - E_1$ will knock the electron into the short-lived state 2. From there the electron quickly falls back to the ground state, releasing a photon with a characteristic energy $E_2 - E_0$. So if such a photon is detected, we know that the unit must have been in the state 1 originally (see Figure 7.3).

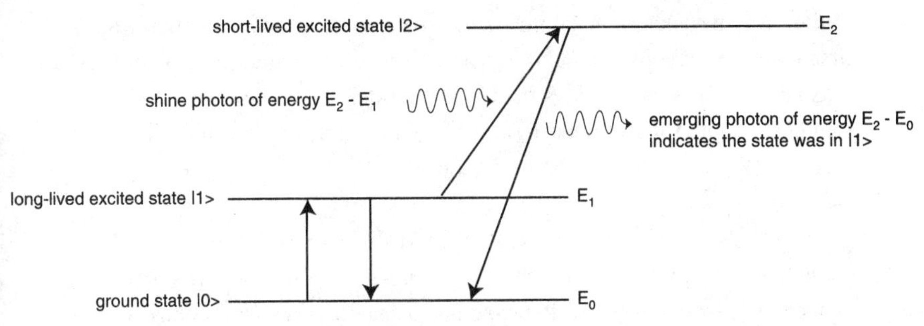

Figure 7.3. Readout from a polymer quantum computer.

Lloyd also pointed out that the short-lived state could be used as a simple method of error correction because it would be possible to do a form of majority voting. If the majority of bits in units ABC was 0, for example, the whole state could be set to 000, and if the majority was 1, the whole state could be set to 111. By running a computation using multiple triplicate copies of qubits, you could reduce the error rate substantially. This form of error correction would not cope with all kinds of errors, but it was at least an attempt to address a potentially serious problem for any practical quantum computer.

Although Lloyd suggested a polymer as a substrate for a quantum machine, he argued that the same kind of scheme could be applied to virtually any system in which there are local interactions. Quantum dots and nuclear spins, as we will see, also offer such possibilities. And as Lloyd's quote at the beginning of this section implied, almost any material will do. It's another example of the "computational friendliness" of physics that David Deutsch attaches so much importance to. But unfortunately that friendliness has its limits because it doesn't necessarily make the task of producing a real working quantum computer easy. Far from it . . .

The Trouble with Decoherence

Although Deutsch's quantum networks and Lloyd's polymer machine helped address one of Rolf Landauer's criticisms of research into quantum computing—the problem of translating mathematical prescriptions into physical reality—there remained other sizeable objections. The most important of these was the fact that all of the quantum models for computation referred to physically perfect machines. Physicists often make assumptions of the "assume a perfect vacuum" kind, so it's not surprising that they would first consider idealized computing machines. But is such an assumption reasonable? Back in the days of the first models of reversible computing, Landauer argued that indeed it was not. Consider, for example, his comments on Fredkin's billiard ball computer:[15]

> This is a system that requires perfection. A slightly misdirected ball will go increasingly further off course with time until eventually the system ceases to work altogether. In fact, this is a chaotic system; collisions between balls will cause an exponential, rather than linear, growth in the error. Thus, any source of error—incorrect initial launching velocity or position, a misaligned mirror, friction or noise—will cause the system to fail. This is therefore a pathological form of computation.

Landauer argued that, by analogy, a rather similar fate would be in store for the quantum computer.

> Thus, a computation that is to yield a result that depends on two or more simultaneous computational trajectories requires a completely phase-coherent quantum mechanical evolution, in the absence of friction. But . . . that in turn requires unrealizable perfection for the parts of the computer.

The obstacle Landauer was referring to was the problem of *decoherence,* a term that came into the language of quantum mechanics sometime early in the 1980s. It derives from the word *coherent,* which is used by physicists to describe certain forms of wave motion. Laser light is the classic example of coherence because in a typical beam the photons travel together locked in phase and frequency with one another. It is this coherence that endows laser light with many of its useful properties, such as being able to travel over large distances without spreading. The ability to create holograms too depends on the interference between coherent laser light directed toward a photographic plate and the same light scattered from an object. The light from an electric lightbulb, on the other hand, is *incoherent* because the photons are emitted with random phases and many different frequencies.

In Young's two-slit experiment, it is the coherence between the waves arriving from the two slits that causes an interference pattern on the screen.

Note that that doesn't mean different photons are coherent with one another, as in the laser. Rather it is the coherence between two different trajectories of an individual photon that gives rise to interference in this case. Furthermore, at most points on the final screen, the phases of the two trajectories are different, but they are still coherent because the phase differences remain constant with time. If you placed two different light sources behind the slits, the phases between different photons arriving would be completely unrelated. The phase relationship would vary rapidly and randomly, causing any interference fringes to be lost in a blur. Thus we see that coherence is crucial to observing interference phenomena.

The idea of decoherence arose as a number of theoreticians attempted to understand the way quantum states are perturbed by the environment or measuring apparatus. As an example, consider what would happen if we tried to follow the route each photon takes in the two-slit experiment. Suppose we were able to do this by using detectors that register photons without destroying them. If such detectors were placed close to each slit, what would we see? Versions of this experiment have been performed and though they involve a slightly more elaborate setup than described here, the findings are clear: we do not witness photons splitting in half and traveling through both slits, as the wavelike phenomenon of interference would suggest. Instead we see each photon taking just one route, as we might expect for a particle. However, we also discover that the interference pattern disappears.

It is as if the intrusion of the measuring apparatus forces the photons to choose which way they are going to travel through the two slits. In the language of decoherence, the superposition of trajectories for each photon decoheres, and it is this decoherence that washes away the interference pattern.

In this situation the decoherence is caused by our deliberate intervention, but as we'll see, it can arise spontaneously in any quantum computing device because of the difficulty of shielding quantum states from the outside world.[16]

We saw how entanglement arises in the EPR experiment when pairs of photons are generated in such a way that the quantum properties of each photon in a pair are closely linked. When a quantum system undergoes decoherence, it is usually because parts of the apparatus or the environment become entangled with the system. As this entanglement becomes more and more widespread, the different quantum states within a superposition involve more and more bits of the environment, making it increasingly difficult or impossible to perform interference experiments. The result is that superpositions "collapse" onto classical states. We will explore this process in more detail in the next chapter when we'll look at how it might be overcome. Suffice it to say for now that decoherence is, without doubt, the biggest problem ahead for anyone hoping to build a quantum computer because it totally disrupts quantum calculations.

Trapping the Atom

In his book *Nano* Ed Regis begins one chapter with the following delightfully tongue-in-cheek sketch of how scientists used to regard the subatomic realm (and perhaps, to a lesser extent, still do):[17]

> Ever since the dawn of quantum mechanics in the 1920s, one thing that everybody "knew" about atoms and elementary particles was that working with them was out of the question. Isolating a single atom absolutely could not be done. In fact, even entertaining the thought of atoms as clearly defined entities was a sign of retrograde thinking, a barbarity, a relic from the olde-tyme Newtonian physics. The very notion reeked of ruff collars and powdered wigs . . .

Regis continues in a similar vein until he introduces the physicist Hans Dehmelt of the University of Washington.

> . . . Dehmelt became the sworn enemy of what he regarded as "a still persisting wave of quantum-mystification in the literature," the notion that atoms, electrons, and suchlike were somehow not fully "real" objects. They were real enough, he decided, if you could isolate them for months at a time, give them names, and make them perform "quantum jump ballets."

For that is precisely what Dehmelt did: He pioneered the isolation and trapping of individual atoms and electrons by confining them within electromagnetic fields. In 1973 he and his colleagues captured a single electron in his apparatus like—as Regis puts it—a fly in a bottle. In 1980 he trapped a single barium ion, which he grew rather fond of and christened Astrid Atom. When illuminated, it could be seen with the naked eye as a tiny blue-white star. Further experimentation led Dehmelt's group, in 1986, to be the first to observe directly a quantum leap, in which a barium ion jumped into a higher energy state—turning invisible in blue light—and then leapt back into its normal state, turning visible again. Dehmelt's team was the first to isolate and detect an individual proton, an individual positron, and small numbers of antiprotons. In short, Dehmelt showed the world how to tame these quantum fuzzballs. And the secret to these successes were devices called ion traps.

Ion traps come in many different types and geometries, but they generally depend on coaxing ionized atoms or other charged particles toward the central region of an evacuated chamber using electric and magnetic fields. One type, known as the Paul trap,[18] dispenses with magnetic fields and relies on a combination of static electric and radio frequency fields to imprison particles within the central region. The traps can hold large numbers of ions—the electric repulsion between like charges ensures that the ions remain at some distance from one another. Alternatively, by letting the more energetic particles escape, the numbers can be whittled down to just a handful or even a single ion. Experimenters can count the number of ions by shining laser light on them, an operation that makes it possible to see even a single atom with the naked eye.

Dehmelt had foreseen in the 1950s that ion traps could be useful for increasing the resolution of atomic frequencies, properties that govern the atomic clocks we use today for keeping global time standards. A typical atomic clock uses lasers and microwaves to excite atoms—usually of the element cesium or rubidium—emitted in a beam. Certain atomic energy levels, known as hyperfine transitions, are used to generate ultrastable frequencies that determine the ticking rate of atomic clocks with an accuracy of up to one second in a million years. The hyperfine transitions arise from the tiny energy difference arising from opposite spin orientations of the outer electron in an atom such as cesium relative to the spin of its nucleus.

Using trapped particles offers the advantage of keeping atoms relatively still and almost perfectly isolated from outside disturbances. Even so, trapped atoms still retain thermal energies that when translated into velocities can be surprisingly high. At room temperatures, for example, the molecules and atoms in the air move at speeds of around 1,000 miles per hour. Even at liquid-helium temperatures (approximately 3 degrees above absolute zero), the atomic motions are around 100 miles per hour, enough to cause significant Doppler shifts, which detune the frequencies of the atomic transitions. To improve the accuracy of time standards, scientists needed a way to reduce the fluctuations in frequencies caused by these motions.

In 1975 Dehmelt, in collaboration with David Wineland, and, independently, Theodor Hänsch and Arthur Shawlow proposed a technique for cooling atoms using lasers. The method depends on bombarding atoms with laser light tuned to slightly less than a resonance frequency of the atoms. Atoms at rest or moving away from the photons are left undisturbed by the light because of the deliberate mistuning, but atoms moving toward the beam see a Doppler shift in the frequency, bringing them into resonance. At resonance, photons are absorbed and then reemitted in random directions. Each time this scattering process happens, the atom receives a kick that slows down its movement toward the laser. The result is that any forward motion among the atoms is quickly reduced.

The idea was first demonstrated using ions rather than atoms in 1975 by Dehmelt and Peter Toschek at Heidelberg, and independently by Wineland and colleagues at the National Institutes of Standards and Technology (NIST) in Boulder, Colorado. In 1985 Stephen Chu and colleagues used a three-dimensional arrangement of laser beams to cool atoms. This system is known as *optical molasses* because at the intersection of the laser beams, the atoms are slowed in every direction as if they were moving in a thick, viscous fluid. This method of cooling brought the temperature of sodium atoms down to 240μK—240 millionths of a degree above absolute zero. Further work by Claude Cohen-Tannoudji's group at the École Normale Supérieure in Paris and William D. Phillips at NIST in Gaithersburg, Maryland, brought further substantial improvements that have reduced the temperature to as low as 0.2μK—less than a millionth of a degree above absolute zero. For these developments, which have produced the coldest known temperatures, Chu, Cohen-Tannoudji, and Phillips were awarded the 1997 Nobel Prize for physics.

Note that laser cooling can't remove all of the atomic motion, because even at absolute zero atoms and subatomic particles have a zero-point energy, an irreducible quantum level of energy. Nevertheless, the technique has opened the way to improved time measurement potentially accurate to one part in 10^{18}, several orders of magnitude better than the current cesium standard. It has also made possible experiments that depend on ultracold conditions, such as the creation of novel states of matter known as Bose-

Einstein condensates. In 1995 scientists at the University of Colorado in Boulder were the first to create such a state when they condensed some 2,000 rubidium atoms into a superatom, a coalescence predicted by quantum mechanics but never before seen because of the record low temperatures required. And in 1999 Lene Hau and colleagues at the Rowland Institute of Harvard University used a Bose-Einstein condensate to slow light to 38 miles per hour—a speed that can be surpassed by a cyclist.

Laser cooling and ion trapping have also led to exciting advances in practical approaches to quantum computation. In 1994 two theoretical physicists, Ignazio Cirac and Peter Zoller, working at the University of Innsbruck, heard a talk given by Artur Ekert about the burgeoning subject of quantum computation. Ekert was encouraging other scientists to apply their minds to the problem of finding a suitable hardware platform for testing the ideas of quantum computation. Cirac and Zoller immediately saw that laser-cooled trapped ions could be a suitable medium for doing quantum logic because, as Dehmelt and others had shown, such ions were isolated very effectively from the environment. They could see how to implement single-qubit logic operations using a laser to excite and de-excite single ions, but the question was, how could quantum operations requiring *interactions* between qubits be achieved. Cirac and Zoller produced an ingenious solution[19] that was quickly recognized as a conceptual breakthrough.

Cirac and Zoller's starting point was a "linear" trap holding a number of ions in a straight line, like pearls on a string, a design first envisaged by Dehmelt. The ions would be laser cooled so that they would all be in their ground state. Each ion could be selectively excited into a relatively long lived state by a pulse of light from a laser beam directed specifically at that ion. The trap would therefore act as a quantum register, with the internal state of each ion playing the role of a qubit.

The clever part of Cirac and Zoller's idea was to show how to exploit *vibrations* between the ions to implement interactions between the qubits—that is, two-bit operations. The ions are strongly coupled because of the combination of the electric repulsion between them, forcing them apart, and the electric fields of the trap, squeezing them together. The coupled system behaves a little like a series of pendulums connected by a wire. When one pendulum swings back and forth, it causes a vibrational wave to travel along the springs, rocking the other pendulums. In the ion trap, these vibrational interactions are quantized and are carried by *phonons,* analogous to photons as the carriers of light. The result is that the vibrations have their own ground and excited states that can encode qubit information. So, using 0 and 1 to denote the ground state and the excited state, respectively, we can represent a single ion in its internal and vibrational ground state as 00.

Now, the vibrational frequencies are much smaller than the frequencies involved in exciting the internal energy states. Cirac and Zoller calculated that by tuning a laser to a frequency equal to the difference between the ion's internal frequency and that of its rocking motion, it would be possible to selectively flip an excited ion into its ground state *while at the same time exciting its vibrational state.* This operation would, in effect, transfer the internal state of an ion onto its vibrational state, thus transforming 10 into 01. The same operation would also perform the reverse action, turning 01 into 10. So this laser pulse is a very useful way to move qubit information between the different quantum aspects of an individual ion.

More significant, the reason the vibrational states are so useful is that they are shared by all of the ions in the trap. So by shining similar kinds of laser pulses consecutively on two different ions, it is possible to transfer qubit information between them, offering a means of shuttling qubits around at will. That still isn't enough, though, to perform a controlled-NOT operation on two qubits. To achieve that, Cirac and Zoller proposed a further laser operation. They suggested how the state 01 (ground state, excited vibrational state) could be selectively rotated via an "auxiliary" energy state, changing its sign (equivalent to a 180-degree phase change), without affecting the other states—00, 10, and 11 (see Figure 7.4).

With three laser pulses—this conditional phase change pulse sandwiched between two swap pulses—you get an interaction that changes the sign of the state, but only if both ions are initially in their excited states. With an additional one-bit rotation on one of the ions, this procedure yields, believe it or not, a controlled-NOT operation.

A huge bonus with this scheme is that because the vibrations influence the whole system, the ions don't have to be adjacent, and it's even possible to realize three-bit gates or higher by using extra laser pulses directed toward other ions. The ion trap system potentially offers enormous flexibility and certainly appears to make possible every operation needed for a quantum computer.

The decoherence time looked promising on paper too, with some ear-

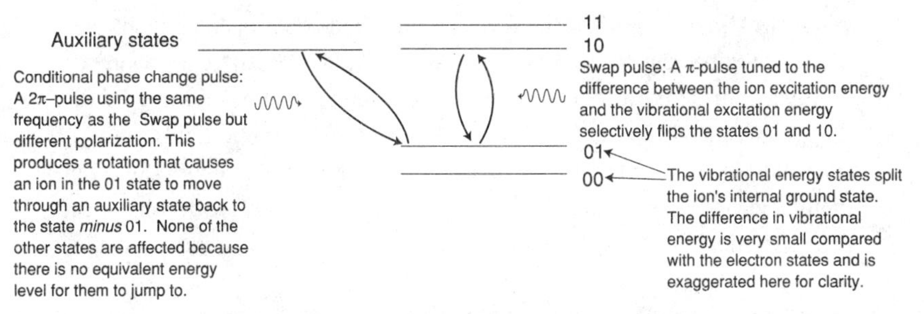

Figure 7.4. Cirac and Zoller's ion trap scheme

Cirac and Zoller's scheme exploits two main types of laser pulse. The swap pulse (*right*) switches an ion's internal state with its vibrational state. The conditional phase change pulse (*left*) flips the phase of the whole quantum state only if it starts in the 01 state. Used in a particular sequence, these pulses produce a controlled-NOT interaction.

lier ion trap experiments putting in running times of many minutes before quantum disturbance—enough time to do many thousands, if not millions, of quantum operations. So here was a truly realistic hardware proposal that had a chance of working. All it needed was somebody to demonstrate it.

David Wineland, now the leader of the ion storage group at the National Institute of Standards (NIST) at Boulder, Colorado, duly took up the challenge with his colleagues Chris Monroe, D. Meekhof, B. King, and W. Itano. By 1995 they had constructed the world's first working quantum gate—a two-bit controlled-NOT gate—using ion trap technology. Although their experiment was based on Cirac and Zoller's scheme, they simplified it by using just one solitary ion. They trapped a beryllium ion, Be^{+}, which has a single electron in its outermost shell. Also, rather than using atomic excitation to a higher energy orbit, they exploited the ion's hyperfine energy levels, which depend on the alignment between the electron spin and the ion's nucleus. They therefore exploited these two internal energy states of the ion for one qubit, and two vibrational modes for the second qubit. By applying a sequence of laser pulses similar to that specified by Cirac and Zoller, they were able to demonstrate the controlled-NOT operation with something like 90 percent reliability.

The results showed that Cirac and Zoller's scheme worked. Here, at last, was the first genuine quantum gate. There were some caveats, though. The measured decoherence time was less than a thousandth of a second, which was long enough to perform a controlled-NOT operation but certainly not sufficient for an extended computation. The short decoherence time was attributed to deficiencies in the laser beams, positioning of the ions relative to the beams, fluctuation of external magnetic fields, and instabilities in the electronics responsible for the trapping fields. So even though ion traps appeared to be a good bet for avoiding decoherence, many practical factors still seemed likely to disrupt any quantum computation.

When I visited Wineland and Monroe in their laboratory, they were optimistic about making substantial improvements to their apparatus. "This experiment was thrown together with a lot of these problems built in," Wineland said. "We know some causes of noise that we can significantly reduce. In some of our clock experiments, where we rely just on the internal qubits [using a different energy transition of beryllium], we've seen coherence times in excess of 10 minutes. So our feeling is that the internal qubit coherence looks very good and the problem without question is the motional coherence. I'm pretty sure that's going to be the dominant thing for a while."

The other big issue was and still is whether the same quantum operations can be made to work in a multi-ion system. There's no problem making multi-ion traps, but manipulating each ion is not going to be so easy. "Cirac and Zoller's scheme relies on the fact that you can address a single ion at a time with a laser beam," Wineland said. "If you run through the formulas for laser beams, that looks just fine, but in real optics there's always some outlying radiation from a typical beam and that could disturb the adjacent ions. Another problem is that as you add more ions to the trap, the number of vibrational modes increases. With even a very small number of ions, selecting the desired mode is going to be very difficult.

"The factorization guys want to have at least a few thousand ions for their calculations. In the long run, if the ion scheme continues to fly, then we're reasonably sure that rather than having one big long quantum register, we'll have to start thinking about multiplexing [joining] ion traps—each containing a manageable number of ions. But that's step two from where we are now. Step one is to address how many ions we can do this logic on. I think if we can get around some of the mundane problems we have right now, we can do *n* ions. What *n* is, is not clear, but I can't believe we won't be able to do two."

"Five!" interjected Monroe. "Five, okay. Maybe six," said Wineland jokingly. "But we strongly feel that the process is going to run out of steam before 2,000 ions. Our goal right now is to characterize what the problems are with *n* ions, when *n* is relatively small, and see how that scales as *n* gets larger."

Scaling is the crunch issue for virtually all of the potential technologies for quantum computing because while it may be possible to implement a few controlled-NOT gates with ion traps or whatever, no one really knows whether a quantum program can be run for long enough over enough qubits to perform a useful calculation.

"It seems like the problem has the quality of a general analog computer," added Monroe. "There's no fundamental reason we can't go to 2,000 ions. It's just that these errors add up in a funny way. Like analog computers, you get an exponential error accumulation."

Errors, scaling, and decoherence: These are the watchwords by which different practical hardware implementations will be judged. Ion traps appear to be the most promising technology for quantum computation to date, which explains why America's premier code-making and code-breaking center, the National Security Agency, is funding some of NIST's work on quantum logic and also similar work at Los Alamos National Laboratory.

That a secretive and powerful organization like the NSA (otherwise known as Never Say Anything) is involved certainly helps give this work a certain mystique. When I gleaned during my visit to Wineland's laboratory that somebody from the NSA had been visiting the same morning, I suddenly felt rather privileged even to be able to talk to the researchers. "If the people who are interested in factoring large numbers thought there was a good chance of this happening soon, we wouldn't be talking now," said Wineland ominously. "We wouldn't be talking?" I queried. Monroe spotted the double entendre: "Yeah, we wouldn't be alive anymore!" We all laughed, but Wineland added hurriedly: "They're very reasonable people. Everybody is on the same wavelength here. It's clear that it's a real long shot. They're taking the attitude that there is a chance, and if it does work they don't wanna be left behind with the technology."

Richard Hughes, who leads similar pioneering ion trap work at Los Alamos National Laboratory, takes pretty much the same view. "We're still in the vacuum-tube era of quantum computing," he said. "I don't think this is going to be the ultimate way quantum computing works. What we're trying to do is find out if it can work in the sense of ever being able to do a useful computation. Can it factor *any* number, let alone a big one?"

Flying Qubits

Another technology that belongs to the vacuum-tube era of quantum computation is known as cavity QED—short for cavity quantum electrodynamics. Quantum electrodynamics, or QED, is the name given to the quantum description of the behavior of light, and "cavity QED" refers to the unusual behavior of light when it travels through a special kind of cavity with highly reflecting walls. It turns out that atoms and photons in small cavities of this kind behave rather differently from those in free space. The reason is that the cavity reinforces certain wavelengths of light and suppresses others. The preferred wavelengths are those that fit precisely within the space of the cavity, just as the musical note of an organ pipe is determined by the wavelength of sound waves that can be accommodated within the length of the pipe.

If an atom is inside a reflecting cavity, its tendency to emit or absorb any photons is constrained by the resonances of thc cavity. It is possible to make atoms emit photons ahead of schedule or remain in an excited state indefinitely by choosing the appropriate cavity size. This ability to modify the quantum properties of atoms has attracted a number of research groups to study cavity QED systems in detail.

One of the leading groups is at the Institute for Quantum Information and Computing run by Jeff Kimble at Caltech. As a sign of the strategic importance of his work, in 1996 Kimble received a $5 million grant from the Defense Advanced Research Projects Agency to investigate the feasibility of quantum computing. His interest began around 1994 when, soon after Shor's algorithm had appeared, he received a visit from Artur Ekert, who once again played the role of catalyst in encouraging new experimental approaches to quantum computation. After exchanging ideas with Ekert and later with Seth Lloyd, Kimble set about applying cavity QED systems to quantum computation. He and his colleagues Quentin Turchette, C. Hood, W. Lange, and H. Mabuchi devised a cavity QED version of a quantum gate that appeared in the same issue of the journal *Physical Review Letters* that carried the report of Monroe and Wineland's ion trap demonstration of the quantum controlled-NOT gate.[20]

Kimble's group demonstrated "conditional phase shifts" using beams of light interacting with an atom inside a cavity. As we saw before, quantum phase shift gates are enough to build any quantum logic circuit when supplemented with one-bit rotations. So, with one or two caveats, this ex-

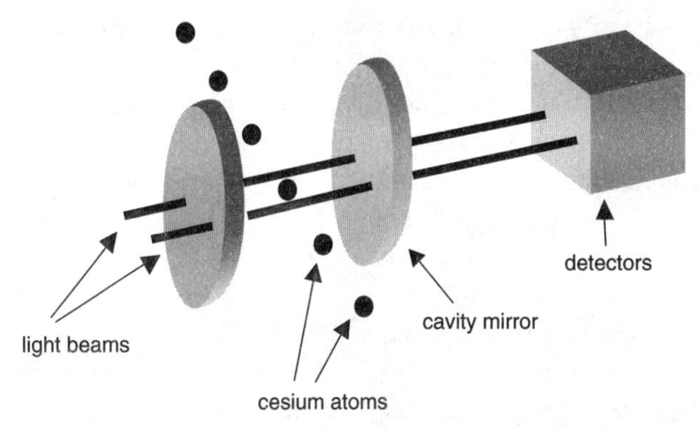

Figure 7.5. Quantum phase gate using cavity QED.

periment put cavity QED in the running alongside ion traps as a potential hardware platform for quantum computation.

The heart of the experimental apparatus is illustrated in Figure 7.5. A beam of excited cesium atoms passes through a cavity $^1/_{20}$ of a millimeter across kept in an ultrahigh vacuum. The beam is sufficiently tenuous so that only one atom is likely to be inside the cavity at any one time. At right angles to the atomic beam, two beams of laser light of slightly different frequencies travel toward the back of one of the mirrors. Although the mirrors are very high quality to ensure ultrahigh reflectivity, they are very slightly leaky and allow a small fraction of photons to enter the cavity. Once inside the cavity, any photons are likely to be captured and rereleased by a passing atom many times as they travel back and forth before finally escaping through the second mirror, which is leakier than the first.

Unlike the ion trap system in which quantum information is encoded in the internal and vibrational states of atoms, the qubits in the cavity QED experiment are carried by the photons in each laser beam. The information carried by these "flying qubits," as Kimble calls them, is encoded in the circular polarization of each photon. After the photons have passed through the cavity, their polarizations are measured.

If we represent the clockwise and counterclockwise polarizations as + and −, then the experiment showed that if either or both the inputs were −, the beams passed through unchanged, but when both were +, the emerging beams were shifted in phase by about 16 degrees, which might not sound like much but is enough to be useful for building logic circuits. So how does this system work?

The point of the experiment is to find a way to get two qubits—in this case photons—to interact *conditionally,* that is, such that something happens only when both of them are in one particular quantum state. This is not so easy to arrange because photons don't interact with one another—they glide through one another as easily as ghosts traveling through solid walls. The reason is that photon propagation is a linear process, so that when two photon wave fronts overlap, they simply add up or subtract from one another as demanded by the principle of superposition. Such linearity is no good for conditionality. But getting the photons to impinge on an atom introduces a nonlinear aspect to their interaction *if* they are of the right frequency to excite the atom. The point of the cavity is to amplify this nonlinearity by extending the amount of time each photon spends there reflecting back and forth between the cavity mirrors. The resonance of the cavity is chosen to coincide with the excitation resonance of the atom, while the frequencies of the photons are chosen to be close to the resonance. They are deliberately detuned slightly away from the resonance peak so as to favor an interaction only when the photons are circularly polarized in a parallel direction to the atom's spin. The atomic spins are aligned in one direction by laser excitation before they enter the cavity so that only if both photons are polarized in one particular direction will they interact strongly with the atom. When this happens the emerging photons are significantly shifted in phase.

When this work appeared in *Physical Review Letters* in 1995 along with the Boulder report on the ion trap quantum gate, it caused quite a stir. At last it looked as though quantum computation was beginning to be an experimental science. It was later pointed out, though, that similar work had already been published by another leading group of cavity QED researchers: Serge Haroche, Jean-Michel Raimond, and colleagues at the École Normal Supérieure in Paris. A year earlier they had demonstrated conditional phase shifts using photons in the microwave regime interacting with highly excited states known as Rydberg atoms. An article in *Physics Today* discussing these new experimental developments went some way to putting the record straight and included the following barbed comment about the Paris work by its leader, Haroche: "That dispersive-cavity QED experiment was already a demonstration of conditional dynamics at the single quantum level . . . at a time when logic gates were not yet fashionable in quantum optics. Whether such experiments should be heralded as first steps toward a large-scale quantum computer, which I consider an impossibility, is a matter of taste."[21]

Haroche's severe pessimism over the prospects for building a quantum computer was spelled out even more fully in an article he and his colleague Jean-Michel Raimond wrote for *Physics Today*[22] not long after. It was forebodingly entitled "Quantum Computing: Dream or Nightmare?" When I went to see Kimble, he was quick to admit that his group's 1995 experiment wasn't the right paradigm for building larger quantum logic circuits. "The kind of interactions that we get have a time scale of 10 nanoseconds. How far does light fly in 10 nanoseconds? It flies a few meters. The photons we're talking about aren't little particles a millimeter long bouncing around—they're things like elephants! And it's kind of hard to think about a compact arrangement to manage the flow of livestock."

The photons could be shrunk down somewhat by using shorter wavelength lasers, but Kimble couldn't envisage making them flea-sized, which would be the sort of scaling you would need to build a reasonable-sized machine with thousands of gates. Another problem was that he was obliged to use weak beams of photons in his apparatus, which entailed looking at large numbers of measurements. For a quantum computing machine, physicists would really like to fire individual photons on command like balls in a pinball machine so that they could know precisely when the photons were due to arrive. As we noted in the quantum cryptography experiments, devising such a trigger isn't easy. "Nobody has learned how to do that in optics—to make photons when you want them, where you want them, and of the shape you want," Kimble said. "The good news/bad news about photons is that indeed they fly. They move at the speed of light, and when they leave your cavity you'd better have thought about what you're gonna do with them. They're not gonna sit around and wait!"

Kimble added that there were promising ideas about how a photon trigger might be built, so perhaps this problem may not prove insuperable. One exciting proposal he and his colleagues were working on was an idea for combining cavity QED and ion trap technologies. Imagine a quantum computing network in which there were lots of multibit ion traps. How are these devices going to talk to one another? In 1997 Cirac, Zoller, Kimble, and Mabuchi outlined how photons could be used to transfer quantum information between trapped atoms. Their solution envisaged the use of cavity QED techniques to exchange quantum information between distant atoms using photons conveyed perhaps by fiber optics. The proposal has the virtue of exploiting the best features of both ion trap and cavity QED technologies: using photons for what they are good at, which is traveling very fast, and using atoms for storing information and local processing. The potential importance of cavity QED is that it offers a method for exchanging quantum information between these two different carriers and therefore a way of being able to scale up atom trap technology.

The Doctors of Spin

Early in 1997 came surprising reports in the popular science press claiming that the answer to the dreams of would-be quantum computer programmers had been sitting right under their noses. It was going to be possible, according to the news stories, to run a quantum program right there on your desktop in a mug of coffee! These somewhat fanciful claims were triggered by a research paper published in the journal *Science* by Ncil Gcrshenfeld of MIT's Media Lab and Isaac Chuang then at the Institute of Theoretical Physics at Santa Barbara. *Science* itself ran a news article about the work under the headline PUTTING A QUANTUM COMPUTER TO WORK IN A CUP OF COFFEE.[23] What's more, another group—David Cory of MIT and Tim Havel and Amr Fahmy of Harvard Medical School—had hit on a very similar idea at more or less the same time.[24]

Both groups had found a way to harness the nuclear spins inside the atoms of liquids as their qubits. Their methods depended on *nuclear magnetic resonance* (NMR) spectroscopy, a technology widely used by physicists, chemists, and biologists to analyze molecular structures and by doctors for medical diagnosis. In its medical setting NMR is better known as *magnetic resonance imaging* (MRI), a name that avoids unnecessarily provoking fears over the dreaded N-word: nuclear. Nuclear magnetic resonance actually doesn't involve radioactivity or any ionizing radiation but instead uses harmless but powerful magnetic fields and pulses of radio waves to probe the atomic nuclei that quietly reside within all materials, chemical and biological.

Like the electron, the nucleus of an atom can be loosely regarded as a tiny spinning ball of charge but with quantum properties. The rotating charge, which acts as a loop of electric current, gives rise to a magnetic field like that of a tiny bar magnet. The orientation of these bar magnets when measured is quantized, so that in an external magnetic field a hydrogen nucleus, which is a single proton of spin ½, will appear to be in one of two possible states: either aligned parallel with the field (spin up) or antiparallel with the field (spin down). Initially the nuclear bar magnets will be randomly oriented in either direction, but a strong external magnetic field causes the spins to feel a small energy difference between the two states. Over time some of the nuclear spins flip to the lower energy state, where they point with the field. The spins are said to become magnetized. However, the process is slow, and because of occasional disturbances to the spins from thermal vibrations, the overall magnetization at equilibrium with even a very powerful magnetic field is rather small—an excess typically of one spin in a million.

Although the nuclear spins appear to point up or down, their spin axes actually point at a small angle away from the field and precess around it like the wobbling motion of a toy gyroscope as it leans away from the vertical. This precession happens at a frequency governed by the energy difference between the spin up and spin down states.

Now, this precessional motion gives rise to a resonance when the spin is bombarded with photons in the radio frequency range. The effect is somewhat similar to the periodic force required to push a child on a swing. If an electromagnetic pulse of the same frequency as the precession is beamed in at 90 degrees to the magnetic field, along the x-axis, say, the nucleus feels a periodic force that nudges its spin axis sideways at the same point during each wobble. Such a pulse tips the spin axis round such that the longer the pulse, the larger the rotation. A 180-degree tipping pulse will flip spins from up to down and vice versa. It's possible and indeed useful to have intermediate tipping pulses. A 90-degree pulse, for example, will orient the spins to point either way along the y-axis (see Figure 7.6). Note that such intermediate orientations are not inconsistent with the idea of quantization: A spin pointing along the y-axis can be regarded as a superposition of spin up and spin down just as light polarized at 45 degrees to the horizontal can be regarded as a superposition of photons polarized vertically and horizontally.

To see these resonance effects, it is necessary to use very powerful magnets and to ensure that the field is very accurately maintained. A typical magnetic field strength of, say, 2 tesla is around 40,000 times stronger than the Earth's magnetic field and strong enough to cause a hammer to fly across the room. Furthermore, someone walking into the room with a metal cigarette case will ruin the whole experiment.

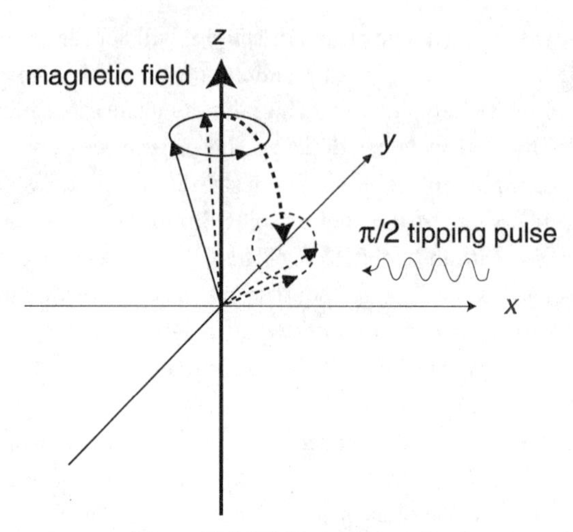

Figure 7.6. NMR tipping pulse

When a material is placed in a powerful magnetic field, some of the nuclear spins align themselves at an angle to the field and precess around it. By applying a radio frequency pulse from the side, it is possible to tip the spins over by an angle determined by the length of the pulse. Such tipping pulses are an important tool in NMR spectroscopy and NMR quantum computation.

For a hydrogen nucleus in a 2-tesla field the resonance frequency is 86 MHz, which is just off the bottom end of the dial on the FM radio band. In an NMR apparatus, pulses of carefully tuned radio waves are beamed in to the sample. This actually has a dual action; it not only causes some spins to flip but also causes the spins to precess in unison like a choreographed array of tiny gyroscopes. (Normally the precessional motion of the spins is incoherent.) Once a pulse of radio waves has finished, the precessing spins continue to move together for a while, generating an oscillating magnetic field that can be detected by coils mounted around the apparatus. The signals are then turned (using Fourier transforms) into a frequency spectrum that reveals details of the nuclear resonances. The precessional motion gradually loses its coherence once the radio pulse has finished, causing the signal to fade away.

A surprising aspect of NMR is the role played by the thermal jostling and tumbling of atoms and molecules in liquids and gases. Although thermal vibrations keep the overall magnetization very low, the thermal tumbling and jostling of molecules in liquids and gases actually helps to *increase* substantially the resolution of spectral lines observed. This happens because of the way the electron clouds in molecules superimpose their own local magnetic effects on the nuclear spins, over and above the external magnetic field. These local fields, which depend on the geometry, structure, and orientation of the molecules, alter the resonance frequencies of the nuclear spins slightly. In a solid the molecules may be frozen in many different orientations, causing the resonance peaks of the nuclear spins to appear as broad plateaus. However, in liquids and gases the rapid tumbling of the molecules averages out these variations in the local magnetic field, resulting in extremely sharp resonance peaks. The sharpness of the spectroscopic lines makes it possible to see many more features.

Take medicine, for example. Most hospital MRI scanners are tuned for detecting protons, the hydrogen nuclei, because they are relatively easy to magnetize and are abundant in biological tissues in the form of water. To gain depth information, the scanner's magnetic fields are arranged to vary in strength across the human body. This causes the resonance frequencies of hydrogen nuclei to change from one side of the body to the other. By sweeping the pulses of radio frequencies across the spectrum, it's possible to work out the three-dimensional distribution of hydrogen in different parts of the body. However, a picture of the hydrogen distribution alone would not be much use because it tends to be present in soft tissues at roughly the same concentration. The power of magnetic resonance imaging comes in its ability to discriminate between hydrogen atoms *in different chemical environments.* We'll see how that works in a moment, but the consequences are that once the MRI scanner receives its signals, they can be translated by a computer into color-coded images of almost unrivaled quality: The pictures can reveal not only gross anatomical features but also many biochemical details, including abnormalities such as those caused by cancer.

The possibility of using NMR for quantum computing arose when scientists realized that nuclear spins had many of the requisite properties for playing the role of qubits and that each molecule could act like a single quantum computer. The immediate attraction of nuclear spins is that they have wonderfully long coherence times: typically many seconds or minutes in liquids and many hours in gases. As Gershenfeld told me, "You can view the nucleus in a molecule as a perfect atom trap. So in that sense the hard part of the fabrication is done inexpensively by nature. What's right now the most severe scaling limit in ordinary computation is the cost of the

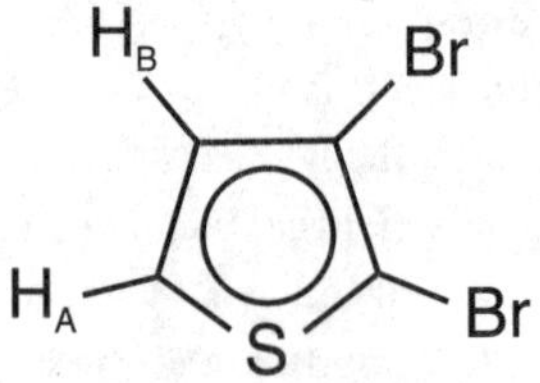

Figure 7.7. 2, 3-dibromothiophene

This chemical can serve as a two-bit quantum processor. The two hydrogen atoms, labeled H_A and H_B, exhibit slightly different NMR resonances because of the differing influence of their neighboring atoms. The different resonance frequencies enable the hydrogen nuclei to be addressed independently.

manufacturing plant." A second consideration is that nuclear spins are easy to load with data and to read out using the techniques of NMR. And last but by no means least is the fact that their properties are influenced by their molecular or chemical environments.

This last property is crucial for doing two-bit (and hence universal) logic. Consider, for example, the two hydrogen atoms in the molecule 2,3-dibromothiophene (see Figure 7.7).

When this molecule is subjected to a magnetic field of 9.4 tesla, the nuclear spins tune into radio frequency pulses of around 400 MHz. However, the precise frequencies of the two nuclei differ by about 130 Hz because of the differing magnetic influences of their neighboring atoms (see Figure 7.8). One hydrogen atom, labeled H_A, has bromine as a neighbor, while the other, H_B, has sulfur. The frequency difference, which is known as a *chemical shift,* makes it possible to address each nuclear spin separately. On top of that, though, the orientation of the proton spins influences the field too, adding a further, smaller shift that splits each spin's resonance into two peaks separated by a mere 6 Hz. If a radio frequency pulse is tuned precisely to one of these resonance peaks, the only molecules that will respond are ones in which both nuclei are aligned in one particular configuration. This phenomenon, which is known as *double resonance* because both nuclei must participate in the resonance, makes it possible to change one spin in a way that is *conditional* on the state of the other, precisely the operation we need to emulate a controlled-NOT gate.

Double resonance phenomena between nuclear spins and electron spins were known in the 1950s when NMR was first being developed. Indeed, in 1956 spectroscopic experiments established that it was possible to do a

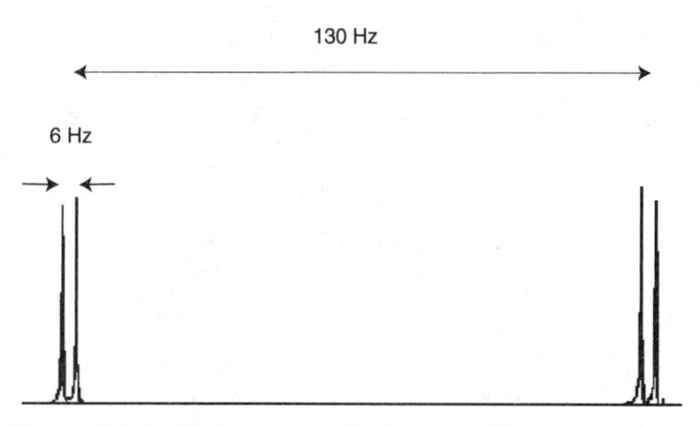

Figure 7.8. NMR spectrum of liquid 2, 3-dibromothiophene

The spectrum produced in a 9.4-tesla field reveals the resonances of the hydrogen nuclei (that is, protons) within each molecule. Each proton gives rise to a doublet separated by 6 Hz, caused by its two possible spin orientations. The doublets are separated by 130 Hz because of the effect of chemical shifts on the nuclei.

kind of controlled-NOT operation on nuclear and electron spins. The effect occurs as a by-product of a procedure known as ENDOR (short for Electron Nucleus DOuble Resonance), which uses two pulses of radio frequencies to transfer the spin state of an electron to a nucleus. The fact that this operation also performs an interesting logical function, XOR or controlled-NOT, was not appreciated at the time.[25]

NMR appears, then, to have all the right things going for it as a hardware platform for doing quantum logic. So what's the catch? Well, the obvious problem is that it inevitably involves huge quantities of molecules: numbers in the region of a hundred billion trillion (10^{23}) molecules if you take your average mugful. What's more, most of the spins are randomly oriented. That means that we have, in effect, a vast array of tiny quantum computers all starting out in different states. These circumstances seem to fly in the face of the idea of computing with a coherent or "pure" quantum system. Indeed, we have in an NMR system a very good example of a mixed quantum state (see page 122, Chapter 4). Mixed states don't sound like a very promising basis for doing quantum computation.

One early suggestion for overcoming this drawback was to build an apparatus that could address individual nuclear spins in much the same way that it is possible to manipulate single ions in a trap. A device known as an atomic force microscope, which can manipulate individual atoms, remains a candidate for such a role, but so far this idea remains untested.

The breakthrough by the two American teams was to find ways to select from a macroscopic quantity of molecules an ensemble of molecules that shared the same quantum state. On the face of it, this looks like a tough call. It would be like trying to discern the song of a few harmonious choristers in a vast and unruly football crowd. However, it turns out that the overwhelming randomness of the spin orientations can be turned to advantage because the NMR signals generated by most of the molecules average out to zero. This happens because oppositely polarized spins precess in opposite directions, so for nearly every spin generating an electrical signal in the detector coils, there is another spin generating the exact opposite.

The result is the cacophony of the crowd is silenced, leaving only the small excess of spins aligned with the magnetic field to sing out like soloists in a huge but silent choir. However, this is not the whole answer because the soloists do not, unfortunately, all sing from the same hymn sheet.

Suppose we had a magnetic field pointing upward and we used a slightly more complicated molecule than before: one with four addressable nuclear spins. If we were able to look at just those molecules that contributed to the final signal, we would certainly find some with four spins pointing up, but there would also be others with only three spins pointing up. These latter molecules are not fully aligned with the field, but they are still present in excess over the general population and therefore make a contribution to the NMR signals. It also turns out that the molecules with three or four spins pointing *down* are slightly *under*represented in the population, and the absence of these molecules adds to the signals too. The only molecules that don't affect the signals are those with two spins up and two spins down. Once you subtract the random background, you're left with what's called a thermal distribution of states rather than a pure quantum state. Including all these molecules in a quantum computation will not be much use because they will start off in different states. So the question is, how can we home in on a group of molecules that are all in the same quantum state?

It turns out there are several different ways of doing this. Gershenfeld and Chuang's method was to use some of the nuclear spins on each molecule as a kind of check for quantum purity. Suppose, for example, they used spins 1 and 2 in a molecule for this purpose and used spins 3 and 4 for doing a quantum computation such as controlled-NOT. What they can do is beam in radio frequency pulses that will only affect nuclear spins 3 and 4 *provided* spins 1 and 2 are aligned with the magnetic field. Using such conditional manipulations, it's possible to ensure that the quantum calculation is conducted only among an ensemble of molecules that all start in the same quantum state.

There's an extra bit of setting up that has to go on first, though. For molecules of four or more spins, the system has to be "prepared" before the quantum calculation starts. The problem is that with four spins, for example, merely insisting that spins 1 and 2 were pointing up would still leave ambiguous the state of spins 3 and 4. However, with an extra sequence of radio pulses, it's possible to redistribute the different ensembles so that spins 1 and 2 do guarantee the purity of spins 3 and 4. (See Appendix H for more details.)

How Useful Is NMR Quantum Computation?

One penalty of Gershenfeld and Chuang's NMR scheme is that to ensure the purity of the quantum state, they need to sacrifice some qubits, reducing the number left over for their calculations. On the plus side, with larger molecules and more spins, the total requirement for these purity spins becomes proportionately less and less. However, going to progressively higher numbers of spins soon becomes a big problem.

The method adopted by Cory and his group for purifying their quantum ensemble was different and goes under the name of *spatial averaging*. Like Gershenfeld and Chuang, they also prepare their "pseudopure" state with a series of radio frequency pulses. However, in the middle of the sequence Cory's group used a special pulse known as a gradient pulse, which delivers a magnetic field of varying strength across the sample. This has the effect of spatially stratifying the sample so that molecules in different layers are in slightly different spin states. It turns out that this manipulation, provided it is performed to just the right extent, together with the rest of the pulse sequence, rearranges the total population of spins in such a way that when you take the average, everything magically cancels out apart from the "pure" state the experiments want for doing their quantum calculations. (If you're not convinced, the only way to believe it is to look at the equations.) This method has the advantage of not expending precious qubits but the disadvantage of being exquisitely sensitive to inaccuracies in the gradient pulse.

Both groups of researchers have demonstrated many simple forms of quantum logic using their methods with small numbers of qubits. Indeed, Gershenfeld and Chuang were the first to demonstrate with quantum logic that one and one make two. Beyond that, the prospects look bright for NMR to outrun all other quantum technologies—for a while, at least. However, the roadblock looming further ahead for NMR is the issue of scaling.

There are several things that go wrong as you add more qubits to the system. The most obvious is that the strength of the NMR signal decreases *exponentially.* The reason for this loss of signal strength is that as you add more spins, the number of possible spin states for the whole molecule increases exponentially. Because the purification process merely *extracts* rather than *amplifies* one of those states (that is, the one with all spins pointing with the field), the number of representative molecules in that state decreases exponentially.

The incentive is to work with molecules that have as many extra spins as possible (that's where coffee might have come in—caffeine molecules have a handful of different spins), because for each qubit you add to the system, the potential power of the computation doubles. But the downside is that for each extra qubit, the signal halves in strength, and no amount of computational power will overcome the problems of a signal that is weaker than the noise. With current NMR methods, ten qubits looks like taking experimenters to the limits of measurability. While ten qubits may be enough to factor the number 15 using Shor's algorithm, they would scarcely amount to the computational powerhouse quantum theorists have been hoping for.

What are the prospects for overcoming this limit? Late in 1997 the journal *Science* published a very negative assessment by Warren S. Warren, a NMR spectroscopist at Princeton University.[26] Chief among his arguments for why he thought NMR quantum computing didn't have much of a future was the problem of scaling. With a 100-spin system at room temperature, he pointed out that for 99.99999999 percent of the time a sample of some 10^{22} molecules would contain *no* molecules in the sought-after pure state—that is, the one in which all spins point with the field. Furthermore, he added, the desired pure state was only 1 percent more probable than the opposite state, in which all the spins point *against* the field. "Thus," he continued, "for every 100 times one molecule accidentally gets in the 'right' initial state, there will be 99 occurrences of the 'wrong' initial state giving exactly the negative of the desired signal." Warren concluded: "In summary, quantum computing might well turn out to be capable someday of solving certain problems better than conventional techniques; but if so, bulk NMR is not likely to play any role in a practical implementation."

In a robust defense published alongside Warren's critique, Gershenfeld and Chuang suggested a number of ways in which standard NMR spectrometers could be optimized for quantum computing. Their longer-term aim was to go well beyond ten qubits. When I spoke to Gershenfeld shortly after this exchange had appeared, he certainly sounded tremendously upbeat. "We've gotten much further than anyone expected in *any* approach to quantum computing even a year ago. We've made circuits, we've implemented quantum algorithms, we're beginning work on a compiler because it's getting too hard to do these [NMR programs] by hand, and we're developing simple tabletop prototypes. There are some significant scaling challenges as to whether this can ever seriously beat classical computation, but there's a very strong argument to be made that if anything can, it has to be this."

So how far did he think the technology could go? "Round about fifty qubits is when you begin to beat classical computers," Gershenfeld said. "What that means is that with custom hardware tuned for computation with spectroscopy, you could just begin to graze the point where classical computers run out."

Fifty qubits is certainly way beyond anything scanning on other people's NMR screens at the moment, but among the improvements Gershenfeld believed could propel NMR systems forward were optical pumping techniques for "cooling," or aligning, the nuclear spins. The basic idea here is that if many more of the nuclear spins can be coerced into aligning with the magnetic field, the population size of the pure state could be boosted, providing a much stronger signal. Optical pumping involves bombarding the sample with light tuned to polarize the electron spins in one direction. The electron polarization can then, in principle, be transferred to the nuclei using double resonance techniques.

David Cory, on the other hand, was not so sanguine about the prospects: "The take-home message is that ensemble quantum computing is a wonderful way of exploring the physics of quantum computing, and it gives us for the first time an experimentally realizable paradigm for looking at these issues. But it is not, in the mode that has been recently suggested, a computationally interesting approach."

Nevertheless, Cory was optimistic that NMR systems with small numbers of qubits could still be very interesting to study as machines to emulate other quantum systems, an idea first proposed by Richard Feynman (see Chapter 3). Another attractive feature of NMR is that it is relatively cheap and widely available. Many universities have their own NMR machines and could, in principle, conduct their own quantum computation experiments. Researchers at Los Alamos did just that in 1997. By using a standard NMR system, they were able to create, for the first time, a very interesting three-qubit entanglement known as the Greenberger-Horne-Zeilinger state (see Chapter 8, page 282). Following this, in 1999 Cory and colleagues demonstrated the first implementation of the quantum Fourier transform while Isaac Chuang and colleagues unveiled the first working version of Grover's algorithm, a search program we'll also examine in the next chapter.

Connecting the Quantum Dots

If present forecasts hold up, by about the year 2015 semiconductor manufacturers are likely to be making chips whose logic elements are close to the atomic scale. But we don't necessarily have to wait until then to explore this regime. Research since the 1980s using specially fabricated semiconductor devices has already yielded many clues about what we can expect at these scales. Among the most interesting devices to have emerged from such studies are *quantum dots,* tiny islands of semiconducting material on a chip.

A typical quantum dot may be a few hundred atoms across, but by applying a negative electric field around one of these little blobs, it's possible to control the number of freely moving electrons inside. A negative voltage will squeeze out some of the electrons and can be adjusted so that only one electron is left inside. Trapped inside what is, in effect, a tiny box, the electron is forced to occupy one of a set of discrete energy levels. This quantization is a consequence of the way the wavelike aspects of the electron have to fit inside the box exactly and is analogous to the discrete energy levels of electrons around atoms. For this reason the quantum dot is sometimes described as an artificial atom; an atom, moreover, whose properties can be made to order.

For these quantum properties to reveal themselves, the whole device typically has to be cooled to liquid helium temperatures. Quantum dots and other nanoscale devices usually also require special manufacturing techniques that are different from the lithographic methods employed in conventional silicon chip manufacturing. The key technology here is that of *molecular beam epitaxy.* Invented in the 1960s by Alfred Cho and John Arthur of Bell Labs, MBE enables researchers to assemble their devices almost as if they were spray-painting with atoms. At the heart of an MBE machine is an ultrahigh-vacuum chamber in which sits a heated semiconductor wafer. Elements such as gallium, arsenic, and aluminum are evaporated in separate chambers, and at appropriate times the vapors are squirted toward the main chamber, where they leave a deposit on the surface of the wafer. The whole process happens slowly, so that different layers of atoms can be grown on the wafer as a single crystal, each layer lattice-matched with the one below.

The materials commonly used for these so-called heterojunction devices are gallium arsenide, which is a semiconductor, and aluminum gallium arsenide, which is an insulator. Conventional microchips use silicon as the semiconductor, and silicon dioxide as the insulator, but because silicon dioxide is amorphous, it won't lattice-match onto silicon. This doesn't matter in conventional devices, but in the quantum regime it's crucial because any imperfections in the crystal cause electrons—the conductors of electrical current—to scatter as they travel through. Such scattering kills off any obvious quantum effects when researchers study the electrical properties of their devices. In short, imperfections cause decoherence. Avoiding thermal disruption also explains why these systems have to be supercooled to liquid helium temperatures.

Recent developments have made it possible to construct large numbers of quantum dots on a single chip using so-called self-assembly techniques. If indium arsenide is grown on a gallium arsenide substrate, it initially forms a perfect, atomically flat monolayer. If more indium atoms are added, however, they begin to clump together in little hillocks separated by valleys. Originally this clumping was considered a nuisance—until researchers realized that the clumps could serve as quantum dots.

Quantum dots are usually surrounded by an electrical insulator such as aluminum gallium arsenide, and electrons can be encouraged to tunnel in and out of them under the control of external metal layers called gates. Each dot behaves like a tiny capacitor, so that every time an electron is added, its potential changes significantly. It's actually possible to "see" each electron flow into and out of a quantum dot by watching the voltage jump.

Such delicate manipulations have opened the way to intensive research into the concept of single-electron transistors, which could be used as memory and logic devices. Computer memories typically depend on the flow of millions of electrons to represent the binary states 0 and 1. The potential advantage of single-electron memory devices is that they could be much smaller and require far less power. If these devices could be produced controllably in large numbers, manufacturers could, in principle, pack colossal amounts of memory onto microchips—way beyond anything possible today. For this reason quantum dots are now the subject of research not only in academic research centers but also at some of the leading Japanese semiconductor manufacturers, such as Fujitsu, Hitachi, Toshiba, and NTT.

But quite apart from all of these more immediate ambitions, a number of quantum theorists have advocated quantum dots as a possible platform for constructing a quantum computer. In 1994, before the advent of quantum computing with atom traps, Adriano Barenco, David Deutsch, and Artur Ekert suggested how quantum dots could be used to implement conditional quantum logic. Suppose, they suggested, we had two single-electron quantum dots closely spaced on a semiconductor chip. The first two energy states of the electron in each quantum dot would serve as representations for 0 and 1. The electrons, like those in conventional atoms, could be excited from the ground state to the first excited state by a π pulse of laser light tuned to the frequency corresponding to the energy difference. Superpositions of 0 and 1 could be produced by using shorter pulses, so that, for example, a $\pi/2$ pulse would put an electron starting from the ground state into a superposition of 0 and 1. Such a pulse would act as a square root of NOT operation. Other single-qubit rotations would be similarly straightforward to implement by varying the pulse length.

The possibility of conditional logic arises from the interactions between quantum dots. Barenco and his Oxford colleagues showed that the state of one quantum dot could modify the resonance frequency of the other one if the two dots were subjected to an electric field. This is possible because in the presence of an external static electric field, the electrons in the ground state tend to move away from the field while electrons in the excited state move toward it. The same effect could arise in two closely spaced quantum dots so that the resonance frequency of one dot, the target qubit, would be influenced by the state of the other dot. So if we engineered the resonance frequencies to differ sufficiently, we could use a π pulse to selectively flip the target qubit state subject to the state of the control qubit.

The system is appealingly simple and, if it could be made to work, would offer all the functionality we need to implement quantum algorithms. However, the overwhelming problem with quantum dots is the decoherence time. Impurities and thermal vibrations in the crystal lattice typically cause decoherence within much less than a nanosecond (one billionth of a second). Fortunately, the laser pulses used for quantum operations could be much shorter, perhaps of the order of a picosecond (one trillionth of a second). That could allow enough time, in principle, for several hundred laser pulses before decoherence. But it's a daunting prospect to realize that a complete quantum algorithm must be completed within a billionth of a second. Again, we are up against the problem of scaling. This kind of technology may make possible some very simple quantum calculations, but it offers little threat to the dominance of conventional computers.

In 1998 Daniel Loss of the University of Basel and David Di Vincenzo of IBM proposed a different way of exploiting quantum dots that might suffer less from decoherence. Their idea was to store the quantum information in the *spin* orientation of each electron in a quantum dot. With two dots closely spaced together, they suggested how the dots could exchange spin states by quantum tunneling, a phenomenon that allows particles and interactions to cross barriers that would otherwise be too high to jump across. By switching the tunneling on and off with an electric field between the two dots it's possible to create logical states to represent 0 and 1. A spin-swap operation could then be achieved by raising the voltage of a metal gate for a set period of time and with a half pulse you would get an operation Loss and Di Vincenzo call "the square root of swap." With this and some simple single-bit rotations, the result would be a quantum XOR gate.

Another variant of this idea is to use superconducting quantum dots. When two superconductors are separated by a very thin insulator and a voltage is applied across them, a tiny oscillating current can tunnel its way from one side to another. This is the principle of the Josephson junction, a device that was much vaunted in the 1970s as a way of building low-energy supercomputers. Although those hopes were dashed, it seems Josephson junctions could be headed back for the big time. In 1999 Hans Mooij and colleagues at Delft University were able to create a quantum superposition of states in a chip that incorporated several Josephson junctions. Shortly afterward, a team at NEC's Fundamental Research Laboratories in Tsukaba, Japan, led by Yasunobu Nakamura, announced that they had electronically manipulated a quantum state—a qubit—on a chip, achieving a very respectable decoherence time of around 2 nanoseconds (2 billionths of second). With many new research papers appearing every month on these solid state approaches it seems likely that quantum dots will soon become a serious candidate for doing real quantum computing experiments.

Runners in the Quantum Race

All the technologies we've examined have their strengths and weaknesses. How can we decide which, if any, is likely to prove the winner in this technological race? One simple calculation we can do is to look at two key parameters that characterize each one: the switching time and the decoherence time. Dividing the first into the second gives an estimate of how many quantum operations can be performed before our calculation is likely to be disrupted. The ratio can be regarded as a rough figure of merit so that the higher the number, the better the prospects for doing an extended computation. See Table 7.1.[27] For good measure I've added some very notional figures for the likely scalability of the different technologies.

If ion traps are the vacuum-tube technology of quantum computing, then quantum dots would seem to be the obvious candidate for the role of quantum microchip. But as we can see from the figure of merit scores in the table, quantum dots don't look promising at the moment. But if they could be made to work, quantum dots would probably be much easier to scale up. Meanwhile, NMR has stolen a march on everything else. Of course, these figures are very approximate, and with these technologies still in their infancy, the whole situation could change perhaps dramatically with the entry

Table 7.1. Competing technologies for quantum computation

Technology	*Switching time*	*Decoherence time*	*Figure of merit*	*Scalability*
Ion traps	10^{-7}	10^{-1}	10^{6}	50 qubits?
Cavity QED	10^{-14}	10^{-5}	10^{9}	2–5 qubits?
NMR	10^{-3}	10^{4}	10^{7}	10–50 qubits?
Quantum dots	10^{-9}	10^{-6}	10^{3}	1,000 qubits?

of a different technology or with new ideas about how to improve the ones above.

What's more, none of these technologies would be worth a candle if it weren't for the fact that quantum theorists have discovered, much to many people's surprise, ways to handle errors in quantum computation. They've also produced some interesting new applications for these machines to chew over should they ever see the light of day. It is to these subjects—quantum error correction and other novel quantum algorithms—that we now turn.

8

Quantum Error Correction and Other Algorithms

I feel that a deep understanding of why quantum algorithms work is still lacking.[1]

JOHN PRESKILL

Processing in the Dark

In the movie *Scent of a Woman* there's an amusing sequence in which Al Pacino, playing a blind military veteran, pressures his chaperon, a young college graduate, into letting him take the wheel of a Ferrari they've just borrowed for a test drive. Initially, Pacino's irascible war hero cruises ever so slowly up a quiet street, but a sudden suicidal craving for excitement takes over and he puts his foot down. As the road nears its end, his terrified passenger has to yell out the precise moment when Pacino must make a turn. Somehow they manage to survive the corner without crashing.

Anyone who has driven a car will instinctively feel a certain heart-stopping terror in that scene. Yet ultimately we know it is unbelievable. Taking a corner calls for sophisticated coordination between the hands, eyes, and brain. You might get some information from other senses, but without vision you'd surely be totally lost. In real life, there's no way a simple instruction "TURN LEFT NOW!" would be good enough to enable a blind person to negotiate a bend at speed. Why not? Quite apart from the problem of the delay in reaction time, there's more to turning than simply deciding when to begin the maneuver. You need to know how fast to turn the wheel, how far to turn it, and when to start rotating it back. Turning a corner is a fairly complicated analog feedback process.

So what? Well, in some ways I think driving a car blind is a good analogy for the idea of getting a quantum computer to calculate *without looking where it's going.* They're both unbelievable and for similar reasons. Take the value of a qubit. We know that it can exist as both 0 and 1 simultaneously and that the strength or amplitude of each state in a superposition can vary continuously. That would seem to imply that quantum information possesses analog properties. We also know that quantum information is extremely fragile because of the problem of decoherence. Yet in a quantum computer we need to be able to process these delicate analog states, push them this way and that, while we remain blind to precisely where they're going. And in a long computation, we need to be able to do this over and over again even though we know that errors will accumulate. Rather like Fredkin's billiard ball computer, the computation is going to be "pathological." In fact, in some ways it's going to be worse, because at least with the billiard ball machine we can see what's happening.[2] With a quantum computer, we're simply not allowed to look—not until the end. It's no wonder people like Rolf Landauer thought the whole idea of quantum computation was unbelievable.

Yet until the early 1990s, many of his colleagues were happy to ignore such worries and blissfully concentrate on the wonders of pure quantum computation unadulterated by noise. In the process, a huge gap opened up between theory and experiment in quantum computation. Experiment, as we have seen, has until recently barely got beyond the stage of building a single quantum gate, yet theorists have raced ahead to fathom what they could do with a perfectly engineered machine endowed with thousands, if not millions, of gates. It was a fun ride, but sooner or later it had to end with a crunch.

Or did it? Few people have been better placed to observe these trends than Richard Hughes, leader of a large team of quantum theorists and quantum experimentalists at the Los Alamos National Laboratory. Largely funded by the NSA, he and his researchers are at the cutting edge of human knowledge in both spheres, but he is the first to recognize how far theory in quantum computation has outstripped experiment. In lectures, he neatly sums the situation up with a slide depicting the Starship *Enterprise* from *Star Trek* warping its way through space. Below appears the label THEORY, while to the right we see a cartoon of a Stone Age man attempting to ride a primitive-looking wheel. Below him we see the word EXPERIMENT.

The reason for this huge discrepancy between theory and practice in quantum computing research comes down to the extreme fragility of coherent quantum states. It's worth remembering that one of the reasons digital computers long ago elbowed their analog brethren out of the way was that digital designs could do calculations with complete accuracy to more or less any number of digits, while analog computers were often hopelessly inaccurate. The robustness of digital information compared with analog also explains why CDs have replaced vinyl records and very likely why life on Earth evolved a digital code to store genetic information. So compelling are the advantages of digital information that the evolutionary biologist Richard Dawkins confidently predicts that should we discover extraterrestrial life, it will almost certainly have, like us, a digital form of inheritance.

The reason digital information is more accurate than analog is that it is much easier to protect against errors. Zeros and ones are either correct or incorrect: if the signals representing them are slightly adrift, they can be amplified to put them back on track. If a large deviation occurs, causing some zeros to turn into ones or vice versa, provided the damage is not too severe proofreading mechanisms can detect such errors and correct them. Such mechanisms protect computers against corrupt programs and data, enable CD players to give us audio recordings free of hissing and crackling, and protect our bodies from falling apart as a consequence of damage to the DNA in our cells. Analog machines are not totally without some form of error correction: watching our maneuver while driving around a corner is, in fact, a form of negative feedback—a mechanism that appears in all kinds of analog systems. But effective though negative feedback is, the correction is never perfect.

On the face of it, then, calculating with quantum states seems to be a step backward into the retro world of analog. Furthermore, we're hampered by the fact that we are not allowed to look at quantum states in midcalculation because of the disturbance measurement can cause. We therefore cannot apply negative feedback.

"This is what had a lot of us stymied when we were first thinking about quantum error correction," John Preskill, a leading quantum computing theorist at Caltech, told me. "Analog computers are notorious for being unstable because of the accumulation of small errors. It didn't seem that the situation would be very different here. What's really remarkable is that it turns out that you can digitize the errors even though, in principle, the errors form a continuum." *Digitize the errors.* Once again, the topsy-turvy world of quantum mechanics sprang a surprise on its learned explorers.

Democracy Among the Qubits

If you've ever hooked a PC up to a bulletin board by telephone, you've probably used a terminal emulator program that dispatches packets of data back and forth as bytes, each encoded with a parity bit that's added to each 7 bits of data. Each parity bit is there to check whether the total number of 1s in a byte is either always odd or always even, depending on the choice of settings you made in the program. If you choose even parity, the program will check each byte to make sure it has an even number of 1s. If one byte is received with an odd number of 1s, the program knows the byte must have been damaged. It can reject that chunk of data and ask for it to be resent.

That's one very simple example of digital error correction, but there are many other much more sophisticated error checking schemes, ones that can correct errors on the spot without the need for asking data to be resent. Such error checking and recovery methods are used on connections to the Internet and within computer components such as hard drives, floppy disks, memory chips, and so on. The common aspect of all these schemes is the idea of redundancy: using extra bits as a check on the integrity of the data.

As we have already noted, though, the robustness of digital data in computers depends not only on redundancy. More fundamentally, digital signals can be amplified so that whenever an input signal wanders slightly off its proper value, the logic gate will push the output back on track. Quantum information cannot be amplified in this way, because of the no-cloning theorem. So we apparently lose one of the great advantages of dealing with binary digits. But the possibility of exploiting redundancy remains.

David Deutsch had appreciated very early on that to answer the criticisms that had been leveled against quantum computation by Landauer and others, it would be necessary to find some way of introducing error correction into quantum systems. The question was, were the difficulties matters of principle or were they "merely" matters of technology. In 1987, at the time he produced his paper on quantum networks, which was motivated by similar concerns, Deutsch produced the first quantum error correction algorithm. He refrained from publishing it because it was very inefficient, and he hoped other people would find better algorithms. To his dismay, not only did this not happen, but the "quantum computers cannot be built because of the error correction problem" objection was cited more and more.

Eventually Deutsch presented the algorithm at a conference in Broadway, England, together with a plea for people to take error correction algorithms seriously. On the way back from that conference, Richard Jozsa worked out a simple way to implement the algorithm and a way to analyze it. There were many more technical problems after that, but it eventually led to a paper that was published with Andre Berthiaume in 1994.[3] The idea required a quantum computer to run simultaneous copies of a computation. If no errors occurred, all of the separate copies would produce the same answer and the overall system would display a certain symmetry. If, however, there were any errors, there would be an asymmetry. The Deutsch algorithm showed how it was possible to perform operations on the whole system, and measurements on a subsystem of the computer, in such a way that the whole machine would be projected back into a symmetric state again, thereby eliminating any errors.

Deutsch admits the algorithm was cumbersome and its efficiency so low that it was most unlikely ever to be useful. Nevertheless it was a start and, more important, a *proof* that quantum error correction was, in principle, possible. Furthermore, in 1999 Susanna Huelga and colleagues showed how the algorithm could be employed to improve frequency standards and the accuracy of atomic clocks.[4]

In 1995, Peter Shor once again demonstrated his cunning ability to pull algorithmic white rabbits from the quantum programmer's hat by discovering a way to correct quantum errors bit by bit.[5] The method convincingly showed that it was possible to overcome the analog nature of quantum information by "digitizing the errors," making it possible to recover from them completely and exactly.

At a Royal Society meeting on quantum computation held in London at the end of 1997, Richard Hughes described Shor's error correction discovery as one of the three major achievements that had transformed quantum computing from an intellectual curiosity into an academic industry, the others being Shor's factoring algorithm and Cirac and Zoller's ion trap proposal.

Although Shor got there first, Andrew Steane of Oxford independently discovered his own error correction scheme, which was submitted for publication not long after Shor's proposal. And following both Shor and Steane, a rash of new schemes appeared, all variations on a theme but offering many enhancements and optimizations.

Shor's original error correction scheme used nine qubits to encode each qubit of information. Shor showed how an error occurring on any single qubit as a result of decoherence could be corrected. What was remarkable was that the original quantum state could be fully restored without *any* inaccuracy. How was this possible?

It turns out that we can get an error correction circuit to measure a quantum system in such a way that it looks only at the noise and not at the signal. This opens the way to removing the noise by subtracting it from the contaminated signal.

Like many classical systems and like the Deutsch algorithm, the general idea was based on the idea of democracy and, as Seth Lloyd puts it, "the tyranny of the majority." What Shor showed was that if one qubit in a group becomes corrupted, then it's possible to take a majority vote and correct the odd man out. That such a system will work on individual qubits in a quantum computer was a big surprise, because at first it looked as though taking a vote would not be possible without corrupting quantum information through making a measurement.

Of course, the value of majority vote systems is easy to appreciate not only in politics but also in safety critical applications. The space shuttle, for example, has five computers on board to control all its flight and operational functions. The outputs of these computers are continually monitored and compared. Should one computer develop a fault, it's possible to see immediately which one has failed. The remaining four will outvote the erring computer and take command. Should two computers fail, there are still enough to win the vote and keep things in order. (That, incidentally, is where the creators of HAL went wrong in *2001: A Space Odyssey.* They only had *one* backup of HAL, which was four hours away by light speed *on Earth!* No wonder they weren't sure what was going on when HAL began to go haywire.)

Although Shor's quantum error correction scheme required nine qubits to encode the state of each qubit of information, it is possible to implement a simpler scheme using just three. Unlike Shor's scheme, this encoding does not deal with all types of decoherence, but it's very instructive to see how it works.

Let's first consider the general plan for what happens to each qubit of information. First, an encoder processes a qubit of information together with two other qubits (referred to as the *ancilla,* from the Latin word for "maidservant") each initially set to 0. The three resulting qubits are then either stored or independently processed in the main quantum program—that is, the program in which the original quantum qubit would have participated. It's during this time we assume that the qubits will be prone to decoherence. After a set interval of time or perhaps after just one step in the quantum program, the three qubits are recombined in a decoder. If none or just one qubit has decohered, the decoder is able to extract the corrected signal, which appears on the first qubit. The ancilla qubits, which now contain error signals, are thrown away (or reset to 0 for subsequent use elsewhere). See Figure 8.1.

Three-Bit Quantum Error Correction

In the three-bit scheme, the encoder and decoder are actually surprisingly simple. For the encoder, the input qubit is entangled with two other qubits using two controlled-NOT gates as shown in the circuit in Figure 8.2.

If the top input qubit state is |0>, the encoder does nothing and so leaves the three qubits each in the state |0>, which we can write as |000>. If the top qubit is |1>, the controlled-NOT gates flip the lower qubits, producing the state |111>. More generally, if the top qubit is in an unknown quantum superposition, we can write its state as *a*|0> + *b*|1>, where *a* and *b* are the amplitudes of the two states representing 0 and 1. The encoder

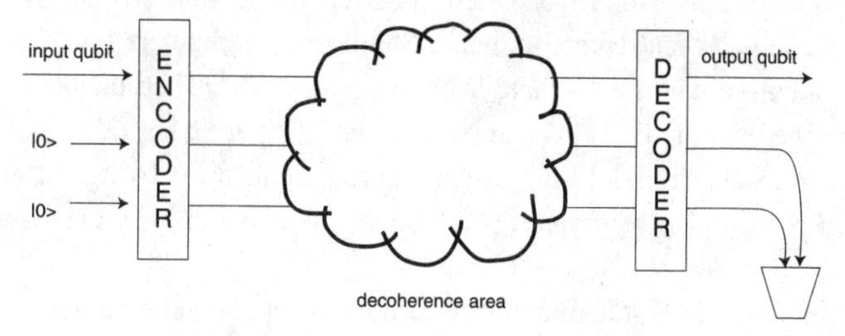

Figure 8.1. Quantum error correction.

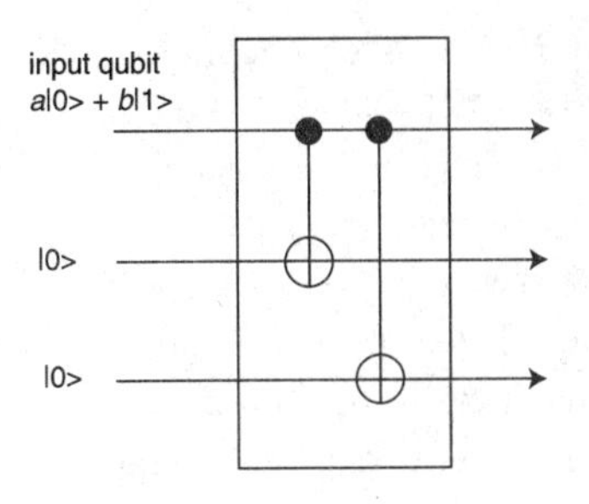

$(a|0\rangle + b|1\rangle)|00\rangle \rightarrow a|000\rangle + b|111\rangle$

Figure 8.2. Three-bit error correction encoder circuit.

translates this state into the entangled state $a|000\rangle + b|111\rangle$. The three qubits can now be processed independently as part of a larger quantum computation, or they can be stored. At some later point we check that the three qubits still agree with one another. However, we can't simply look at the state of each qubit, because a superposition of states will collapse if disturbed by measurement. Instead, the qubits are decoded by the following circuit.

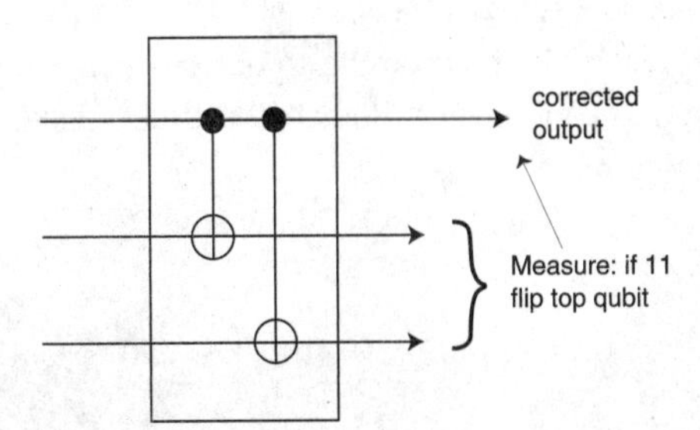

Figure 8.3. Three-bit error correction decoder circuit.

If you think you're suffering from déjà vu, that's because this circuit is virtually the same as the encoder. The only difference is that we measure the output from the ancilla qubits and perform a bit flip on the top qubit if the output reads 11. Let's see what this does.

If the encoded state is $|000\rangle$, nothing much happens. We get $|000\rangle$, which yields 00 on measurement of the ancilla qubits. If the encoded state is $|111\rangle$, the ancilla qubits are flipped by the controlled-NOT gates to give $|100\rangle$. Again the ancilla reads 00 on measurement. So we see that the encoded state $a|000> + b|111>$ is decoded into $a|000> + b|100>$, which we can rewrite as $(a|0> + b|1>)|00>$. On measurement the ancilla reads 00, and the top qubit (which we don't measure) is left intact as the superposition $a|0> + b|1>$, which is the original unencoded state.

So far so good. Now we consider what happens if one of the encoded qubits becomes corrupted. Decoherence can actually cause only one of two types of errors: amplitude errors, in which a qubit is flipped (for example, $|0> \rightarrow |1>$), and phase errors, in which a qubit is reversed in sign (for example, $|0> \rightarrow -|0>$). I'll explain why this is so in a moment. This particular encoder and decoder only protects against amplitude decoherence—that is, errors involving bit flips, but we'll see how its action can be generalized.

If the second qubit flips, we get the corrupted encoded state $a|010> + b|101>$, which is decoded into $a|010> + b|110$ or $(a|0> + b|1>)|10>$. The ancilla qubits read 10, and the first qubit is in the state $a|0> + b|1>$. Likewise, if the third qubit flips, the decoded state becomes $(a|0> + b|1>)|01>$, so the ancilla qubits read 01, and the top qubit is in the state $a|0> + b|1>$. The only time we need to do anything special is if the top qubit has flipped: in this case the encoded qubits become $a|100> + b|011>$, which decodes into $a|111> + b|011>$ or $(a|1> + b|0>)|11>$. The ancilla now reads 11, so according to our procedure we must flip the top qubit, giving the correct state $a|0> + b|1>$ once again. Voilá—quantum error correction!

The summary of these translations follows:

No errors

$a|000> + b|111> \rightarrow a|000> + b|100> = (a|0> + b|1>)|00>$

Top qubit flipped

$a|100> + b|011> \rightarrow a|111> + b|101> = (a|1> + b|0>)|11>$

Flip top qubit

$\rightarrow a|0> + b|1>$

Middle qubit flipped

$a|010> + b|101> \rightarrow a|010> + b|110> = (a|0> + b|1>)|10>$

Bottom qubit flipped

$a|001> + b|110> \rightarrow a|001> + b|101> = (a|0> + b|1>)|01>$

Although the decoder circuit calls for measurements on the ancilla qubits and a flipping action on the top qubit if the result is 11, the same effect can

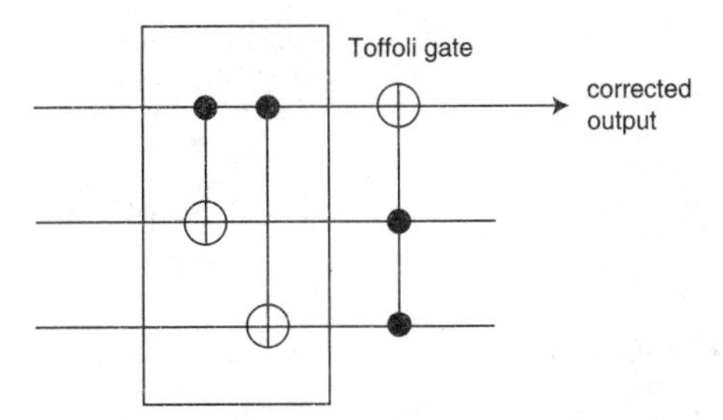

Figure 8.4. Decoder circuit without measurement.

be brought about without measurement by using a quantum Toffoli gate (or controlled-controlled-NOT gate). This flips the target qubit if the other two are both 1. So we can draw a combined decoder circuit as in Figure 8.4.

As we noted before, this three-bit error correction scheme only copes with one kind of decoherence, amplitude errors. It doesn't correct phase errors—sign changes such as $|0> \rightarrow -|0>$. However, if we add an H gate ($|0> \rightarrow |0> + |1>$, $|1> \rightarrow |0> - |1>$) to each output of the encoder and add another to each input of the decoder, the circuits will now correct phase errors instead of amplitude errors. Shor's error correction scheme cleverly combined these ideas using $3 \times 3 = 9$ qubits to protect against both kinds of errors.

As I also mentioned beforc, given that superpositions are effectively analog, you may wonder why we need only consider digital errors—that is, complete flips or sign reversals. After all, if quantum states can take on a continuum of different values, why can't the errors too?

In the first version of the decoder circuit, where we make measurements on the ancilla, the measurements project the errors onto one state or the other. That, after all, is what quantum mechanics is all about: it quantizes things. In this case, it quantizes the errors, leaving the original information intact.

In the second version of the decoder, using the Toffoli gate, the answer is not quite so easy to understand because there are no classical measurements. Nevertheless, the method still works because all errors can be expressed as a sum or linear combination of the intact state and the states with complete phase and amplitude errors. So a qubit that has done a partial flip (such as a small-angled rotation or phase change) can be regarded as a superposition of a complete flip and no flip. It turns out that the error correction circuits perform the appropriate corrections for each part of the superposition (in different universes, if you like) so they still cope, amazingly enough, with partial errors.

How Does Quantum Error Correction Scale?

Peter Shor's nine-qubit method of correcting quantum errors turned out not to be the most efficient coding scheme. Andrew Steane's system needed seven qubits, but the number of qubits necessary for perfect error correction was later squeezed down to a minimum of five. Even so, these methods can only deal with single errors—that is, one qubit going off track in each representative group. To cope with more than one error at a time, it's necessary to use larger groups of qubits and codes that are more elaborate. To deal with three simultaneous errors requires a code that uses as many as twenty-three qubits for each original qubit. For five errors we need fifty-five qubits, and for seven errors we need eighty-seven. The drawback of these schemes is that in order to render our quantum computer immune to errors, it has to grow in size quite a bit. After all, eighty-seven seems like an awful lot of qubits just to do the work of one. However, these increases pay big dividends: For each extra qubit the code is able to check, we get an exponential increase in accuracy.

Let's consider an interesting long-term goal. To factor a 130-digit number, the sort of task that takes months on today's classical computers, it's reckoned that you'd need a quantum machine with around 2,000 qubits, and you'd need to perform around 20 billion quantum operations. And that's without error correction. For such a machine to be successful in a reasonable number of runs, the error rate would have to be around 1 in 10^{13} gate operations. Current gates based on ion traps and NMR are not much better than about 10 percent accurate, so we're off by a factor of a trillion (10^{12}). Such a huge improvement in accuracy is probably impossible: Estimates of the best likely decoherence rates for any practical gates point to a limit of around 1 error in 1 million operations (that is, 1 in 10^6). That would still be 10 million times too high to do the factorization.

With error correction, though, it's a different story. Using an eighty-seven-qubit code, it is theoretically possible to reduce the decoherence rate by the necessary seven orders of magnitude. So, in principle, error correction could make large factorizations feasible—at the expense of increasing the size of the quantum computer by about a factor of 100. That means a machine with 200,000 qubits—totally outlandish by today's standards, but if qubits and their processing environments could be miniaturized like transistors on a microchip, then it's not such an impossible dream.

What's truly encouraging is to see how these figures scale beyond the classical horizons. If we wanted to factor a number twice as long as the previous example—that is, one with 260 digits—we'd find that with the best of the known classical algorithms the running time would increase by a factor of 1 million. With a quantum machine, however, the length of the calculation increases only by a factor of 8—that's because the number of gate operations scales roughly as L^3 where L is the number of digits. Meanwhile, to accommodate the larger numbers we would clearly have to double the number of qubits, but that's no big deal considering we're doubling the length of the number to be factored. With longer running times we might also have to step up to a more powerful error checking code, but as we have just seen, the impact on the number of qubits required is likely to be modest. If we're already using a fifty-five-qubit code, the increase to eighty-seven is only 60 percent. So it looks as though we won't have to worry about an unreasonable increase in the demand for resources.

However, one rather crucial aspect we've ignored in this discussion of error correction so far is that the error correction circuits themselves will make mistakes. They're using the same technology as the quantum computer itself, so inevitably they will be subject to the same problems of decoherence. If left unchecked, the error correction circuits could cause more damage than they prevent. Again, it was Peter Shor who first produced a complete answer to this problem. He outlined a method of *fault-tolerant recovery,* which used extra ancilla qubits and some extra circuitry to double-check the diagnosis of errors.

In addition, Shor dealt with the need for *fault-tolerant computation.* With the simple three-bit scheme we examined in the last section, this isn't a problem: We merely use copies of whatever gate is required, one for each qubit. So if we wanted to perform a NOT operation as part of our original quantum algorithm, we simply apply a NOT gate *bit-wise*—that is, one NOT gate to each encoded qubit. However, this prescription won't work with more complex error correction schemes because in many, the encoding tends to scramble the states |0> and |1> into complicated superpositions of states. In Steane's seven-qubit scheme, for example, |0> and |1> are translated into:

$$|0\rangle_{code} = |0000000\rangle + |0001111\rangle + |0110011\rangle + |0111100\rangle + |1010101\rangle + |1011010\rangle + |1100110\rangle + |1101001\rangle$$

$$|1\rangle_{code} = |1111111\rangle + |1110000\rangle + |1001100\rangle + |1000011\rangle + |0101010\rangle + |0100101\rangle + |0011001\rangle + |0010110\rangle$$

Performing a simple quantum gate operation now has to be translated into a form that works with these peculiar encoded states. Not so straightforward. Fortunately, some of the gates in this code remain virtually unchanged. If you apply, for example, a NOT operation bit-wise to the state $|0\rangle_{code}$ shown above, you actually get $|1\rangle_{code}$ and vice versa. The same is true for the Hadamard rotation and the controlled-NOT gate. However, life is not so easy when we look at gates necessary to complete the set for universality.

Single-bit phase changes, which are very handy for implementing one elegantly simple version of the Fourier transform in Shor's factoring algorithm (see page 234), are no longer available in single-bit or single-gate form. Instead, fault-tolerant fixed-angle versions (of angle 53.13 degrees) have to be combined to approximate whatever angle is required. Alarmingly, the circuit for each fixed single-bit rotation involves two three-bit Toffoli gates, two Hadamard gates, and a phase-flip gate. What's more, the implementation of each Toffoli gate involves an elaborate circuit requiring the participation of some sixty-three qubits! So much for elegant simplicity.

Part of the problem is that the fault-tolerant gates form a discrete set, unlike idealized quantum gates, which form a continuum. That, in a sense, is the price we have to pay for expecting perfection from imperfect materials. Nevertheless, even with the apparently quite substantial overhead required for fault-tolerant computation, it's not thought that the resources will scale unreasonably with increasingly large quantum calculations. Indeed, one of the surprising aspects of the Fourier transform required for factoring is that it works even if the phase changes are horribly inaccurate.

Crossing the Error Threshold

In the previous section we saw that by using error correction codes that dealt with more mistakes per block we could substantially reduce the error rate. Taken to its logical conclusion, the implication would seem to be that provided we used a sufficiently complex code, we could maintain quantum information in storage for as long as we wanted. Such nonvolatile quantum memory would make possible long-range quantum teleportation and many other intriguing phenomena that require stockpiles of entangled particles. It would also mean that we could run quantum programs indefinitely.

The problem with this argument is that it ignores the escalating costs of the increasingly complex recovery procedures required. For a start, the bigger the error code, the larger our quantum computer has to be. That makes the performance of a quantum computer rather different from that of an ordinary computer: Imagine that the longer you wanted to run your word-processing software, the bigger your machine had to be. However, this particular resource problem isn't too severe because the growth in size of the computer with respect to running time is quite moderate. A more serious problem is that as we attempt to rectify more errors, the complexity and length of the recovery procedure rapidly increase, which means there's more chance of things going wrong during the correction process. Analysis of the costs and benefits of using larger error correction codes shows that there is a cutoff point at which no further gains in accuracy can be made.

However, researchers found that this limitation could be overcome by using a special kind of code called a *concatenated* code. Imagine we were using Steane's seven-qubit scheme for encoding each qubit. A concatenated version of the code replaces *each* of the seven qubits with a *block* of qubits. Now, if we zoomed in on any one of these blocks we would see seven subblocks within the block. If we looked closer still, at one of these subblocks, we would see it too consisted of seven smaller subblocks. We could repeat this process over and over until we reached the deepest layer, in which the smallest blocks consisted of just seven qubits. So if there were just two levels within the concatenated code, the total number of qubits would be $7^2 = 49$. For three levels, it would be $7^3 = 343$ qubits, and for N levels, it would be 7^N qubits. The overall structure of the concatenated code is like a rapidly branching tree—each branch splits seven ways, and the leaves represent the qubits.

Although the concatenated code explodes in size the more levels we have, we also get an exponential increase in the accuracy of our code. In addition, the process of error recovery within each tier of the hierarchy remains almost as simple as that of the original seven-qubit code. This multilayered approach works by a divide-and-conquer process in which the commonest errors are dealt with at the leafiest end of the tree, secondary errors are dealt with on the twigs, tertiary errors on the branches, and so on down to the root of the tree.

When researchers analyzed the performance of these concatenated codes, they discovered a very interesting threshold effect: If the error rate of individual gates is *higher* than a certain threshold amount, adding extra layers to the correction code causes more damage than it cures. But if the error rate is *below* the threshold, then, in principle, we can increase the accuracy of our scheme without limit.

What then is the magical threshold number? Unfortunately, it's not quite as straightforward as that. A lot depends on the specifics of the type of code used, the type of errors, and whether errors occur more often in qubit storage or in gate processing. However, according to John Preskill the results can be crudely summarized as follows: If storage errors are negligible, the threshold rate for gate errors is about 1 in 10^4, and if storage errors dominate, the threshold error rate is about 1 in 10^5 per time step, where a time step is the amount of time required for each gate operation.

The reassuring thing about these figures is that they offer a realistic target for experimenters to aim for. If the figures had been much smaller, say, 1 in 10^{12}, the future of quantum computing would have been very limited. Instead, the numbers are tantalizing, such that while they won't be easy to reach, they don't look impossible.

All in all, recent progress in quantum error correction has exceeded everyone's expectations and has transformed the prospects for making quantum computing a practical reality. Even the most hardened skeptics have been impressed. "I had been too pessimistic—it's that simple," Rolf Landauer told the *New York Times*[6] early in 1997. But there remained plenty of doubt over the practicability and economic viability of building a quantum computer. "I'm not ready to put my money in quantum computing," Landauer added.

Creating the GHZ State

Given that we are unlikely to see a large-scale quantum computing machine any time soon, what can we look forward to? As we saw in the last chapter, NMR techniques probably offer the greatest potential in the near future to run small quantum programs requiring up to ten qubits. Although that's nowhere near enough to beat classical computers at tasks like factoring, it is sufficient to do some uniquely quantum tricks.

One example of this was the work of researchers at Los Alamos who in 1997 became the first to create a Greenberger-Home-Zeilinger state. The

GHZ state is a very interesting quantum entity involving three entangled qubits. It is closely related to the Einstein-Podolsky-Rosen phenomenon (discussed in Chapter 3), in which apparently separated particles appear to exhibit "spooky" correlations. What's particularly interesting about the GHZ state, though, is that these correlations defy classical logic even more strongly than does the EPR effect.

Recall that the distinction between quantum and classical behavior in the EPR phenomenon is defined mathematically by the Bell inequality. However, it's not an all-or-nothing effect. To prove that quantum mechanics violates the Bell inequality, it's necessary to carry out an extended run of measurements. From these we can calculate the frequencies with which pairs of measurements agree with one another. Only then can we convince ourselves that the results are inexplicable classically. No single measurement on its own is enough to convince us.

The way we demonstrate the EPR effect is a litte like the way experiments that attempt to prove the existence of telepathic powers depend on statistical samples. A typical setup is to get one subject in a room to look at a sequence of pictures while a second person, sitting in another room, is asked to guess what the other person is looking at. The results in favor of telepathy stand or fall on the strength of the correlations. Sometimes experimenters claim to have results that show something interesting is going on but usually the numbers are within the margins of statistical noise. Either that or somebody cheated. But even the occasional statistically significant correlations offer little practical value. If purveyors of psychic powers really want to impress the skeptics, what they need is a surefire effect that works 100 percent of the time rather than some small "effect" buried within the noise of statistical chance.

Of course, the EPR effect is not really in the same position; it *has* been convincingly demonstrated with statistically significant results. But, psychologically, wouldn't it be all the more impressive to have a demonstration of nonlocal effects that happens in one shot every time? Well, that's where the GHZ state comes in.

The idea, originally a thought experiment, was first proposed by Daniel Greenberger, Michael Horne, and Anton Zeilinger. The GHZ experiment is basically an extension of the EPR experiment with three correlated particles instead of two. We saw in Chapter 4 that an EPR pair can consist of two polarization-entangled photons in a superposition such as $|00\rangle + |11\rangle$. If 0 refers to horizontal polarization and 1 to vertical, this tells us that if we measure one particle's polarization and it comes out horizontally polarized, so will the other's. Likewise, if the first particle comes out vertically polarized, so will the other. But that in itself does not prove anything nonclassical is going on. It's only when we make slightly different measurements on each particle that we get "spooky" correlations.

Now, in the case of the GHZ experiment, rather than using photon polarizations, we'll first consider using electrons, which are particles with spin $^1/_2$. And rather than representing up by 1 and down by 0, we'll switch to the labels 1 for up and -1 for down. This is just a matter of notational convenience and doesn't refer to any physical differences. The particular state we're interested in for the GHZ experiment is the entangled superposition $|1,1,1> - |-1,-1,-1>$. (The state $|1,1,1> + |-1,-1,-1>$ is also a GHZ state and will also work, but some of the numbers come out the other way round.) It turns out that there is a particular measurement that we can make on this GHZ state that will defy classical logic at a stroke. First of all, though, we'll need to examine the experimental setup, which is taken from a refined version of the GHZ experiment[7] proposed by David Mermin of Cornell.

It's possible to measure the orientation of a particular electron using a device known as a Stern-Gerlach apparatus, which consists of magnets with nonuniform magnetic fields. As an electron passes through the Stern-Gerlach device, the magnetic field pushes it one way or the other, depending on which way the electron spin is pointing (see Figure 8.5). Now, as we have seen before, whenever the spin of a particle such as an electron is measured, it is observed to point only in one of two directions: up or down with respect to the measuring apparatus. However, we can point the measuring device any way we want. Indeed, the advantage of using electrons

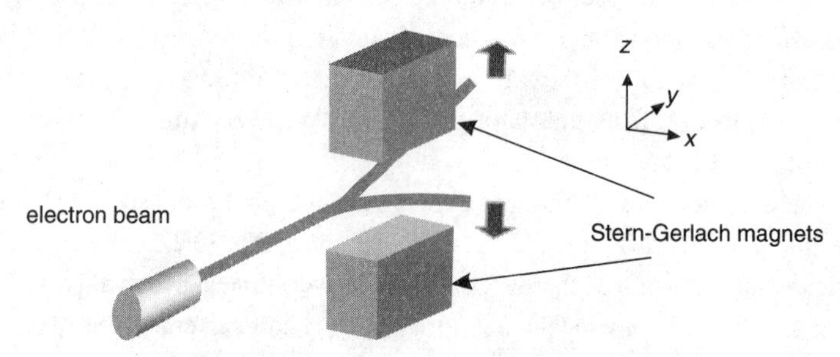

Figure 8.5. Stern-Gerlach magnets oriented for separating up and down spins.

rather than photons is that we can orient the measuring devices along any of three axes, *x*, *y*, and *z*. (With photons it's not quite so obvious how we get polarizations pointing in all three dimensions.)

Now, imagine that our GHZ state $|1,1,1> - |-1,-1,-1>$ consists of three electrons that shoot off in different directions toward three different Stern-Gerlach devices. In this state each -1 refers to spin down and each 1 refers to spin up when measured along the *z*-axis. So if all three Stern-Gerlach devices are set for measuring along the *z*-axis, we'll get perfect correlation between all three measurements. The result of measuring the first electron will be random because the overall state is in a superposition of up and down. But the measurements on the other electrons will agree with the first.

That's the basic setup for this experiment. Now, suppose we point the three Stern-Gerlach devices not along the *z*-axis but instead along a combination of the *x* and *y* axes. What results should we get? One very useful result that can be derived from the quantum mechanical equations is that whenever we measure one spin along the *x*-axis and the other two along the *y*-axes and multiply the answers, we should always get the result $+1$. So either all three spins point forward with their axes or two point backward and one points forward, giving $-1 \times -1 \times 1 = +1$. If we represent the *x*-spins of the three electrons as s_x^1, s_x^2, s_x^3 and the *y*-spins as s_y^1, s_y^2, s_y^3, we can write:

$$s_x^1 s_y^2 s_y^3 = s_y^1 s_x^2 s_y^3 = s_y^1 s_y^2 s_x^3 = +1 \qquad (1)$$

Now, it also turns out that if we measure all of the spins along only the *x*-axis, we ought to get the answer -1. So either all of the spins point backward or two point forward and one points backward giving $1 \times 1 \times -1 = -1$. We can write:

$$s_x^1 s_x^2 s_x^3 = -1 \qquad (2)$$

Classically, however, we will show that this last product "really" ought to be $+1$, which means *either* all three *x*-spins point up *or* two point down and one points up. So the crucial clash between quantum and classical reality can, in principle, be decided in one measurement.

What is the classical argument for claiming that the product ought to be $+1$? First, we'll pretend we don't know the answer to what the product in (2) should be and write the equation as $s_x^1 s_x^2 s_x^3 = m$, where *m* is the mystery number (either $+1$ or -1) we want to find. According to the original EPR argument, *if* it is possible to infer the state of a particle with certainty without disturbing it, then its state must already exist as an "element of reality," that is, as a definite value. In this particular situation we can predict with certainty the x-spin of one electron by measuring the y-spins of the other two because we know the final answer must be $+1$. So classically, the x-spin must be well defined even *before* the measurements.

By a similar argument, the x-spins of the other two electrons must already exist too. Furthermore, the y-spins must also exist as elements of reality before any measurements because we can infer the y-spin of an electron by measuring one x-spin and one y-spin of the other electrons. So from this classical perspective, we believe that the x-spins and y-spins must all exist as elements of reality even before we make any measurements. We can therefore claim that the equations in (1) must all be true before any measurements:

$$s_x^1 s_y^2 s_y^3 = +1, \; s_y^1 s_x^2 s_y^3 = +1, \; s_y^1 s_y^2 s_x^3 = +1$$

Now comes the clever part. If we multiply all three of the above equations with our original mystery equation, $s_x^1 s_x^2 s_x^3 = m$, we get:

$$s_x^1 s_y^2 s_y^3 \times s_y^1 s_x^2 s_y^3 \times s_y^1 s_y^2 s_x^3 \times s_x^1 s_x^2 s_x^3 = m$$

If you look carefully at each symbol in the above equation, you'll see every spin state appears twice. So every x-spin value and y-spin value is *squared.* As each value must be either $+1$ or -1, the square of each will always be $+1$ and the whole product must be $+1$. Therefore m must be $+1$. So, by this classical argument, we deduce that $s_x^1 s_x^2 s_x^3 = +1$. But according to quantum mechanics it should be -1. So in the GHZ experiment, then, we need only measure the x-spins of all of the electrons to differentiate between the classical claim and the quantum claim.

Of course, in practice we wouldn't rely on just one measurement to demonstrate this phenomenon. Indeed, to demonstrate nonlocality it would be necessary to do something similar to Aspect's EPR experiment, in which he switched the detectors between different orientations at high speed. Because the detectors were separated by several meters, he was able to eliminate the possibility of any signals traveling between them.

In 1997 Raymond Laflamme Manny Knill, Woijcech Zurek, and coworkers at Los Alamos National Laboratory created the world's first demonstration of the GHZ state. They used nuclear spins instead of electron spins and applied the techniques of nuclear magnetic resonance spectroscopy to manipulate their sample. They manipulated the spins of one hydrogen nucleus and two carbon nuclei in the chemical trichloroethylene, better known as an anesthetic and a common solvent, although for NMR purposes the chemical they used was labeled with the isotope carbon 13. The nucleus of an ordinary carbon 12 atom doesn't have a net spin because the equal number of protons and neutrons cancel one another's effect.

As we saw in the last chapter, NMR quantum computing relies on measuring signals from a thermalized mixture of molecules whose nuclear spins are weakly polarized by a magnetic field. So, unlike the EPR effect, which has been cleanly demonstrated with individual photons, the Los Alamos GHZ state was observed as the aggregated signal from what was, in effect, a huge ensemble of quantum computers. To create the GHZ state, the researchers beamed a series of NMR pulses into their sample to effect the following circuit:

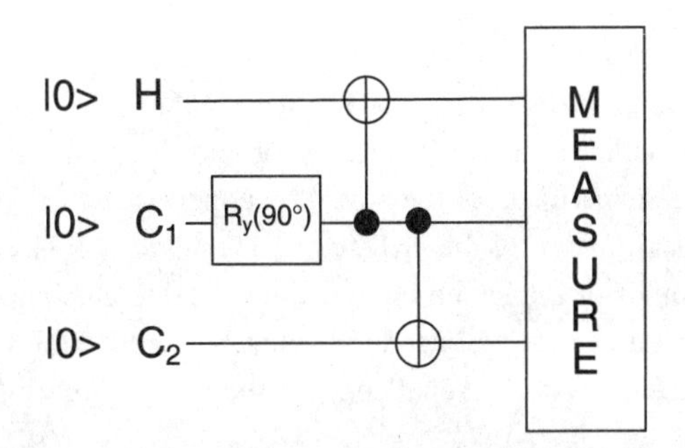

Figure 8.6. NMR circuit for implementing the GHZ state

Creating the GHZ state requires only a modest amount of quantum computation. One spin is tipped 90 degrees about the *y*-axis and then used as the control input on two controlled-NOT gates applied to the other two spins. The measurements reveal the state of the spins along the *x*-axis.

The Los Alamos team used a novel approach to select their pseudopure initial state |000>. Rather than "prepare" their sample, they relied on their ability to resolve the complete quantum state of the mixture of polarized spins in the final readout using a special form of measurement known as tomography. Like computerized axial tomography—CAT scans—the technique scans the state of different planes within the sample to build up a picture of the internal state of the sample. The box marked $R_y(90)$ refers to a rotation about the y-axis. What this does is to put one of the carbon nuclei into a superposition of up and down (that is, |0> + |1>) that is then copied across to the other nuclei using the two controlled-NOT gates to produce the GHZ state |000> +|111>.

Although the Los Alamos scientists were the first to produce a GHZ state, they weren't able to measure its weird quantum correlations. This was first achieved at MIT in 1999 by Richard Nelson, David Cory, and Seth Lloyd, again using NMR and a specially labeled three-qubit molecule—this time of the chemical alanine. Independently, Lloyd also showed how it should, in theory, be possible to use a four-qubit molecule to reveal these correlations without the need even to select a pseudo-pure state in the NMR experiment. His proposal was to perform a computation on which the answer to the question of whether $s_x^1 s_x^2 s_x^3 = -1$ is revealed on the fourth qubit. The rather amazing aspect of this idea is that the computation should lead to the same spin orientation on the fourth qubit regardless of the initial state of each molecule.

However, it won't be possible to test the nonlocal aspect of the GHZ experiment using NMR techniques. The three qubits, based as they are on nuclear spins within a molecule, are far too close together to allow for a delayed-choice experiment of the kind Aspect performed on EPR pairs. To see the nonlocal feature of the GHZ experiment, we will have to wait for the other methods of implementing quantum gates to catch up with NMR. Even then experimenters will need to be able to separate their qubits by at least a few meters—a perfect challenge for the flying qubits of cavity QED perhaps.

Take a Ride on the Universal Quantum Simulator

Beyond creating unusual quantum states like the GHZ state, what are we likely to be able to do with a quantum computer with only a small number of qubits? One possibility that's been attracting increasing attention is the idea, originally proposed by Richard Feynman, of constructing a quantum simulator. In 1982 he showed how classical computers had difficulty simulating quantum systems because as the size of the system increases, the number of variables the computer has to track grows exponentially fast. Feynman conjectured that a machine that worked by quantum means would make a much more efficient simulator than a classical machine.

Feynman acknowledged that his thinking was partly motivated by the idea of a cellular automaton, in which everything happens locally. As we saw in Chapter 2, cellular automata employ simple local rules to govern the behavior of cells and their immediate neighborhood. But applying these rules across the whole automaton can produce rich and complex dynamical evolution. Physical systems appear to evolve in a similar way. The motion of the molecules in a gas, for example, is determined by collisions with nearest neighbors rather than long-range forces exerted by molecules far away. So Feynman envisaged a quantum computer that worked by simple local interactions rather than an enormous computer with arbitrary interconnections.

Feynman also recognized that a quantum simulation would have to reduce everything that happened in a specified volume of space and time to a finite number of logical operations. "The present theory of physics is not that way, apparently," he said. "It allows space to go down into infinitesimal distances, wavelengths to get infinitely great, terms to be summed to infinite order, and so forth; and therefore, if this proposition is right, physical law is wrong. So good, we already have a suggestion of how we might modify physical law, and that is the kind of reason I like to study this sort of problem."[8]

What that meant was that a simulation would have to make space and time discrete. Discrete space-time lattices are commonly used for classical simulations of physics problems, but they typically require huge computational resources.

In 1996 Seth Lloyd fleshed out Feynman's idea by showing how the logical operations available on a quantum computer could be marshaled to simulate the behavior of more or less any quantum system provided its dynamics were determined by local interactions. As an example of a quantum simulator's superiority over a classical computer, he considered a system of forty spin-½ particles. On a classical computer such a state would require 2^{40} (roughly 10^{12}) numbers in memory. Even worse, to calculate its evolution in time would require the manipulation of a matrix of numbers involving $2^{40} \times 2^{40}$ (almost 10^{24}) elements. Yet such gargantuan classical calculations could be handled by a mere forty qubits in a quantum computer.

A quantum simulator would, in effect, be a laboratory in which physicists and chemists could study models of quantum phenomena that are too difficult or impossible to simulate classically. The range of possible applications extends across some of the most difficult problems in physics and molecular chemistry. Examples include the dynamical evolution of chemical reactions (which would be of great interest to pharmaceutical companies), the behavior of liquids and gases, lattice-gauge theories, highly correlated many-body systems such as ferromagnetic materials and superconductors, and even that last stop on the way to the unified theory of everything, quantum gravity.

Lattice-gauge theories include quantum chromodynamics (QCD), the theory of the strong nuclear force, which is the force responsible for binding nuclear particles together within nuclei, the fusion energy we get from the sun, and the destructive power of nuclear weapons. QCD forms part of what is known as the standard model of particle physics, the best model we have of the fundamental forces and particles of nature. But unlike other aspects of the standard model, our understanding of QCD is still far from complete. Physicists and computer scientists currently use some of the most powerful supercomputing facilities in the world to simulate the behavior of quarks and gluons, the component parts of neutrons and protons. In particular, they want to understand more about why quarks are confined within nuclear particles and about how these nuclear constituents ought to behave in high-energy collisions within particle accelerators. The problem with such simulations, though, is that they often turn out to require almost limitless amounts of computational resources. Even with the added power of a quantum computer, Lloyd estimates that lattice-gauge models may require hundreds to thousands of qubits because of the need to simulate continuous variables. However, as Lloyd has also pointed out, it may be possible to take advantage of a greater range of quantum hardware. For example, decoherence, instead of being a liability, could be turned to advantage in a simulator, he says, because it could be used to mimic the interaction of the simulated system with its environment.

Superconductivity is another phenomenon for which a theoretical understanding is quite limited, particularly in regard to high-temperature superconductors. Researchers are looking for superconducting materials that operate at still higher temperatures, but in the absence of an adequate theory to explain why some materials are better than others, a quantum computer simulation could offer a whole new approach to the subject.

At the moment it is too early to say how difficult it will be to translate these difficult problems into a suitable form for a quantum computer. But to paraphrase Tom Toffoli, as the first simple forms of quantum computational hardware begin to emerge, it's likely that physicists will start trying to hitch a ride on their quantum simulators.

Searching a Quantum Phone Directory

Despite the undeniable power of Shor's algorithm, two years after its discovery people were beginning to ask whether quantum computers were just a one-hit wonder. If all they were going to do was factor large numbers (and perform quantum simulations as a party trick), was that really enough to justify the heady claims of a quantum information revolution?

Fortunately, in May 1996 a new quantum algorithm turned up just in time to revivify any flagging spirits. This was the remarkable discovery by another employee of Bell Labs, Lov Grover. His algorithm is mathematically not quite as spectacular as Shor's in that it doesn't offer an exponential boost in speed over classical computers. However, it is likely to be far more important in the long run because it offers a very worthwhile increase for what is a much more common computational task: searching for information. Or, as Grover put it in the title of his paper in *Physical Review Letters:* "Quantum Mechanics Helps in Searching for a Needle in a Haystack."

Consider the way we typically use a phone book. If you know someone's name, getting his or her number is simple and fast, provided, of course, the person is listed in the book. You just look for the alphabetical entry. But if you only have a number and want to know who it belongs to, it's quite a different matter. If the directory is available electronically as a simple list, a computer would have to look through, on average, 50 percent of the entries before finding the correct name. For a typical-size phone book, such a search would be no real problem on a computer. But what if you wanted to search through a much larger database. For example, suppose you were given a short piece of text and you wanted to find the book in which it had been originally published. If you had access to an electronic version of all the books in the Library of Congress, the search probably would take quite a long time because the computer would have to read, on average, half the library to find your quote.

The problem gets even worse if you consider a search on a "virtual phone directory," which could be so large that it would be impossible to fit inside all of the world's computers put together.[9] Examples of this are offered by our old friend, cryptography. In Chapter 5, I mentioned one of the most widely used methods of encrypting information, the Data Encryption Standard (DES). With this system, messages are encrypted and decrypted using a 56-bit key that must be secretly exchanged between the participants ahead of time.

A common scenario in examining the strength of secret codes is to assume that a potential eavesdropper is able to collect at least one example of a message in both its plaintext (unencoded) and ciphertext (encoded) forms. Suppose we're trying to discover somebody's key and we somehow uncover a message in which the plaintext was "attackatdawn" and the ciphertext was "zvbegscixyfe." We now want to discover the key that relates the two.

We can imagine a virtual phone directory in which each name corresponds to a different key and each corresponding telephone number is the encoded form of our plaintext "attackatdawn" using that key. To find the correct key, we need to plow through the phone directory until we find the correct ciphertext "zvbegscixyfe." The corresponding name will be the key we desire. For the DES virtual phone book, this turns out to be a big task. The procedure we've described is, in fact, an exhaustive search and entails reading an average of 2^{55} (about 3.6×10^{16}) entries in our virtual directory before we are likely to find the right one.

However, Grover's quantum algorithm would be able to solve the problem, on average, after reading only 185 million entries in the directory, making the search more than 100 million times more efficient. More generally, when an exhaustive search of N database entries takes on average $N/2$ tries, Grover's algorithm takes on the order of $\sqrt{N}$ tries. This is a "quadratic" increase in speed rather than exponential, but it would nevertheless be tremendously useful in a wide variety of practical applications.

Of course, it wouldn't be much use for *literally* searching through conventional databases so long as classical bits remained cheaper than qubits for long-term storage. Where it would come into its own would be for *algorithmic* searches, such as chess playing and code breaking (as in the DES example).

The quantum algorithm takes for granted that the database is available in a quantum form because it needs to make its search using quantum superposition and entanglement. That, after all, is how the algorithm gets its boost in speed: because it is able, in effect, to "sense" every entry in the database by making a single access. The reason the algorithm then has to make additional accesses to the database is that it needs a way to boost the amplitude, which is initially very small, of the term within the superposition it's looking for.

Let's see how the algorithm works for cracking a secret code. In our DES example, rather than consulting a separate quantum database, we can get the quantum computer to do all of the work. (That is most likely how the program would be used anyhow.) In the following description we'll be using three quantum registers, although the third register might only need to be a single qubit.

The algorithm starts by setting the first quantum register to a superposition of all possible DES keys—from 0 to $2^{56}-1$. The quantum computer then calculates all possible ciphertexts from the plaintext "attackatdawn" by applying the DES encryption procedure with quantum logic to the first quantum register, which holds the superposition of keys. This allows us to calculate every possible encoding efficiently using quantum parallelism. The answers would be placed as a superposition in the second register. The two registers would now be in an entangled state such that each key was tied to its ciphertext. The superposition would thus contain 2^{56} terms, one of which was the desired key paired with its ciphertext "zvbegscixyfe." Each term on its own would have a vanishingly small amplitude so that if we now measured the registers, the chances of finding the right key would be as small as the chances of finding it by picking one out at random.

Now comes the procedure for boosting the amplitude of the desired term, which Gilles Brassard describes as a kind of "quantum shake."[10] The shake exploits quantum interference to increase the amplitude of the target pair of key and ciphertext. The way it does this is to perform two program steps. The first selectively inverts the phase of the term corresponding to the target pair. The second inverts all of the terms about the mean—that is, it subtracts the amplitude of each pair from the average amplitude. The shake is repeated until the desired term dominates the superposition (see Figure 8.7).

Each shake slightly increases the amplitude of the desired term within the superposition while making the other terms shrink. After repeated shakes the amplitude steadily increases until it reaches a maximum after around $\pi/4\sqrt{N}$ shakes. At this point we can read off the answer by measuring both quantum registers. If instead we continue shaking, the amplitude begins to sink.

This process of quantum searching has been likened to cooking a soufflé. You put the initial superposition in a "quantum oven" and wait for the

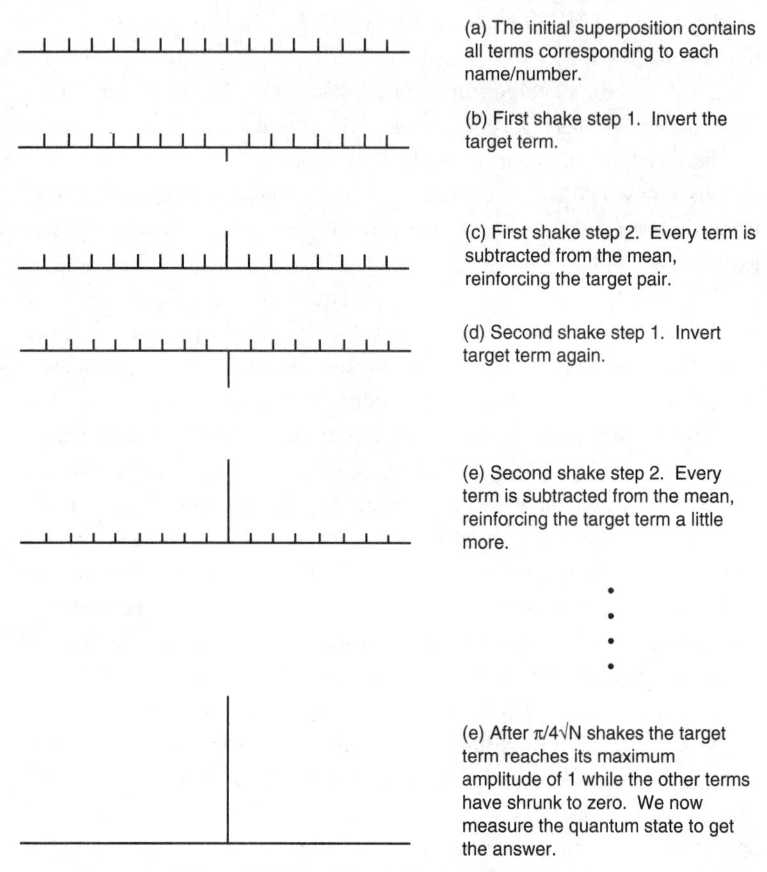

Figure 8.7. How Grover's algorithm works.

desired answer to rise. Provided you open the oven door at just the right moment, success is virtually guaranteed. Too early or too late, and the desired quantum state will have either not fully risen or started to collapse—either way, there's an increased chance of getting the wrong answer.

The first step of each shake can seem rather mysterious at first because it appears to act selectively on one term—the very one we're trying to find in the superposition. How do we manage that? Remember that the second register contains a superposition of all the ciphertexts and we have a target value for the ciphertext—"zvbegscixyfe," in our example. All we need to do is evaluate whether the second register is the same as our target value.

We could do this by subtracting one from the other, producing 0 only for our target term in the superposition. A little more logic would convert this to 1 and all of the nonzero values to 0. We then perform a conditional flip on that part of the entangled quantum state for which the number in the third register is 1. We can do this by using a controlled-NOT gate acting on a qubit set to $|0\rangle - |1\rangle$ (see the circuit in Figure 8.11). This produces a kickback phase flip if the control input is 1, and leaves things unchanged if it is 0, which achieves the desired effect of inverting the target term relative to the rest.

The second step, subtracting from the mean, is actually the more complex part of the process, but it can be broken down into a relatively simple series of Hadamard transforms and conditional phase shifts.

All together, Grover's algorithm is a very appealing discovery because of its wide variety of possible applications and because it may well be considerably easier to implement than Shor's algorithm. Also, we see once again that anyone who manages to build a sufficiently powerful quantum computer will be able to blow away the security of another of the world's most widely used cryptographic systems, DES. Grover's algorithm also has implications for quantum complexity theory.

Amadeus and the Quantum Complexity Puzzle

Assessing the potential applications for quantum computing has some of the character of a parlor game; as always with a new technology, if we knew what was there to be discovered, it would already be discovered. Nevertheless, it is possible to make some general comments about the potential limitations of quantum computers.

The exciting feature of the Shor algorithm is that it gives a quantum computer an exponential advantage over a classical computer. Such an increase in performance is particularly interesting apropos the complexity classes P and NP because of the apparent intractability of certain problems, particularly the NP-complete problems. However, as we saw in Chapter 5, factoring has not been shown to be NP-complete nor has it been proven to be intractable (that is, not in P) on a classical computer. It's just that nobody has found an algorithm that makes it tractable. So until someone finds a proof that it is intractable, a faint question mark hangs over the theoretical advantage of the quantum computer in this task.

Is it possible that a quantum computer could solve an NP-complete problem efficiently? If so, then by the definition of the NP-complete class, a quantum computer would be able to crack all NP problems. The discovery of an efficient NP-complete quantum algorithm would be truly sensational. It would mean that all the difficult problems found within the NP class, such as the traveling salesman, network optimization, and bin-packing problems, could be solved using quantum algorithms. Another of the NP-complete problems, known as *satisfiability,* would be solvable too, with interesting if not amazing repercussions. One form of satisfiability concerns the problem of looking at a logic circuit and evaluating whether it can produce a particular set of outputs. The obvious and only known method to find out is to try every possible combination of inputs and evaluate whether the circuit produces the desired outputs. This is easy enough on a simple circuit but as more and more inputs are added, the number of evaluations needed grows exponentially. So why would finding a fast way to satisfy a circuit be such a big deal?

In one of my conversations with David Deutsch, he dramatized the idea as follows. "There are a lot more people who can appreciate Mozart than there are who can compose like Mozart. If you had a way of solving NP-complete problems, then you could translate your way of appreciating Mozart into a way of generating new Mozart. That's the level of the breakthrough this would be."

Generating new Mozart? How does Deutsch leap from the rather prosaic task of "satisfying a circuit" to the possibility of *composing like Mozart?* Think of it like this: Suppose we were able to produce a, no doubt, very complicated circuit that could evaluate a piece of music and tell you whether it sounded like Mozart. Such a circuit is not beyond the realm of possibility. Some AI researchers have found ways to get computers to analyze English prose to test the authenticity of newly discovered works supposedly by famous writers. One could do something similar for music. Once we had a Mozart evaluation circuit, provided it worked efficiently (that is, didn't take too long), then if we had a way of solving NP problems quickly, we could produce a novel piece of Mozart, even finish his famous *Requiem,* by using our fast algorithm for solving satisfiability. New Mozart at the flick of a switch. We could do the same for writers too, so we'd soon be producing new sonnets and plays by Shakespeare.

If it all sounds too good to be true, then you're in good company: most theorists, including Deutsch, are quite sure that quantum computers will never, in fact, be able to solve NP-complete problems efficiently. Grover's algorithm is a pointer here because the problem we just examined of cracking DES is actually a good example of the satisfiability problem. By searching for the key, we are trying to satisfy the demand that the DES encryption circuit produce the desired ciphertext. As we noted before, Grover's algorithm gives a quadratic improvement over classical methods rather than an exponential boost, so it won't put NP-complete problems into the QP (quantum polynomial) class. Furthermore, recent work has suggested that the quadratic improvement may be the best we can get out of a quantum computer in these kinds of tasks.

In particular, if the DES circuit or circuitry relating to another NP-complete problem is treated as a black box (or oracle), then according to a paper by Charles Bennett, Ethan Bernstein, Gilles Brassard, and Umesh Vazirani,[11] there is little that can be done to improve on the speed of Grover's algorithm. They say that this does not completely rule out the possibility that a quantum computer could perform NP-complete problems efficiently. But it seems unlikely.

The Shape of Quantum Circuits to Come

Another aspect of quantum algorithms that become apparent with the unifying work of Richard Cleve, Artur Ekert, and colleagues was that they all have a rather similar structure.[12] Recall the circuit used for David Deutsch's first quantum program:

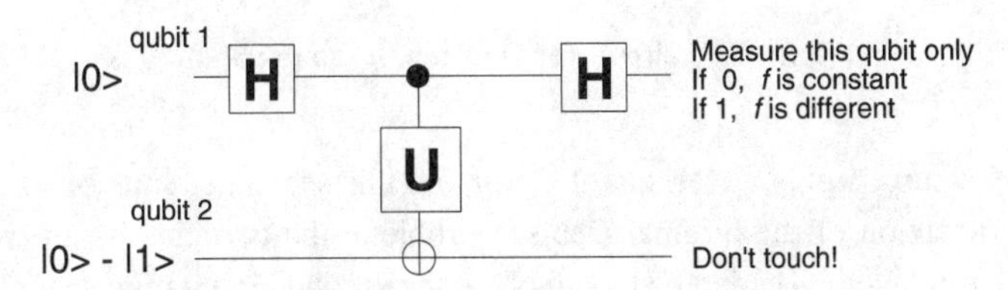

Figure 8.8. Circuit for Deutsch's stock market program.

As a reminder, let's review how this circuit works in the light of everything we now know. The circuit presents a superposition |0> + |1> to the quantum circuitry **U,** which evaluates (using a unitary operation) some mathematical function *f*. **U** attempts to place the answer on the second qubit via a controlled-NOT gate, but because this qubit has been set to the superposition |0> − |1>, a strange thing happens. If the output of **U** is 0, then there's no change, and if the output is 1, then qubit 2 is flipped, giving |1> − |0>, which is the same as the initial state apart from a phase factor of −1. Because qubit 2 is now entangled with qubit 1, the phase change is actually shared by the whole state. So there's a kickback effect that changes the phase of qubit 1.

Now comes an additional Hadamard transform. What this does is *erase* any information about which number was presented to **U.** This is similar to the way it is necessary to ensure in Young's two-slit experiment that we don't know anything about which path a photon takes. As long as we don't make any attempt to see which slit a photon passes through, we see the interference fringes on the final screen. The same applies in this circuit. If we measured qubit 1 before the last Hadamard transform, we would collapse the superposition onto 0 or 1. **U** would then perform only one type of evaluation, and we would lose any interference effects.

Now let's examine the circuit for the Deutsch-Josza problem:

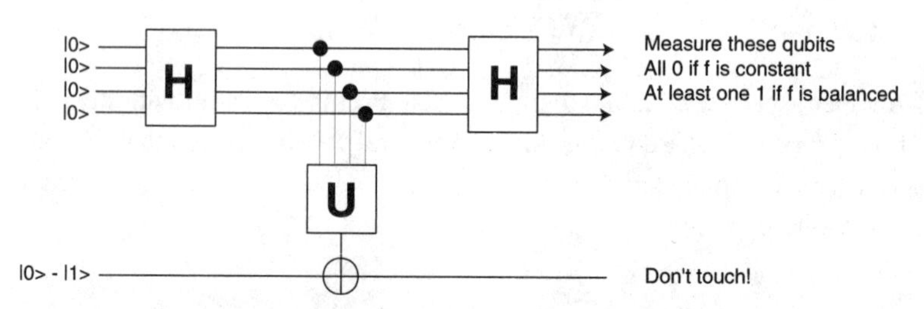

Figure 8.9. Circuit for Deutsch-Josza problem.

Not surprisingly, given that the problem this second circuit solves is a generalization of the original Deutsch problem, the two circuits are very similar in layout. The large **H** symbols in the second circuit represent clusters of Hadamard transforms so that each qubit is subjected to one Hadamard transform before the function evaluation circuitry **U.** Again we start by creating a superposition of numbers that is then evaluated by **U.** However, in this case we begin to see the power of quantum computation, because with this arrangement we are able to get **U** to work on an exponential number of different numbers. Then again we have another set of

Hadamard transforms, which erase the "which way" information and allow us to see any interference effects.

Now, let's examine a generic circuit for Shor's algorithm:

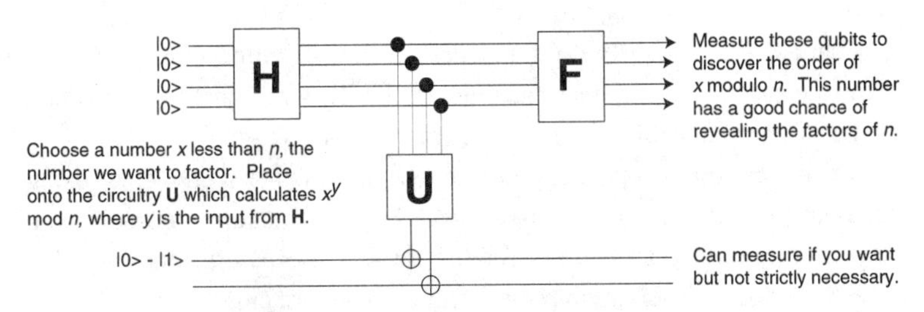

Figure 8.10. Circuit for Shor's algorithm.

Although there appears to be more going on in this circuit than before, the overall sandwich-type structure **HUH** is the same except that the final **H** is replaced by **F,** a circuit for performing Fourier transforms. First we set up a superposition of numbers using the first block of Hadamard transforms. These are presented to the circuit **U,** which calculates numbers of the form x^y mod *n,* where *x* is a randomly chosen number and *y* is the input. Then we have **F,** which is responsible for extracting any periodicities in the incoming qubits. As it turns out, the Hadamard transform is a special case of a Fourier transform, so mathematically they take similar forms. And like **H, F** erases any "which way" information about numbers presented to **U.** The reason it doesn't matter whether we measure the bottom qubits in this case is that they only provide limited "which way" information. Measuring them partially collapses the superposition but doesn't lose the important periodic information extracted by **F.**

So now, I think, we begin to get the picture. These quantum algorithms all exploit the same bag of tricks. How about Grover's algorithm? This time the circuit is more like a multilayered sandwich.

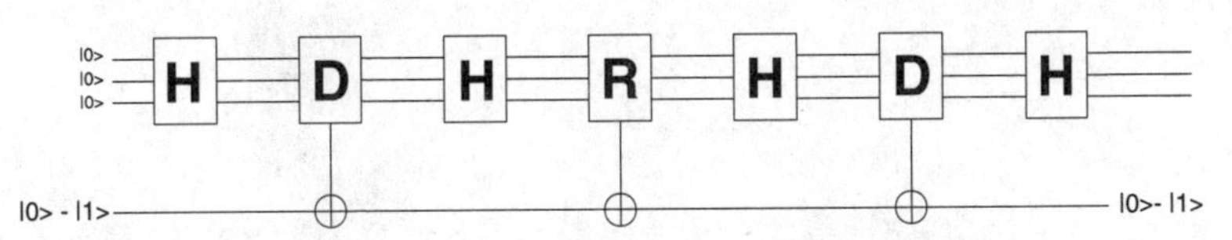

Figure 8.11. Circuit for Grover's algorithm.

In the circuit in Figure 8.11, the boxes marked **D** refer to circuits that access the quantum database and return the value 0 if the input doesn't match the target (the item we're looking for in the database) and 1 if it does. So for each **D** the circuit consults the database once, corresponding to the first step of each quantum shake involved in Grover's algorithm. The box marked **R,** together with the Hadamard transform on either side performs the second step of each quantum shake. **R** is reasonably simple and can be assembled from a collection of controlled-NOT gates.

So all these circuits take the form of various kinds of sandwiches, using Hadamard transforms and occasionally a Fourier transform as the bread, and specific unitary transforms for the filling. The natural question that arises from these observations is whether all quantum circuits will take this form. No one is absolutely sure, but it seems hard to think of any other radically different approaches. Initially, it might seem that the structure offers only a rather limited range of opportunities. If so, quantum computers might seem rather less inspiring than was first thought. But who knows? There could still be some big new ideas just around the corner.

9

Visions of the Quantum Age

> The nineteenth century was known as the machine age, the twentieth century will go down in history as the information age. I believe the twenty-first century will be the quantum age.[1]
>
> PAUL DAVIES

A Quantum Computing Road Map

In this last chapter we're going to take a rather more speculative look at where quantum computing and some issues closely related to it are headed in the coming years. Already we've seen the development of some demonstrations of individual quantum gates and some simple circuits in which several gates work together. Given the crucial role error correction must play in constructing larger circuits, an early target will be to demonstrate simple types of error correction. Already simple kinds of error correction requiring three qubits have been realized using nuclear magnetic resonance techniques along with simple implementations of Grover's algorithm, the quantum Fourier transform, and the Deutsch-Jozsa problem. More elaborate schemes will, no doubt, follow.

Once experimenters have progressed to around ten qubits, we may see the first attempts at factorizations of the numbers 4 and 15. After that we'll probably begin to see simulations of more sophisticated algorithmic searches using Grover's method. From there, the next milestone may be to produce a quantum repeater station for quantum cryptography followed by more complex algorithmic searches offering real-world applications and useful simulations of quantum systems. Beyond that lie the ultimate goals of a factorization engine capable of factoring numbers of more than 150 digits and, finally, a full-blown quantum computer, the ultimate universal computer.

Of course, it may take decades to reach these more ambitious goals—if they can be reached at all. But what would be the significance of achieving them? Andrew Steane of Oxford offers this sober assessment:

> The idea of "Quantum Computing" has fired many imaginations simply because the words themselves suggest something strange but powerful, as if the physicists have come up with a second revolution in information processing to herald the next millennium. This is a false impression. Quantum computing will not replace classical computing for similar reasons that quantum physics does not replace classical physics: No one ever consulted Heisenberg in order to design a house, and no one takes their car to be mended by a quantum mechanic. If large quantum computers are ever made, they will be used to address just those special tasks which benefit from quantum information processing. A more lasting reason to be excited about quantum computing is that it is a new and insightful way to think about the fundamental laws of physics.[2]

I doubt whether many people would argue with Steane's general outlook on the practical consequences of quantum computing in the short term. His view that the biggest motivation for studying quantum computation is for the insights it offers into fundamental physics reiterates David Deutsch's view expressed in the opening chapter. Indeed, a good example of such progress made before the discovery of the code-breaking applications can be found in a paper on time travel published by Deutsch in 1991.[3]

In this paper he was able to give a precise and general answer to the question of what would happen if it were possible to create "closed timelike curves." These are the physicist's jargon for loops in space-time that would make it possible to travel backward in time. Though such things sound like the stuff of science fiction, it seems that there are some special circumstances under which time travel might be possible—at least according to the equations of general relativity. Such possibilities cause physicists to sleep uneasily in their beds because they could potentially open up causal paradoxes in which people go back in time and change the past in a way that is inconsistent with the present. The typical example of this is somebody going back and shooting his grandmother.

Previous analyses of these ideas required horrendous calculations in general-relativistic quantum field theory, in which one tried to follow the behavior of matter in exotic space-times. Heavy approximations were inevitable, making the work far from convincing. It was also hard to separate the question about what would happen in the presence of these closed timelike curves from the question of whether they could form at all. However, by using ideas from quantum computation, Deutsch was able to attack these issues afresh and reduce them to a relatively simple question about the behavior of a quantum computational network with a "negative time delay" to represent the closed timelike curve.

His results fell out easily and exactly. The problem of any causal paradoxes was elegantly obviated by the fact that if you were to travel back and change the past, the result would not conflict with the present because, in fact, such travel would take you to a different part of the multiverse. It would, in effect, take you to a different universe.

Of course, such ideas may sound very fanciful compared with the rather more down-to-earth issues of code breaking. Yet I have heard it argued that the effort to build quantum computers capable of large factorizations could ultimately prove pointless unless the fact was kept secret. That's because as soon as it became public knowledge that such a machine existed, everyone would stop using RSA-based cryptography and use another form of cryptography that couldn't be broken so easily. That kind of argument, though, is a little like saying there's no point in training to become a doctor because if doctors were successful in their mission to stop people becoming unwell, they would have nothing to do. It's far from clear that there are alternative classical cryptographic techniques that would remain safe against organizations equipped with powerful quantum computers. Even if there were, anyone with a machine capable of cracking RSA codes would still be able to read past communications using RSA encryption. Considering that some secrets are meant to be kept for 50 to 100 years, quantum computers would be, without doubt, in huge demand by the intelligence agencies, not to mention underground organizations.

There is yet a wider aspect of the quantum computing revolution that could bring about enormous change not only in the way we think about the world but in the way we live. I'm referring to the development of nanotechnology, a parallel revolution that is likely to be instrumental in the development of quantum computation, a point David Deutsch strongly endorsed in my conversations with him.

"The milestones of nanotechnology are closely linked with the milestones of quantum computers," he said. "I think a fully fledged quantum computer will involve the same sort of technology as self-replicating nanomachines. We may well see one of those two things within months of the other. They are the same sort of thing. And at that point I think nanotechnology will take off and revolutionize the whole of human life. Whether quantum computers will do that depends on what further abilities they have that we don't know about. If they only have the present set of capabilities, then they're very small beer compared with nanotechnology as far as affecting everyday life goes. Nevertheless, at a fundamental level they are still much more important."

Nanotechnology and the Singularity

To understand the role of nanotechnology in the coming quantum age, let's take stock for a moment of how the subject developed. Once again, we're presented with the pioneering work of Richard Feynman. In 1959 Feynman gave an after-dinner talk to the annual meeting of the American Physical Society, an event that holds a key place in Feynman folklore. Under the title "There's Plenty of Room at the Bottom," Feynman set out to describe a totally new kind of technology—one we now call nanotechnology[4] because of the nanometer scales on which it is based. He explained that if you were able to print letters using individual atoms, you could write the entire contents of the *Encyclopedia Britannica* on the head of a pin. He went further, describing how information could be encoded in three dimensions rather than merely on a two-dimensional surface. He envisaged storing a separate bit of information in a cube of atoms 5 on a side, representing a total of 125 atoms per bit. With such a technology, he calculated, you could cram the whole of the contents of the Caltech Library as it then was—some 24 million volumes encompassing most of the world's most important books—into a space smaller than a grain of sand. To quote from his talk:

> It turns out that all of the information that man has carefully accumulated in all the books in the world can be written in this form in a cube of material one two-hundredth of an inch wide, which is the barest piece of dust that can be made out by the human eye. So there is plenty of room at the bottom! Don't tell me about microfilm!

Almost as a portent of his later work on quantum computers, Feynman discussed the possibility of building miniature computers in which wires would be 10 to 100 atoms thick and circuits would be several thousand atoms wide. This was long before people started thinking about quantum effects in the computational process. Feynman's rationale then was to imagine a computer powerful enough to do the kind of intelligent tasks we humans do easily, such as recognizing faces.

Although Feynman was the first to alert the world to the extraordinary possibilities of constructing things on the atomic scale, the idea of nanotechnology was ahead of its time and didn't really take off until the 1980s. The man who rekindled interest in the idea was Eric Drexler, who, not knowing about Feynman's talk, started thinking about the possibility of building molecular computers while still an undergraduate at MIT in the late 1970s. As he developed his ideas, he realized that once engineers had sufficiently precise control over the atomic domain, they would be able to construct more or less anything that was physically possible.

Drexler wrote up a popular account of his ideas in his book *Engines of Creation.* His message was truly visionary. Nanotechnology, he predicted, would change the world because it would make possible tiny machines that could construct virtually anything that the laws of nature allow to exist. Such machines could be programmed to enter human cells to fight disease, cure cancer, and even prevent aging. They could build spaceships from superstrong, superlight diamond-based materials, making possible spaceflight to the stars. They could even resurrect people from the dead, provided their bodies had been suitably preserved, an idea that spawned the California fashion for having one's head frozen at death. All this and more would be made possible by constructing tiny little nanomachines that could take materials apart and reconstruct them atom by atom.

Not surprisingly, some people found it difficult to take Drexler's ideas seriously. Although he and his colleagues produced detailed schemes for molecular components such as atomic bearings and gears, some scientists cast doubt on whether even these relatively simple ideas could work in practice. How, for example, could we be sure that a molecular bearing wouldn't weld itself together because of the redistribution of electrons and forces that so often occurs within complex molecules?

The simple answer, of course, is to build these molecules and see how they behave in practice. Unfortunately, constructing molecular bearings is still beyond current capability. Another possibility is to simulate such molecules either on a classical computer or, even better, on a quantum computer. Scientists working at NASA Ames Research Center in Palo Alto have produced some physically semirealistic animations of molecular gears based on (classical) computer simulations. But even if these components did work, there is still a huge gap between building them and constructing the sophisticated nanorobots that will be required to realize the more far-reaching ideas envisioned by Drexler and his supporters.

Nevertheless, progress has been faster than many people had expected. One of the most striking developments was the famous atomic IBM logo, first published in the journal *Nature* on April 5, 1990. The logo appeared in an extraordinary photograph taken by two researchers at the IBM Almaden Research Center, Donald Eigler and Erhard Schweizer. Using a scanning tunneling microscope, they pushed thirty-five atoms of xenon around on a nickel surface, like pieces on a chessboard, to spell out the letters IBM. It was a defining moment in the development of nanotechnology.

The scanning tunneling microscope, or STM, that made this feat possible was in itself a landmark achievement in the history of nanotechnology. Invented in 1981 by Gerd Binnig and Heinrich Rohrer of IBM's Zurich Research Laboratory, the STM enabled scientists to "see" atoms on the surface of crystals in unprecedented detail. For their work, Binnig and Rohrer duly won the Nobel Prize for physics.

The STM uses an ultrafine probe that is brought very close to the surface of the sample crystal. An electrical current is passed through the sample, and electrons hop across the gap between the probe and the sample by a process of quantum tunneling. The measurements are normally performed in a vacuum chamber at liquid helium temperatures to minimize disturbance from air molecules and thermal vibrations. The tunneling current is very sensitive to the size of the gap, so any slight variations on the surface reveal themselves in the current flow through the probe. A scanning mechanism moves the probe back and forth across the surface and at the same time adjusts the height to keep the tunneling current constant. The surface undulations can thus be read like braille to build up a "picture" of the atomic surface.

The ability to move individual atoms around came when Binnig and Rohrer discovered that their STM probe tip could sometimes pick up stray atoms and move them from place to place on the surface. Within a few years, competing laboratories announced a series of structures made using atom-by-atom control, culminating in the IBM logo.

Since that time other researchers have produced a range of nanopictures, including a version of the famous photograph of Albert Einstein sticking his tongue out, an atomic gingerbread man, and the world's smallest map of the Western Hemisphere.[5] More prosaically but of potentially much greater importance, Eigler and Schweizer went on to produce some of the first nanocircuits, including an atomic switch.

Drawing atomic pictures and building a nanoswitch were still obviously a far cry from the structures Drexler had envisaged, but they were close to the sort of thing Feynman had anticipated. They were certainly more in the spirit of his thinking than the famous tiny motor for which Feynman was obliged to shell out a thousand dollars in 1960. That was the result of offering two prizes in his talk, one for the first person to make a motor smaller than a 1/64-inch cube. Within a year a former Caltech engineer responded to the challenge by building a working motor that satisfied Feynman's conditions. Feynman paid up, but the motor, although smaller than the period that ends this sentence, was still millions of times larger than the atomic regime. It wasn't build atom-by-atom and hence wasn't within the purists' definition of nanotechnology. The IBM logo, on the other hand, clearly was.

In 1989 Drexler and like-minded scientists set up the Foresight Institute, a research organization in Palo Alto dedicated to furthering the study of nanotechnology. And in 1996 the Foresight Institute announced the Feynman Grand Prize of $250,000 for the first person to design and build two crucial nanotechnology devices: a nanoscale robotic arm and a nanocomputing device.

The importance of these devices would lie in their use in constructing a nanoassembler, a tiny robot we could program to build whatever we wanted by snapping atoms together like Lego blocks. The nanoassembler would be a *universal constructor,* a machine first advocated by mathematician and polymath John von Neumann in his studies of cellular automata. Von Neumann used the concept of a universal constructor, in conjunction with a universal computer, as the core component in a two-dimensional self-replicating system.[6] A nanoassembler would be a three-dimensional universal constructor whose function would be not only to churn out atomically crafted materials but, more significantly, to produce copies of itself. Once scientists build such a machine, the possibilities for nanotechnology become almost limitless. Self-replication means the universal constructors will cost next to nothing to make, which in turn will mean that these miniature manufacturing factories will be able to produce virtually any product we desire for practically zero cost. The only limits will be the supply of raw materials and energy and the laws of physics. But given the ability to process materials atom-by-atom, the universal constructors will be able to transform the dullest forms of garbage and waste into pristine shiny products.

At about the time these nanoassemblers become available, the nanotech visionaries predict that society will witness a period of very dramatic change of a kind that is almost impossible to imagine. The prospect of unlimited wealth and material goods for all is only part of the story. Control over the atomic domain will enable scientists to construct atomically perfect silicon chips with nanocircuits of unprecedented computational power. At around this point, it's possible computers will overtake the overall computational power of the human brain. If software developers can figure out what is required to produce genuine artifical intelligence, given enough processing power, computers could become smarter than human beings, and at this point they could begin to redesign themselves.

The evolution of computers could thus be driven by the computers themselves, leading to a process of positive feedback that again will result in a period of extremely rapid change. This coming revolution has been dubbed the *Singularity* by an Internet-connected virtual community of futurists who call themselves *Transhumanists.* To them the Singularity is "a moment in time when things are changing so fast it's impossible to predict what will happen next" or "a developmental discontinuity, an ultimate event horizon beyond which predictability breaks down totally."[7]

When is all this likely to happen? Estimates vary, but a widely touted date is around the year 2035, although Drexler believes it could come much sooner, at around 2020.

Some scientists scoff at these claims and write them off as the utopian ravings of a bunch of California dreamers. After all, they make even the wildest predictions of the quantum computing researchers look tame by comparison. *Scientific American* attempted to capture the essence of this negative vibe when, in April 1996, it published a skeptical article about the prospects for the nanotechnology revolution. The article, by one of the magazine's in-house science writers, Gary Stix, suggested that the grand claims of the Foresight Institute amounted to little more than a science-fiction fantasy. It provoked quite a storm on the Internet. Ralph Merkle, a nanotechnologist at the Xerox Palo Alto Research Center and one of Drexler's closest associates, published a devastating line-by-line critique of the article's contents on the Foresight Institute's Web site. *Scientific American* eventually replied with what initially seemed like a robust defense of its article, but which led to another incisive response from the Foresight Institute. I won't go into all the details (you can read an entertaining account of this saga on the Foresight Institute's Web site at *http://www.foresight.org/*), but the upshot was that none of the *Scientific American* article's technical criticisms—of which there weren't many, anyway—stood up to detailed scrutiny.

Nevertheless, it was clear that *Scientific American* had correctly tapped the mood of more than a few scientists. I produced a radio program for the BBC in 1997 in which two British nanotechnologists discussed their views on where their subject was headed. They said that there was a big cultural difference between the views of researchers in America and researchers in Britain. The Americans, they claimed, were far too gung ho about the possibilities. During the program, they heard extracts from interviews we had recorded with Ralph Merkle, who talked about being able to make superstrong materials using reinforced diamond, and Al Globus and Leon Crevitt at NASA Ames, who talked about making new kinds of rocket materials full of nanomachines and the possibility of building a space elevator that would take you from the surface of the Earth up into outer space without the need for expensive chemical rockets. When our British nanotechnologists were asked what *they* saw coming out of the field, the best they could come up with was a mobile phone that you could wear on your wrist.

As far as quantum computing is concerned, as Deutsch said, its milestones are likely to be closely linked with those of nanotechnology. Of course, there's no reason the computer that controls the nanoassembler has to be a quantum computer. Indeed, Merkle has suggested that initial designs might concentrate on building mechanical computers almost like nanoversions of Babbage's Difference Engine, the mechanical precursor of the modern computer. Nevertheless, nanotechnology will open the way to the design of atomically perfect nanocircuits that not only could display quantum properties but might also be made sufficiently robust to withstand the insidious problem of decoherence. If there's a chance of building a full-fledged quantum computer, molecular nanotechnology surely looks like a good bet.

DNA Computing

One of the rebuttals against skeptics who claim that nano is a no-no is to point out that there is already an existence proof that it can work: life on Earth, after all, depends on an extraordinary collection of nanomachines. These include, in particular, the DNA molecules and the protein molecules that coordinate and mediate processes necessary for survival and self-replication.

In recent years computing scientists and mathematicians have become increasingly interested in the information processing aspects of biological systems. Charles Bennett, as we saw, drew attention in 1973 to the computational aspects of the biological mechanisms responsible for processing genetic information in DNA. A crucial development that catapulted this idea back into the forefront of research more recently was the work of the mathematician Len Adleman on DNA computing. In 1994 Adleman, one of the coinventors of the RSA algorithm (see Chapter 5), published a paper in *Science* describing the first implementation of a DNA-based computer, a wonderful piece of interdisciplinary science that launched the subject of molecular computation.[8]

Everybody knew that DNA could store information, but the idea of computing with it, Bennett's insight notwithstanding, was quite new. Adleman beautifully described how he hit on the idea in a fascinating interview published by the University of Southern California in its magazine, *Networker*.[9]

> I was sitting in bed reading [James] Watson's *Molecular Biology of the Gene*. It described the action of polymerase. Polymerase is a protein that produces complementary strands of DNA. DNA is like a long necklace with beads, and the beads can be any of four colors. The polymerase works like a machine. It comes along, it latches on to the DNA, and it slides down the strand. As it moves down, the polymerase creates a new strand. Now, in the 1930s, logicians had started to investigate the idea of computation. They made computational models, and one of the

> famous models was Alan Turing's machine. Turing had described a machine running along a tape of digital information; and now here was this polymerase, which runs down strands of DNA information. In the middle of reading this, I thought: "Wow! This is like a computer. This looks like it could compute." So I got out of bed and started to think about it.

Why should anyone want to get DNA to compute? The obvious reason, apart from the sheer novelty of the idea, is that molecules are very small and therefore potentially offer a cheap form of massive parallelism. One estimate was that a pound of DNA molecules suspended in 1,000 quarts of fluid would represent more memory than that of all the computers ever made. (The human body contains about half a pound of DNA.) If there was a way of getting molecules to compute, even a small test-tube solution of DNA containing some 10^{20} molecules would represent a fantastic number of processors.

Using well-established techniques for synthesizing and replicating DNA molecules, Adleman found a way to solve a variation of the traveling salesman problem known as the "directed Hamiltonian path" problem. In this task, you are given a map like a network of airline routes and told to find an itinerary that takes you from a specific starting point to a specific end point, traveling via a number of cities such that you pass through each city just once.

At first sight this conundrum seems easier than the traveling salesman problem because you don't have to find the shortest route; you simply have to find any route that works. However, like the traveling salesman problem, this problem is NP-complete. What makes the problem potentially tricky is that some cities are not directly linked to one another or are only served by connections in one particular direction. Consider the map in Figure 9.1. You are asked to find a route from A to G. The only solution that works is to travel from A to B to C and so on up to G. Finding an answer in this case is easy enough, but finding a route generally is an NP-complete problem and becomes rapidly more difficult as the size of the network increases.

Adleman implemented a search for an itinerary among the routes between seven cities shown in Figure 9.1. His approach proceeded along the following steps:

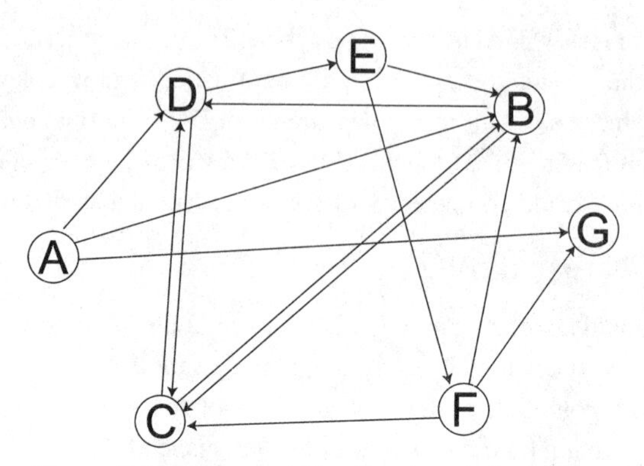

Figure 9.1. A directed Hamiltonian path problem

The problem is to find a route starting at A and finishing at G such that you pass through every letter only once. In this case there is only one solution: A-B-C-D-E-F-G. The general problem is NP-complete.

1. Generate random paths.
2. Keep only those paths that begin with A and finish with G.
3. Keep only those paths with seven cities.
4. Keep only those paths that go via all of the cities.
5. What remains are the answers.

DNA normally takes the form of a double-stranded molecule like a spiral ladder whose rungs consist of pairs of molecules known as bases. The bases come in four types that always pair up the same way: adenine with thymine, and guanine with cytosine. The base sequences of each strand, which determine the information content of genes, are said to be complementary, in that one is a kind of mirror image of the other.

Adleman encoded the Hamiltonian path problem in DNA as follows. First he randomly chose a unique sequence of twenty bases to represent each city. He then ordered from a gene vendor short pieces of single-stranded DNA consisting of the complementary sequence for each city. In addition, he ordered sequences representing each flight connection between two cities. Each of these consisted of the last half of one city's sequence and the first half of the other city's sequence. So a flight between A and B would be represented by a single strand of DNA with twenty bases, ten from A and ten from B.

Step 1 of the algorithm simply involved mixing the DNA sequences of the flight connections and the cities (in their complementary form). Match-

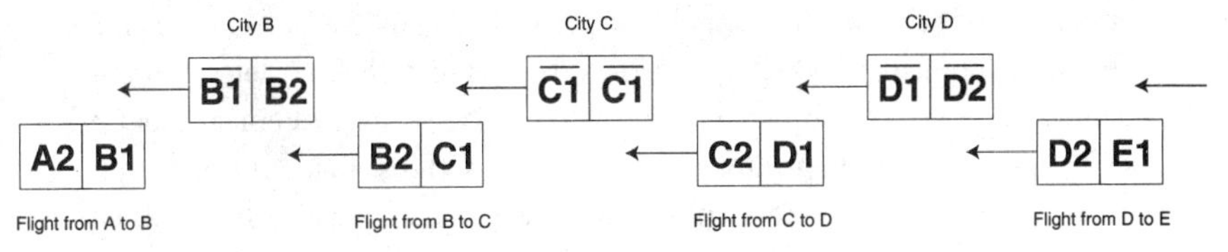

Figure 9.2. DNA computing

The natural tendency of matching complementary sequences of DNA to stick together enables specially made blocks of DNA to discover a route through the Hamiltonian path problem. Each square represents a sequence of DNA consisting of ten bases. Each city and flight is represented by two squares consisting of a unique sequence of twenty bases. Flight sequences are chosen to be complementary with the second half of the sequence of the departure city and the first half of that of the city of arrival. City and flight DNA spontaneously pair up to form extended chains of DNA representing possible flight itineraries.

ing complementary sequences have a natural tendency to stick together (to form double-stranded DNA), so the result was a collection of DNA strands representing random paths through the flight route network (see Figure 9.2).

The next steps involved amplifying and purifying the correct answers. Step 2 exploited PCR, or polymerase chain reaction, the standard technique for duplicating copies of DNA. Adleman used PCR to copy specifically all sequences that started with A and ended with G. For Step 3 he filtered the sequences by passing the products through a gel by electrophoresis. DNA of different lengths traveled at different rates through the gel, allowing the sequences with just seven cities to be separated from the rest. The product was then amplified by PCR and gel-purified several more times. Step 4 involved selecting only those strands that contained every city. This was done one city at a time using special pieces of DNA attached to magnetic beads. The complementary DNA sequence for city A, for example, would be replaced by DNA attached to a magnetic bead. The sample would then be placed in a magnetic field, ensuring that only DNA sequences including city A remained in the test tube while the rest were washed away. The process was then repeated for the other cities, B to G. At the end, all that was left were DNA sequences representing possible routes from A to G.

After a little thought, it's evident that the actual computation is performed entirely in the first step. The molecules of DNA compute many different itineraries in parallel when they are first mixed. This process takes only about a second. The problem, though, is finding the correct answer among the myriad random sequences formed. This is the objective of the rest of the procedure, which has been compared to fishing for letter combinations in alphabet soup. This part of the process took Adleman about a week. For a toy problem involving only seven cities, that's not very impressive. Genetic computers, like quantum computers, are clearly not about to replace silicon machines in everyday number-crunching operations. But how well can a genetic computer cope with a big computational task that would take conventional computers months or years?

Adleman calculated that a DNA computer could perform up to 10^{18} operations per second, some million times faster than the best supercomputers. Furthermore, the efficiency of a DNA computer was estimated to be around 2×10^{19} operations per joule,[10] making it vastly more efficient than electronic computers and only fifteen times less efficient than the Landauer limit of $kT \log 2$. Since Adleman's work, there's been a rash of DNA algorithms showing how to compute other NP-complete problems, including searching for satisfiability, doing matrix multiplication, optimizing graph connectivity, packing knapsacks, and coloring road maps. There's also been a proposal showing how to crack the Data Encryption Standard using DNA. It was estimated that, given a ciphertext and plaintext pair, this approach would take about four months of laboratory work to crack a DES key. So potentially, DNA computation looks very powerful.

However, there are several caveats. At the moment a lot of these ideas are theoretical. Just as in quantum computation, carrying out computationally interesting problems experimentally is likely to be complicated by the need for error correction techniques. Furthermore, although a DNA computer can apparently tackle quite large NP-complete problems, it is still only a form of classical, albeit highly parallelized, computation. Its resources are finite, so there is a cutoff point beyond which it will not go. The reason a DNA computer can solve NP-complete problems at all is that it is able to explore all possible paths in parallel provided the problem isn't too large.

According to one early estimate based on Adleman's original approach,[11] to factor a 1,000-bit number would require $10^{200,000}$ liters of DNA solution! Remembering that the universe only has some 10^{80} atoms, we can see how hopelessly impractical DNA factoring would be. Of course, there may be a better way to implement factoring on a DNA computer that would improve on those numbers, but it seems that classical methods can always be thwarted by using sufficiently long keys. DNA computation, unlike quantum computation, doesn't offer the promise of unlimited exponential resources. Nevertheless, it clearly offers some interesting possibilities, not least because of its conceptual similarities with other forms of parallelism—including quantum computation.

Adleman acknowledged the difficulties ahead in his *Networker* interview: "If computers were twenty years behind where they are now, I would probably be optimistic that we'll be able to beat them at certain tasks. We have a certain window of opportunity to overtake them, but it's a short one."

Clones, Consciousness, and the Indivisible Soul

During a visit to IBM's Thomas J. Watson Research Center in Yorktown Heights in 1996, I was fortunate enough to have lunch with some of IBM's leading lights in the science of computation, among them Charles Bennett. Sitting in their rather splendid staff cafeteria, we were idly musing over some of the more recent developments in quantum computation when Bennett made a characteristically provocative comment: "Some people think that the fact that quantum particles cannot be cloned might explain the indivisibility of the human soul." Bennett's interjection, which was said more in jest than anything else, caused our lunchtime conversation to turn to controversial ideas proposed by the distinguished Oxford mathematician and physicist Roger Penrose.

Penrose, highly respected for his work on black holes and quantum cosmology, had gained a certain notoriety for ideas expounded in two books: *The Emperor's New Mind* and *Shadows of the Mind.* In these two remarkable tomes, he offered an audacious scheme to explain the mysteries of both quantum mechanics and the human mind. His claims can be divided roughly into two main ideas. First, computers as presently conceived will never be able to think as human beings can because there are acts of human intuition that can be proven to be "noncomputable." Second, this ability to outwit computers depends on a form of quantum processing that goes beyond currently known physics.

At first sight, the notion of the brain working as a quantum computer has a certain appeal because of what it could imply for consciousness. Scientists and philosophers have long scratched their heads over consciousness: What is it, how does it work, and how can it be studied? Despite many advances in neuroscience, psychiatry, psychology, cognitive science, and numerous other disciplines that impinge on the study of the brain, a significant number of scientists claim that little or no progress has been made in answering what the philosopher David Chalmers calls the hard problem of consciousness: to explain the existence of qualia, the raw sensations of awareness such as pain, happiness, or the recognition of the color red. Chalmers has asked the related question of why we aren't all zombies: people who act exactly like conscious human beings while being entirely unconscious. Such questions pose problems for the promoters of "strong AI," who contend that the main obstacle to classical computers becoming conscious is simply having enough computational power.

Consciousness also has a holistic quality that raises difficult questions about what would happen if a person's brain could be cloned or downloaded onto an advanced artificial intelligence machine. Imagine if the programs running in your brain were exactly simulated on a computer: Where would the "real you" reside—in your brain, in the machine, or in both simultaneously? Or suppose you were teleported from one place to another but something went wrong with the dematerialization process so that you were copied at the destination but not destroyed at the starting place: Which one of the resulting two copies would you think you were?

Different people have different answers. Some would say you and your clone would each start with a separate identity because consciousness depends not only on the software but also on the hardware of the brain. Some would argue that the two "yous" would indeed momentarily, at least, experience the same identity, although they would quickly split and diverge from one another as you and your twin experienced different sensory inputs. Other people would say you could not be copied in the first place because there is more to the mind than the mechanics of the brain. None of these answers, though, seems entirely convincing, least of all the last, which would be the view of the mystic. How pleasing it would be to cut through the fog of incomprehension and attribute the indivisible gestalt of one's self-identity to the fact that quantum states cannot be cloned.

Although Penrose says such arguments are suggestive, he's been driven more by the idea that human beings can transcend the limits of computation. He doesn't mean this in the obvious sense that computers are still very stupid and can in no way compete with human beings in matters of common sense. He means it *in principle,* so that no matter how smart conventional computers became, they could never reason in the way humans can.

His argument for this claim is based on Gödel's famous incompleteness theorem, which we briefly examined in Chapter 3. This theorem shows that within any consistent mathematical framework there are always mathematical truths that cannot be proved. From this fact, Penrose starts with a simple system of computational rules and produces an obviously true statement that cannot be proved from the rules. Somehow we always seem to be able see further than the rules would allow. Penrose claims, therefore, that human thinking includes a noncomputable element.

Penrose's arguments have been the subject of numerous criticisms by scientists and philosophers. For a start, his Gödel-like argument wasn't particularly new. "Emperor's Old Hat" was how one reviewer described Penrose's first book, pointing out that the Oxford philosopher John Lucas had advanced much the same argument in 1961. Indeed, Douglas Hofstadter brought Lucas's contribution and some of the counterarguments to wider attention in his hugely successful and influential book *Gödel, Escher, Bach.*

Hofstadter cites[12] the work of another philosopher, C. H. Whitely, who proposed the self-referential sentence: "Lucas cannot consistently assert this sentence." The point here is that Lucas would neither be able to assert the sentence nor would he be able to deny it because either way he would face a contradiction. Yet we as outside observers can see that the sentence is manifestly true. This argument is similar to that proposed by Penrose when he claims that there are statements whose truth a computer cannot decide but which we can. The inability to see logical truth in certain situations is therefore *not* confined to computers, because, we human beings can be put into precisely the same predicament.

Daniel Dennett, the philosopher and author of several books on consciousness, also offered an elegant counterargument that goes like this. Imagine an advanced chess-playing machine X, which is superbly capable of achieving checkmate. Now, no one knows any algorithm that is guaranteed to achieve checkmate so by analogy with Penrose's reasoning we can deduce that X does not owe its power to achieve checkmate to an algorithm.

But the argument is clearly fallacious, because there is such an X that does use algorithms—one only has to think of IBM's Deep Blue computer that beat the world chess champion, Gary Kasparov, in 1997. By similar reasoning, even if there is no algorithm guaranteed to discover mathematical truth, it seems dubious to conclude that mathematicians cannot be executing algorithms in the brain to make their discoveries.

These arguments and others would seem to cast considerable doubt on

Penrose's case for claiming that thinking depends, in part, on noncomputable physics.

Quantum Gravity and the Measurement Problem

Like many scientists dissatisfied with the Copenhagen interpretation of quantum mechanics, Roger Penrose is troubled by the distinction between the unitary evolution of isolated quantum systems and the wave-function collapse, or state-vector reduction, caused by measurement or decoherence. For brevity, Penrose labels these two types of behavior U for unitary and R for reduction. As Von Neumann pointed out in 1932, it ought to be possible to understand R processes in terms of U processes by taking into account the quantum state not only of the system under scrutiny but also of the measuring apparatus. However, this approach runs into an infinite regress where superpositions on a quantum scale get passed on to the measuring apparatus, which in turn must be measured by something else, which must be measured by something else, and so on, and so on . . . Von Neumann's solution, as we saw in Chapter 3, was to invoke consciousness as the cutoff point at which superpositions were collapsed back onto sensible classical states. Although consciousness plays a major role in Penrose's world view, he doesn't believe the measurement problem is solved by introducing consciousness into the description of state-vector reduction. To quote:

> It is my view that solving the quantum measurement problem is a prerequisite for an understanding of mind and not at all that they are the same problem. The problem of mind is a much more difficult problem than the measurement problem![13]

Instead of invoking consciousness to explain the measurement problem, Penrose opts for new physics. In particular, he believes that the answer lies in quantum gravity. Einstein's general theory of relativity, developed in 1915, transformed the scientific understanding of the force of gravity by reinterpreting it as a distortion in the geometry of space and time. From this perspective, the reason objects fall to the ground on Earth is not because there is a gravitational force pulling them but because the geometry of space and time around the Earth is warped.

General relativity has been experimentally verified to great precision, but physicists are dissatisfied that gravity remains the odd man out among the four forces of nature. Gravity is, in fact, the only force for which physicists have been unable to offer a quantum mechanical description. It is widely believed that a fully unified theory of the laws of physics will have to combine all the forces, including gravity, with quantum mechanics. There has been some progress in developing candidate theories of quantum gravity, but the problem remains far from resolved. So if there is to be new physics, quantum gravity certainly looks like a good bet.

More to the point for Penrose's argument, though, is that there are suggestions that quantum gravity may involve noncomputable physics. In 1986 Robert Geroch and James Hartle found that in trying to formulate a theory of quantum gravity they had to tackle a computationally unsolvable problem. The problem arose because their analysis involved superpositions of different space-time geometries. To do this they needed to be able to count different types of space-time geometries without overcounting ones that were equivalent. In a two-dimensional universe we could imagine space-time taking different shapes, such as the surface of a sphere, the surface of a ring doughnut, or the surface of a coffee mug (with a handle). From a topological point of view, a sphere is different from a doughnut, but a doughnut is the same as a coffee mug because the latter contain one hole each. It turns out that counting different geometries or topologies for two-dimensional closed surfaces is no problem because they can all be easily classified. Space-time, however, is four-dimensional, and it has been proven that there is no general procedure for demonstrating the equivalence of four-dimensional topologies. So the problem of counting them is uncomputable.

Geroch and Hartle's approach to quantum gravity is by no means definitive. Nevertheless, it and some other ideas open a window in physics through which noncomputable operations might enter. Putting this notion together with the claim that human beings are capable of noncomputable reasoning, Penrose has seen a way to link quantum gravity with the measurement problem and with the problem of understanding consciousness.

Penrose believes that the problem of understanding wave-function collapse or state-vector reduction will depend on finding the correct theory of quantum gravity. Although he doesn't claim to have the full answer—between writing *The Emperor's New Mind* and *Shadows of the Mind,* he changed his mind about some of the details—he has proposed a model of wave-function collapse that he calls objective reduction, or OR. This says that a superposition of quantum states collapses onto a single state not as the result of measurement or decoherence caused by the environment but as a self-generated collapse process rather like the radioactive decay of a uranium atom. The rate of collapse, Penrose argues, depends on the gravitational energy difference between the different quantum states within a superposition, while time and energy are related by the uncertainty principle according to the equation energy × time = $h/2\pi$, where h is Planck's constant. For an object of the size a microns (1 micron equals one-millionth of a meter) of roughly the same density as water, the collapse time works out to approximately $1/20a^5$, so that an atom would take more than a million years to collapse, a speck of water 1 micron across would take a twentieth of a second, and a cat would take 10^{-32} seconds, explaining why we would never see a cat in a live/dead superposition. Such calculations assume the object is perfectly isolated from the environment. If the environment is taken into account, the decay times are likely to be even shorter.

Penrose's OR scheme borrows from an idea known as GRW theory, proposed in 1986 by Giancarlo Ghirardi, Alberto Rimini, and Tullio Weber. They added a small nonlinear term to the equations of quantum mechanics that causes the wave function of an object to collapse spontaneously. Their model is much more ad hoc, though, because they chose the parameters of their theory merely so that the theory wouldn't conflict with any known experimental numbers. According to their idea, a quantum particle would see a spontaneous localization once every 100 million years, while a cat would experience it within 10^{-11} seconds. Even though the two schemes predict very different decay times for large objects, both appear to be beyond the limits of experimental tests at the moment. Nevertheless, it is not impossible that these schemes could be subjected to practical tests in the future.

Is the Brain a Quantum Computer?

In *The Emperor's New Mind,* Penrose speculated that to explain alleged noncomputability in human reasoning, quantum processes must occur within the brain. However, at the time he didn't have a specific model of how or where these quantum processes took place. Stuart Hameroff, an anesthesiologist at the University of Arizona in Tucson who had had a long-standing research interest in consciousness, read Penrose's book and was immediately taken by the way Penrose's ideas fitted in with his own. After making contact with one another, Penrose and Hameroff collaborated on a more detailed model of consciousness that became the subject of

Penrose's next book, *Shadows of the Mind.* They developed a model known as Orch OR, short for "orchestrated objective reduction."

Hameroff and Penrose equate coherent quantum computing with preconscious processing, and self-collapse with a discrete conscious event, so that sequences of OR events give rise to a stream of consciousness. Quantum processing is carried out, according to their model, by a lattice of *microtubules* within the nerve cells of the brain. Microtubules are hollow tube structures made from protein polymers called tubulins that make up each cell's cytoskeleton, a kind of scaffolding found in all eukaryotic cells, the kind of cells found in all animals and plants. Hameroff and Penrose believe that microtubules form a miniature information processing system within each cell, adding a whole new layer of complexity to the conventional view of the brain as a large and sophisticated neural network. They also contend that microtubule-associated proteins—MAPs—that form connections between different microtubules are able to "tune" the quantum oscillations of coherent states: hence the name orchestrated OR.

What do we actually know about microtubules? They certainly appear to be extraordinarily versatile in their function: they are involved in structural support, they are responsible for the whiplike motion of the cilia used by bacterium to swim, they pull apart the chromosomes in the process of cell division known as *mitosis,* they transport proteins, they are thought to play a role in regulating the strength of the connections between nerve cells, and they are also thought to be involved in orienting cells with respect to gravity and detecting photons of light in the rods and cones of the retina.

Versatile though they may be, the idea that microtubules act like miniature quantum computers is enough to trigger the crank detectors in many scientists. I've heard more than a few dismiss Penrose and Hameroff's model as "complete nonsense." The problem is that their ideas are bold in conception but short on experimental support. Indeed, it's tempting to say that they're based more on supposition than superposition. Quantum superpositions, after all, are extremely fragile. To maintain coherence even for very short periods of time, quantum systems require extreme conditions within the laboratory, such as liquid helium temperatures and an almost perfect vacuum to ensure isolation from neighboring atoms and particles. The circumstances within the hot sticky fluids of all cells would seem to be completely at odds with the notion of maintaining and manipulating coherent quantum states.

If decoherence throws a wrench in the works of ordinary quantum computation, where we are, at least, able to operate in a more favorable environment, we would expect it to utterly destroy any quantum coherence in the brain. Nevertheless, Penrose and Hameroff think that nature may have found a way to preserve coherence within nerve cells by exploiting special molecular features of microtubules. Penrose points to the example of high-temperature superconductivity, which demonstrates that coherent quantum effects can withstand temperatures of up to 125 degrees Kelvin (−112 degrees centigrade), which though still very cold from a biological point of view are far higher than anyone had expected before the discovery of this phenomenon in 1986. There's also the work of the physicist Herbert Fröhlich, who in 1968 proposed that sets of proteins within living cells could vibrate coherently with frequencies in the microwave or gigahertz region of the electromagnetic spectrum. But although there are published reports of such excitations in biological systems, their interpretation remains controversial.

Hameroff was responsible for initiating a series of scientific conferences on consciousness held every two years at the University of Arizona in Tucson. These have attracted leading scientists and philosophers from around the world and have been widely welcomed as one of the first serious attempts to put consciousness research onto the scientific map. That, however, didn't stop Penrose and Hameroff from coming under fire at the first conference, held in 1994, where they were accused of piling speculation upon speculation. At the consciousness conferences held in 1996 and 1998, they got a softer ride, perhaps because people felt the arguments had already been rehearsed and, lacking any new hard experimental data, there was little more to say for or against their claims. Nevertheless, Daniel Dennett gave an amusing talk at the 1996 Tuscon conference in which he asked the following question: Why, given that all tissues within the human body contain vast numbers of microtubules, would not a severed arm lying on a table feel pain in the same way that its owner would.

Afterward, I asked Hameroff for his response to this very question. "It's the same question as why isn't your big toe or your earlobe conscious," he said. "The reason is, you need a critical mass of quantum superposed tubulins in a situation where they can have coherence. The brain is specifically organized to have microtubules arranged in a highly parallelized local geometry. The microtubules in other cells are radially oriented, whereas in neurons alone they are all parallel. So the case in the brain is quite distinct from other tissues, including the severed arm."

Putting aside doubts about Hameroff and Penrose's claims, what would be the implications of their work for researchers in artificial intelligence hoping to build a conscious machine? Could we expect a sufficiently powerful quantum computer to be capable of consciousness? "I think that has a better chance than the standard AI approaches of just hoping complexity reproduces it," Hameroff said. "However, the main problem is, in our view, that for consciousness you need a self-collapse as opposed to the collapse in quantum computing, which as I understand it is introduced from outside. So it would have to be some kind of self-collapsing computer. And then the problem would be that since it's probably going to be electrons that are in superposed states, it would take an enormous number of electrons a very long time before they reached the quantum gravity threshold for self-collapse. The advantage of the microtubules is that we're assuming that the proteins go into superposition, so the mass involved is much larger. So I don't know whether it will be possible or not."

Back at IBM Research in Yorktown Heights, Charles Bennett told me the answer to his own question about the possible link between the inability to clone quantum states and the indivisibility of the soul. "I think that the entire sensation of one's soul or personality or consciousness could be entirely a classical effect coming out of the complexity of the process that's going on in the brain," Bennett said, placing himself firmly in the strong AI camp. "The high temperature and the degree of strong decohering effects in the brain are such that it's not a good place to look for quantum effects. It's a very inhospitable environment. I think that the ineffable, indivisible, you-can't-quite-put-it-into-words feature of human life is a metaphor for the fragility and ineffability of quantum information. I think there's no reason to believe that it's anything more than a metaphor. So if you made a replica of a person, randomizing many elements at the quantum level but keeping the atoms at more or less the same places, and asked them what they were thinking, I think they would be thinking the same thing. I think their thoughts would have more to do with the average sodium concentration in all their neurons rather than the phase relation among some particles. That's because when we look at the neurological processes, we see things that seem to be pretty classical—chemistry and electrochemistry—not quantum effects."

Bennett has a ready response for the oft-raised argument that if quantum computation were potentially so powerful, nature would have found a way of exploiting it. "There's lots of things that evolution has not discovered how to do. As far as I understand, the only animals that have discovered radio are people, and they don't make it biologically. Another example is that animals have an efficient way of storing fuel like fat, but they have terribly inefficient ways of storing oxidizers. The hemoglobin molecule, which carries four oxygen molecules, is tremendously big. We don't have anything in our bodies that is efficient in the same way as ammonium nitrate is in a bomb. There's nothing in chemistry that prohibits organisms from having a metabolism in which they store the oxidizer more efficiently than the animals we have now. If so, you could be a marathon runner who did all your breathing as well as all your eating before the race. So there are all sorts of niches of technology that have not been discovered by the biological world."

Even David Deutsch, who, as we have seen, does not shy away from an idea just because it is unconventional, does not accept the idea of the brain as a quantum computer. "There is no evidence of that," he said, "and there is every theoretical reason to doubt it. My own view of the brain is that it is a classical computer and no more. What is remarkable about the brain is not its hardware but its software. We don't have the remotest clue about what consciousness is—or how knowledge is created in the brain, which is actually a simpler problem but still very mysterious. The advocates of artificial intelligence always say that we're going to have it within twenty-five years, and the reason they cite is that computers are becoming more powerful and that soon they will be as powerful as the brain—as if there were programmers sitting there saying, 'I've got this wonderful artificial intelligence program, but my hardware is too slow to run it'! That's just not true, that's not the problem. It's not even that we don't know how to program the functionality. The problem is, we don't know what the functionality is. We don't even know *in principle* what consciousness is, and we don't know how such a thing as consciousness could fit into our world view."

Yet although Deutsch thinks that the brain is a classical computer, he's convinced that quantum computers will be important to understanding consciousness because of the new way we should think of ourselves given the existence of quantum computers. Why is that?

His argument is based on history. It's unlikely that Darwin's theory of evolution, with its explanation of the tree of life, could have been produced in the absence of the Newtonian revolution in physics that occurred two centuries earlier. That's not because Darwin exploited any direct reference to Newton's work, but rather because Newtonian physics opened up a completely new world view, in which people saw that even the most mysterious aspects of the universe as it was then understood, including the movement of the planets and the stars, could be explained in terms of mechanisms and dynamical laws. Without such thinking, it's hard to see how anyone could have arrived at Darwinian theory. By analogy, Deutsch thinks that the new world view opened up by quantum computation will be what's needed to fathom the mysteries of the human brain.

It would be easy to dismiss these ideas as rather tenuous. After all, how do we know for sure Darwin wouldn't have succeeded in the absence of Newton? Well, we don't and, equally well, we can't be sure that quantum computation will bring about the new world view in which consciousness may suddenly become explicable. History isn't always a good predictor of the future. Yet we already know that quantum computation offers numerous insights into the nature of knowledge, the universe, and the laws of physics—things that go far beyond the heady delights of producing faster algorithms and superfast computers. It would therefore not seem unreasonable to expect that it could take us much further than ever before in understanding our place in the universe.

Why Is the Universe Comprehensible?

Many of Penrose's ideas, as we saw, are predicated on the assumption that noncomputability plays a crucial role in human intuition and creativity. In Chapter 3 we saw how Deutsch had elevated the Church-Turing conjecture to a physical principle, the Turing principle, which says that anything that can be computed by a physical object can be computed by a universal computer.

If noncomputability were admitted into physics, where would that leave the Turing principle? Penrose, it seems, is happy to sacrifice it. Noncomputability, for him, offers a means for humans to gain access to mathematical truths that exist in some Platonic realm. His reality has at its base the world of abstract Platonic forms from which emerge the physical and then the mental worlds. Awareness, he suggests, is a direct bridge between the mental world and the world of Platonic absolutes.[14]

Like Penrose, Deutsch waxes philosophical about the relationship between mathematics and physical reality, but he comes to a very different conclusion. Deutsch regards computability as a fundamental property of nature and sees mathematics as nothing more than a branch of physics. Mathematical proof, he contends, far from being something pure and abstract, is a physical process.

Although Deutsch does not accept Penrose's argument for the existence of noncomputability in human intuition, he does acknowledge that the evidence for noncomputability in quantum gravity, albeit tentative, is plausible.[15] But such noncomputability, he points out, is only noncomputable in the Turing sense—that is, not computable on a Turing machine. If it turned out that quantum gravity involved a process that was beyond the capabilities of a Turing machine and even a quantum computer, then one could still make it a computable operation by adding a quantum gravity module to the quantum computer. Thus the "noncomputability" of quantum gravity would become a computable operation on the newly enhanced quantum computer.

In 1960 Eugene Wigner gave a famous lecture in which he pondered "The Unreasonable Effectiveness of Mathematics in the Natural Sciences." The point in question was why the laws of physics are so easily describable by the rules of mathematics. If the laws of physics were noncomputable, then presumably we would not be able to comprehend the universe using mathematics. In a talk on this theme given to a meeting on complexity in Santa Fe in 1989, Paul Davies put the problem as follows:

> We seem to have encountered a logical loop here. The laws of physics define the allowed mechanical operations that occur in the physical universe, and thence the possible activities of a Turing machine. These mechanical operations thus determine which mathematical operations are computable and define what might be called simple solvable mathematics (like addition). For some reason, those same laws of physics can be expressed in terms of this simple mathematics. There is thus a self-consistency in that the laws generate the very mathematics that makes those laws computable and simple.[16]

The puzzle for Platonists such as Penrose would seem to be why this apparent conspiracy between physics and mathematics arises in the first place. From Deutsch's perspective, though, the "conspiracy" is the consequence of the physical nature of mathematics and the Turing principle. It takes very special laws of physics, he says, to permit the existence of a machine that can simulate all other machines and encode the laws of physics themselves. According to Deutsch, then, it is Turing's principle that seems to lie at the bottom of questions about why the world is comprehensible. Or, as he puts it:

> The laws of physics, by conforming to the Turing principle, make it physically possible for those same laws to become known to physical objects. Thus, the laws of physics may be said to mandate their own comprehensibility.[17]

Of course, this leaves open the question of why the Turing principle should be true in the first place. Is it inevitable, is it just a stroke of good luck, or is it a God-given aspect of reality?

Trading Histories for Universes

In the 1980s a number of theoreticians developed approaches to understanding quantum mechanics that attempted to go beyond Everett's theory of many universes. These travel under a variety of names, although the common denominator is that they use the word "histories" instead of "universes." One of these ideas, known as the consistent histories approach, was produced by Robert Griffiths in 1984 and has been extended by Roland Omnes. An independent but closely related idea, called many histories, or decoherent histories, was advanced by Murray Gell-Mann and James Hartle in 1986.

The aim of these approaches, as in Everett's idea, has been to make sense of quantum systems without the need for external observers. The use of the word "histories" goes back to Richard Feynman, who showed that to compute the outcome of a quantum mechanical experiment, one could assume that the system evolved along every different trajectory. Each trajectory has a certain amplitude or strength (related to its probability of occurrence), and the outcome is determined by the "sum over histories"—that is, a sum of these amplitudes. In Young's two-slit experiment, for example, each photon when it arrives at the screen behaves as if it had two main histories: in each, the photon traveled by one or the other of the two slits.

The sum-over-histories approach was enormously successful and gave

Feynman a powerful tool for making quantum mechanical calculations. But the question of what physical significance to give each of these histories was something even Feynman was not sure about, as he indicated in his 1981 talk on simulating physics with computers.

> Somebody mumbled something about a many-world picture, and that many-world picture says that the wave function is what's real, and damn the torpedos if there are so many variables. . . . All these different worlds and every arrangement of configurations are all there . . . we just happen to be sitting in this one. It's possible, but I'm not very happy with it. So, I would like to see if there's some other way out.[18]

The many-histories approach of Gell-Mann and Hartle might appear to be one way out. Certainly, the fact that Gell-Mann, who won the Nobel Prize for his work on the quark structure of matter, is associated with it has given the theory fashionable status. Gell-Mann, who used to work just down the corridor from Feynman's office at Caltech, has a formidable reputation. Famously described as having "five brains, and each one smarter than yours,"[19] Gell-Mann is renowned for his prodigious knowledge not only of particle physics but also of modern literature, human history, archaeology, ecology, and languages (he speaks thirteen). But as John Horgan memorably put it in his provocative and amusing book *The End of Science,* "Gell-Mann is unquestionably one of this century's most brilliant scientists. He may also be one of the most annoying."[20] Horgan, a science writer for *Scientific American* who's not afraid of lampooning some of the scientists he writes about, added, "Virtually everyone who knows Gell-Mann has a story about his compulsion to tout his own talents and to belittle those of others."

I witnessed an example of the sort of thing Horgan had in mind when I attended one of the first meetings on complexity held at the Santa Fe Institute in New Mexico back in 1989. The institute was largely the result of Gell-Mann's efforts to set up a center devoted to interdisciplinary research in complex systems. The 1989 meeting was quite an occasion, attended by many scientific luminaries. In one session a speaker had been discussing the nature of quantum uncertainty, which, after a few technical questions from the audience, prompted the query: "I wonder what Einstein would have made of this issue were he alive today?" The question seemed incongrously simpleminded after the previous discussions. Spotting his opportunity, Gell-Mann suddenly cut in with: "If Einstein were alive today, he'd be over 110 years of age. I think it's unlikely he'd have anything sensible to say!"

So what's the deal with many histories? In his book *The Quark and the Jaguar* Gell-Mann offers a glimpse of how he and his colleague Jim Hartle see the theory:

> We consider Everett's work to be useful and important but we believe that there is much more to be done. In some cases too, his choice of vocabulary and that of subsequent commentators on his work have created confusion. For example, his interpretation is often described in terms of "many worlds," whereas we believe that "many alternative histories of the universe" is what is really meant. Furthermore, the many worlds are described as being "all equally real," whereas we believe it is less confusing to speak of "many histories, all treated alike by the theory except for their different probabilities."[21]

One might conclude that the essential difference between the many-universes interpretation as proposed by Everett and the many-histories approach boils down to a matter of words. I say "universes" and you say "histories" could be the slogan. But the difference runs deeper. Supporters of the many-universes idea take a realist's view: Those alternative universes actually exist. Supporters of many histories, on the other hand, take more of a positivist's view: The world is only what we measure it to be. We are only aware of one universe, but it behaves as if it had many histories.

In his book *The Fabric of Reality,* David Deutsch does a masterly job of undermining, if not demolishing, the positivist's position. Take this passage, for example:

> A more exotic variant of what is essentially the same idea is the following . . . "quantum theory is about the interaction of the real with the possible." This, at least, sounds suitably profound. But unfortunately the people who take either of these views—including some eminent scientists who ought to know better—invariably lapse into mumbo-jumbo at that point.[22]

I don't think Deutsch had Gell-Mann specifically in mind when he wrote that, and, equally, Gell-Mann didn't have Deutsch in mind when he said to me, "The world is probabilistic. That's the way it is, and if somebody doesn't like it, they can always move to another universe." Nevertheless, there appears to be an interesting dialectic going on here.

Are Decoherent Histories the Answer?

Where the many-histories approach perhaps has made progress is in illuminating the mechanism of decoherence. In particular, Gell-Mann and Hartle make a distinction between "coarse grained" and "fine grained" histories. Coarse-grained histories arise whenever we choose to ignore certain information about a quantum system.

Imagine, for example, that you walked through a crowded street. On your way you would be jostled and buffeted by other people, causing your path to deviate from a straight line. If you were to describe your journey, you might only be interested in the starting point and the destination: "I started at the south end of the street and finished at the north end," for example. Such a description would be a coarse-grained history. However, you might want a more detailed description that would include the exact course you took through the market. This would entail having to describe every little twist and turn. Such a description would be a fine-grained history.

Now, in quantum mechanics it is usually impossible to know the precise trajectory of a particle. All we can do is measure its initial and final state. Between these two points we have to assume the particle follows all possible journeys that reach the fixed destination on time, according to the sum-over-histories idea. We are therefore often forced to deal with coarse-grained histories, each of which is a sum-over-fine-grained histories.

The interesting point in this exercise is to consider what happens to the interference effects in coarse-grained histories. The way this works is that you have to add up all of the interference terms that arise between the fine-grained histories. It turns out that the interference terms between fine-grained histories will sometimes be positive and sometimes negative. The result when you add up all the terms is usually a lot of cancellation and an answer that is either very small or zero.

Coarse-graining can thus wash away any interference effects. However, this is not always the case. When a photon reaches the screen in the two-slit experiment, it travels along two main paths. Each of these paths is a coarse-grained history because it is actually the sum of an infinite number of fine-grained histories, each of which involves the photon traveling along some alternative route—perhaps close to the main path but perhaps not. Though we end up with two coarse-grained histories, interference still occurs between them. Coarse-graining, in itself, is not necessarily enough to destroy interference.

However, if an object interacts with the environment on its journey, rather as we are jostled by other people as we wander along a crowded street, the interference terms almost certainly do disappear. The reason is that each time the object is struck by another particle (a photon, molecule, dust particle, or whatever), the two objects become quantum mechanically entangled. We are unable and uninterested in following the subsequent trajectories of all the environmental particles and have to perform a sum-over-histories calculation so as to ignore their contribution. Because such environmental jostling is essentially random (unlike the controlled setup of the two-slit experiment) the coarse-graining here ensures that the interference is washed away. This is the origin of the phenomenon of decoherence.

The idea helps explain why macroscopic objects aren't normally found in superpositions. In *The Quark and the Jaguar,* Gell-Mann cites a question Enrico Fermi put to him in the 1950s.

> Since quantum mechanics is correct, why is the planet Mars not all spread out in its orbit? An old answer, that Mars is in a different place at each time because human beings are looking at it, was familiar to both of us, but seemed just as silly to him as to me. The actual explanation came long after his death, with the work of such theorists as Dieter Zeh, Erich Joos and Wojtek Zurrek on mechanisms of decoherence such as the one involving the photons of the background radiation. Photons from the Sun that scatter off Mars are also summed over, contributing to the decoherence of different positions of the planet.[22]

The same argument concerning decoherence is applicable to the measurement problem and to the problem of Schrödinger's cat. In the latter, as soon as the quantum state of the radioactive atoms is coupled to the macroscopic state of the cat, the system will undergo decoherence because of all the interactions between the cat and the environment (such as air molecules, light, heat, and so on). It is, therefore, silly even to think about the cat being in a mixture of dead and live states according to Gell-Mann and supporters of the idea of decoherent histories.

So the question is, does decoherence solve the conceptual problems of quantum mechanics? One of the more recent popular books on this issue, David Lindley's *Where Does the Weirdness Go?,* dismisses more or less all of the alternative interpretations of quantum mechanics and plumps for the idea of decoherence. Yet Lindley admits that it does not provide the whole answer because it does not solve the problem of quantum cosmology—that is, how we should interpret the wave function of the entire universe when there can be no external observers. It also cannot choose between different outcomes of a quantum measurement, a drawback Lindley expresses thus:

> Decoherence guarantees that a chain of events rather than a continuously ill-defined stream of quantum possibilities actually takes place. But it doesn't tell us which chain of events is going to happen. Probability has not been erased; measurements can have several different outcomes, and we cannot predict which.[24]

David Deutsch, on the other hand, says that the problems facing the proponents of decoherent histories are even more severe than they admit. "From the point of view of the interpretation of quantum mechanics, I think decoherence is almost completely unimportant," he says. "That's because decoherence is a quantitative matter. The interference phenomena never completely vanish; they only decrease exponentially until you can't be bothered to measure them anymore. The question of what the [interference] terms mean is still there, even if the coefficient in front of them is very small. It's like being a little pregnant. Those terms, however small, raise the same problem. If the argument is supposed to be that superpositions occur at a microscopic level but not to macroscopic objects, that's a bit like saying that you believe your bank is honest at the level of pennies but is cheating you at the level of pounds. It just doesn't make sense. It can't be that there are multiple universes at the level of atoms but only a single universe at the level of cats."

The Quantum Universe and the Omega Point

If there's one theme running through this book, it is, of course, the intimate connection between physics and computation. In Chapter 2 we saw how the development of reversible classical computation was driven, in part, by Ed Fredkin's belief in the idea that the universe was, in effect, a gigantic cellular automaton in which the laws of physics are somehow programmed.

Then, in Chapter 3, we saw how the development of quantum computation was driven, in part, by David Deutsch's conviction in the idea of many universes. What are we to make of these different conceptions of the universe? And what light do they throw on the question of why physics and computation should be connected?

On the face of it, Fredkin's and Deutsch's world views would appear to have an important feature in common. Both treat the phenomenon of computational universality—that is, Turing's principle—as a fundamental property of the universe. According to Fredkin, the fact that the laws of physics allow us to build universal computers is evidence that the universe itself is a based on a computational process. But this raises a host of questions: On what is this computation being run? Who or what is responsible for the program? How do we know nobody will pull the plug during the computation? What is the purpose of the computation? Fredkin addressed some of these questions in a paper he presented at a physics of computation workshop held in 1992. The following extract gives a flavor of his thinking:

> If space and time and matter and energy are all a consequence of the informational process running on the Ultimate Computer then everything in our universe is represented by that informational process. The place where the computer is, the engine that runs that process, we choose to call *"Other."* Where did *Other* come from? This question is actually quite easy to fence with. The nature of systems of laws that can support computation is very much broader than the nature of systems that are limited to the physics of our universe. . . . There is no need for a space with three dimensions; computation can do just fine in spaces of any number of dimensions! . . . *Other* is that place that has such structure and laws as to not raise the question of its origins, as origins are a concept peculiar to our world.[25]

Fredkin's view of the universe, which he calls Digital Mechanics because in it everything is discrete, is highly idiosyncratic if not heretical. Though some physicists express sympathy with his ideas, I have yet to come across anyone who fully endorses them. Fredkin has been criticized, for example, for holding what is essentially a classical view of the universe. The "ultimate computer," according to Fredkin's view, is a classical cellular automaton. Nevertheless, he claims to be able to account for quantum phenomena within this classical framework. The EPR phenomenon, in which measurements have nonlocal effects, can be accommodated, he says, because his digital universe is completely deterministic:

> The whole universe has spent its whole history arranging for whatever is going on to happen in each and every cell. . . . The question can best be framed by asking whether causing a particular cell to turn on and off will cause another cell at another place and time to blink. The answer depends on the states of lots of other cells. Strangely enough, you would have to be God to arbitrarily turn on or off just one particular cell. . . .[26]

In other words, nonlocal effects arise naturally within a classical cellular automaton because the state of each cell ultimately depends on the state of all the cells if you extrapolate far enough back into the automaton's history. This would imply that when you decided to measure an EPR photon polarization vertically as opposed to some other angle, the universe already "knew" you were going to do that so that it could arrange for the twin photon to exhibit the appropriate correlations. Such determinism would deny the existence of free will with a vengeance. While many scientists are not uncomfortable with the notion that free will is an illusion, it strains credulity to imagine that the universe can anticipate everything that is going to happen in such a deliberate way.

Many physicists, I suspect, will argue that Fredkin has more or less ruled himself out of serious consideration by clinging to a classical picture of reality. Nevertheless, as metaphysical scenarios go, his surely has some merit. It offers a possible reason for why there is something rather than nothing. (Somebody or something decided to run a cosmic computer program.) It offers a possible explanation for why the laws of physics are what they are. (Somebody or something chose them.) It also offers a possible explanation for why and how the universe started in a state of extreme order. This is often phrased in terms of why the *initial conditions* of the universe were so special. To most physicists this question is a great mystery. However, if the universe were a program running on a computer, the question has a simple and natural interpretation. Anyone who has ever written a computer program knows that you usually have to *initialize* certain variables with special values before the program starts running. So special initial conditions are exactly the sort of thing one expects in computation. The same, we might expect, could be true for the great cosmic program.

Of course, anyone of a theological persuasion will object that Fredkin has merely cast God in the role of cosmic programmer, while atheists will complain that invoking the existence of an intelligence beyond the universe does nothing to solve the problem of explaining ultimate causes; it merely passes the buck onto somebody else. These are cogent arguments, but Fredkin doesn't, at least, invoke an external creator who interferes with the running of the universe. Computational universality is no excuse for believing in religious miracles.

David Deutsch likes the idea of regarding physics as a computational process, but in *The Fabric of Reality* he dismisses the universe-is-a-computer idea in one paragraph as being "too narrow" a perspective of reality. I asked him to explain his reasons more fully. "If the laws of physics as we see them are just aspects of some universal computer program," he said, "then by definition we would be prevented from finding out anything about the hardware of that computer. That is the very nature of computing: the power of computing comes from the fact that the computer is a universal machine. If we're just a program, the program cannot obtain information about the machine on which it is running. So there would be an underlying physics responsible for this computer, and we would never be able to find out what that physics is. Furthermore, that underlying physics would *not* be a program running on a computer, unless you want to postulate an infinite regress. Either way, the hypothesis does not solve anything."

The problem therefore, as Deutsch sees it, is that regarding the universe as the result of a program run on a computer does not explain anything and would represent a dead end for physics. The more fruitful idea, according to Deutsch, is to view physical processes *within* the universe as computational processes. As Deutsch put it: "Viewing the universe as a computer doesn't show any promise because it hardly rules out anything. Whereas requiring that the universe be able to *contain* computers that can perform complex computations places a very strong restriction on what the laws of physics can be."

One of the strengths of Deutsch's conception as presented in his book *The Fabric of Reality* is in the way it embraces so many different deep issues, including evolution, computation, knowledge, the physics of time travel, the limits of virtual reality, and even the ultimate fate of the universe. Take knowledge, for example. In classical physics, and in other interpretations of quantum mechanics, it's actually quite hard to produce a satisfactory definition of what constitutes knowledge that distinguishes it from random noise. But if we were able to look across different universes within the multiverse, Deutsch suggests that knowledge would stand out like a crystal in a sea of randomness. The more tried and tested a piece of knowledge, the more it will appear the same in closely related universes. Hence his comments in the opening pages of this book about the similarity between the silicon chips made by intelligent dinosaurs and our silicon chips.

But Deutsch's world view is hardly representative of physicists in general. It's part of the scientific instinct to search for the simplest and most economical solutions to problems, so it's not surprising that many physicists balk at what they see as the unimaginable extravagance of the multiverse. A good example of this line of thinking is offered again by David Lindley in *Where Does the Weirdness Go?* He considers the journey of a photon that strikes your eye, letting you know that the sky is blue.

> That photon has battled its way from the center of the sun to your retina, colliding along the way with innumerable . . . atoms and electrons. . . . The number of possible paths that the photon could have taken in getting from the center of the sun to you is literally infinite. Any number of possible collisions and trajectories could have delivered the photon from the center of the sun to your eye, and along every such path, each individual scattering event has to be accounted for with the appropriate splitting of universes.[27]

Actually, in the modern multiverse theory, it's wrong to imply that universes split at each quantum event. They merely differentiate from one another, but nevertheless, to take account of this for every activity right down to the choice of path taken by a single photon, it is necessary to contemplate a stupendous number of universes embarking on different evolutionary journeys. For many scientists who see nature as being beautifully parsimonious, such profligacy seems quite monstrous. Yet perhaps even more worrying is not what the photons do but the thought of there being unimaginable numbers of copies of us all doing every conceivable and inconceivable thing. The thought of what Adolf Hitler, Josef Stalin, Saddam Hussein, and Slobodan Milosevic have done in this world is scary enough, let alone the misery they may have inflicted in other universes. And for each of us personally, the whole idea that we take every possible path through the forked garden of life may also seem distinctly disturbing. Of course, not all paths will be equally represented in the multiverse. Some will be like well-trodden byways, while others will remain virtually virgin territory. But nevertheless, it's not very comforting to learn that somewhere in the multiverse you may be perpetrating every wicked act that you've ever thought about.

Such emotional reactions, though, carry little weight in the face of scientific truth. Special relativity, after all, seems very strange when you discover how moving close to the speed of light can severely distort the flow of time, yet no one, apart from a few people who specialize in writing with green ink, seriously questions the theory. David Deutsch pointed out to me that though he found it hard to "really" think there were Australians walking around under his feet, such feelings didn't stop him from believing that the Earth was round. If the evidence incontrovertibly points toward the existence of the multiverse, then so be it, we'll just have to accept it whatever its less palatable consequences.

What, then, are the other objections to the interpretation? There's one that can probably be quickly dismissed as a red herring. It's sometimes said that because the many-universes viewpoint refers to other universes, by definition these other universes are inaccessible to us, and if they are inaccessible, we are unable to say anything about them. Why are they inaccessible? Because otherwise they would be part of our universe and not some other universe.

This argument doesn't stand up to close scrutiny, particularly in the light of quantum computation, where we obviously can access the Everett worlds, albeit in limited ways, via entanglement and interference phenomena. The problem is partly in the name "many universes" because the notion of separate and independent universes is only an approximation to the quantum reality. If the bulk of the universes that move apart in some quantum event are carefully shielded from the event's debris, then they can be brought back into contact. The crucial feature is to keep the universes almost identical. That is the nature of interference. When there's decoherence, small differences between the universes become hopelessly entangled with the environments, wrecking any chance of getting the universes to overlap and interfere. In these circumstances the universes do become independent and inaccessible.

In his book *Shadows of the Mind,* Roger Penrose argues that the many-universes interpretation does not really provide a solution to the "measurement problem," the very problem it was originally intended to solve. How, he asks, does the "illusion" of the R process, the wave-function collapse, come about? When we make measurements on ensembles of quantum systems, we observe with great precision the probabilities predicted by the R process, yet in the many-universes interpretation there's no such thing as wave-function collapse or an R process. Instead, the universe is supposed to split into multiple copies in which you, the observer, see different things in different universes. The interpretation says nothing about different probabilities. To quote Penrose:

> . . . the many-worlds viewpoint provides no explanation for the extremely accurate wonderful rule whereby the squared moduli of the complex-number weighting factors miraculously become relative probabilities.[28]

Strangely enough, in 1999 David Deutsch finally released for publication a paper he'd written many years before that purported to show that the probabilities of quantum mechanics could be *derived* from the formalism of the many-universes interpretation. The paper, if correct, amounts to something of a conceptual bombshell—not only because it could answer Penrose's point but also because of the repercussions it must have more generally for philosophy. To quote one line:

> Thus we see that quantum theory permits what philosophy would hitherto have regarded as a formal impossibility, akin to "deriving an ought from an is," namely deriving a probability statement from a factual statement. This could be called deriving a "tends to" from a "does."[29]

If Deutsch's analysis holds up, then he has certainly found a powerful new argument in favor of the many-universes view simply because it does away with what had previously looked like a God-given axiom of nature. But critics are unlikely to be persuaded any time soon. Consider another point that Penrose makes concerning the way our minds are split by observing quantum events:

> Without a theory of how a "perceiving being" would divide the world up into orthogonal alternatives, we have no reason to expect that such a being could not be aware of linear superpositions of golf balls or elephants in totally different positions. . . .[30]

Again we appear to be back with the issue of Schrödinger's cat and consciousness, which may or may not be a problem, depending on your view. But I can't help feeling Penrose has got a point if we consider the following simpler situation. Imagine you measure the horizontal polarization of a diagonally polarized photon. According to the many-universes interpretation, the act of measurement will cause the universes to differentiate into two copies. You too will divide into two copies, one of whom sees a horizontally polarized photon, while the second sees a vertically polarized photon. The question that arises is this: If I perform this experiment and see a horizontally polarized photon, I would like to know what actually decided *which* of the two copies I experienced. Perhaps the other copy of me, who saw the vertically polarized photon, will be asking the same question in the other universe, but the fact is I only experience one "me," and I feel rather uneasy about the seemingly arbitrary way I lose contact with the other "me" in the other universe. What aspect of my consciousness decided that I would stay with this "me" rather than the other "me"? Also, shouldn't I feel a little sad about losing touch forever with my twin in the blink of an eye?

As you can see, it's hard to put some of these ideas into words without walking into an intellectual quagmire. Nevertheless, the fact that there is something rather odd about these issues is reinforced by Deutsch's admission that consciousness, though probably not a quantum phenomenon, is still very mysterious. Furthermore, even one of Deutsch's most loyal supporters, Artur Ekert, hesitated when I asked him whether he believed the other universes of the many-worlds theory really exist. "There's been a certain type of etiquette among physicists that said, 'Don't talk about the many-worlds interpretation if you don't want to make a fool of yourself.' If you remain agnostic on these issues that's easy; you won't offend anyone, but you also don't learn anything. Somehow younger physicists are different. They don't care whether they make fools of themselves. They see that the many-universes interpretation is more consistent than the alternatives. No collapse, only unitary evolution, the Schrödinger equation applies all the time. Now, *how* you interpret those branches of the wave function, people have slightly different views. Whether I would say each world was as real as this one, here I would just say I don't know. It's not the sort of question I have a good answer to."

Seth Lloyd, like Ekert, could also be counted among the younger generation of physicists, but he's not to be found among the supporters of many worlds. "If we consider the evolution of the universe as a whole, which is the sort of problem this theory is often applied to, it starts out with everything being completely uniform. And then you get a little quantum fluctuation and in one place it says, 'Galaxy, you form here,' and in another place it says, 'Galaxy, you form there.' From that point onward the universe has split into two different worlds, in this parlance of many worlds. In one world our galaxy is over here, and in the other world the galaxy is over there. Now I would say there's a slight abuse of the word 'world' here, because if you ask where is the galaxy *really*—is it really over here and over there at the same time—you're in trouble. I would like real things to be things I can get information about so that I can verify whether they are here or there. These other worlds *aren't* real."

On the question of why the universe is computational, Lloyd also offered a slightly different perspective: "It's very important that it *is* computational because we wouldn't be here unless matter were capable of processing information. Our lives depend to a huge degree on the ability to process information, and I'm not just referring to thinking. For example, when an enzyme breaks down sugars in the body, it's in some sense performing a computation. The enzyme provides a piece of information that's necessary to unlock the energy in the sugar. You can think of reproduction as being a computation the species performs in order to adapt to its surroundings. What I'm saying is that it's not surprising that the universe is computational given that it is *quantum mechanical.* I can't make any claims as to why the universe is quantum mechanical, but given that it is, then I would claim that we ought *not* to be surprised that it is, one, capable of computation and, two, doing something interesting."

So, according to Lloyd, the question of why the universe is quantum mechanical precedes the question of why it is computational. Explain one and you'll have the answer to the other. The problem is that as we unpeel nature's layers of explanation, it seems inevitable that we will always be left with an unfathomable core. For many years scientists saw this as the ultimate problem in explaining the origin of the universe. Given the laws of physics and the initial Big Bang, it's quite possible that we could explain its evolution. But how can we account for the initial event?

In the 1980s cosmologists began to think they might have the answer when they hit on the idea of inflation, a period of superrapid expansion in the early evolution of the universe. Suddenly it became fashionable to think of the universe as originating in a quantum fluctuation from absolutely nothing. This tiny fluctuation was then blown up astronomically by inflation within the first cosmological instant. Afterward the early universe settled down into the much steadier rate of expansion astronomers see today. The attractive feature of this scenario was that by invoking quantum processes in the initial creation event it was possible to break the chain of cause and effect on which traditional scientific explanations normally rely. Quantum fluctuations, after all, happen without rhyme or reason.

Appealing though this idea is, it unfortunately leaves unexplained why the universe is quantum mechanical. The British physicist and noted Christian Russell Stannard argues that these attempts to rule God out of the equation in the origin of the universe inevitably fail because of the lack of an explanation for quantum mechanics. If "God" is supposed to be the answer, though, I don't see how it in any way adds to the plausibility of the Christian notion of God—or for that matter most other religious interpretations of "God." If we imagine God as the cosmic programmer responsible for setting up our universe, the idea that he/she somehow interferes directly with the great cosmic program as it is running seems distinctly inelegant. The evidence for any such interference in the form of biblical stories and "divine revelation" is hopelessly open to question.

After all, over the ages, there have been so many cases of individuals deluding others and themselves into believing they are God's chosen emissaries, it's hard to see why we should give such claims any credence. A little knowledge of human psychology ought to teach us that all claims to divine knowledge or empowerment should be treated with the utmost skepticism.

Let's not get hung up on this issue of science versus religion, because there's an altogether more interesting issue lurking within the question of where quantum mechanics comes from. Einstein talked metaphorically when he referred to God not playing dice with the universe. Today most scientists accept that Einstein was wrong and that God (metaphorically or otherwise) does indeed play dice with the universe. The outcome of every quantum event we perceive is determined by the roll of the quantum dice.

In 1998 Sandu Popescu, a researcher in quantum computing at Bristol University in England, gave a fascinating lecture at a symposium on cosmology and computation in which he pointed out that the interesting question was no longer *whether* God plays dice, but *why* he plays dice. Popescu offered an intriguing answer. The reason God plays dice with the universe, he said, was that this offered the only way in which it was possible to maintain both the phenomenon of nonlocality and the rule that no signal can travel faster than light.

Let's unravel that a little. Imagine you're God and you want to design a universe in which interesting things can happen without the need for any intervention because, being a lazy-minded deity, you want the show to run totally hands off. Now, given the structure of space-time, if signals could travel faster than light, causality would be in deep trouble because, as we have seen before, events would be able to precede their causes. So you feel more or less obliged to accept some kind of speed restriction on signals as a necessary constraint for your designer universe. You therefore decide to set a speed limit within the framework of what we understand as Einstein's special theory of relativity.

But being a God who moves in mysterious ways, you also decide that you want the universe to incorporate some element of nonlocality. How do you go about that? The problem here is that nonlocality involves some kind of instantaneous connection between distant particles, which seems to fly in the face of any ultimate speed limits. Nonlocality and special relativity would seem to be mutually incompatible. But the remarkable feature of *quantum nonlocality,* as we have seen, is that it doesn't offer any way of breaking the speed rules because the instantaneous connections are cleverly disguised by the throw of the quantum dice. And it seems that this might be the only way you could incorporate nonlocality without breaking the speed rule. So you say, "Bring in the quantum dice."

So now, as observers within this cunning scheme, we must ask ourselves not why God plays dice but why God chose to endow the universe with nonlocality. Is nonlocality a vitally important feature of the universe? If it isn't, then why is it there?

Popescu didn't have any immediate answers. I wouldn't be surprised if people with a religious turn of mind claimed that nonlocality was God's way of ensuring that she can keep in contact with all of her children in the universe instantly without contravening the laws of physics. But as we've seen, quantum nonlocality allows only very limited forms of long-range contact.

A further twist to this whole argument about nonlocality comes from the work of David Deutsch and Patrick Hayden, who, as mentioned in Chapter 3, published an article in 1999 showing how nonlocal phenomena in quantum mechanics could be explained in terms of purely local effects.

They offered an analogy for how this could happen: Whenever you download a picture on the Internet, it is almost always in some specially compressed format known as a gif or jpeg file. Such compression ensures that the files are much smaller than they would be normally, thereby allowing the pictures to be propagated across the Web relatively quickly. However, odd things can happen if you change the compressed data in any way. Indeed, a tiny modification in the data in one place can bring about changes across the whole picture. The compression, in a sense, causes nonlocal effects. What Deutsch and Hayden argued was that the nonlocal effects of quantum mechanics could be attributed to the way we look at them. If we sliced the picture in a different way, they claimed, the nonlocality would disappear.

How then do they account for the EPR phenomenon, in which entangled particles exhibit correlations that go beyond anything we can do classically? Their answer is that quantum particles can carry *locally inaccessible* information. This information may be present in a qubit, for example, but not available to any experiments carried out on it locally. Nevertheless, the information can become available if the qubit interacts with other qubits. Remarkably, this information according to Deutsch and Hayden can even survive decoherence and travel via classical means. So the strange features of the EPR effect depend not on instantaneous transmission of information but instead on the information flow required to compare the results of measurements made on the entangled particles. During such communication—even if it's classical—locally inaccessible information can reemerge to impress itself on the results we see. Similar reasoning, the authors showed, applied to quantum teleportation. The locally inaccessible information carried in one qubit is teleported to the other simply via the classical channel.

Deutsch and Hayden's work certainly puts nonlocality in a new light, though it is very unlikely to satisfy everyone. Nevertheless, it is another good example of the way research into quantum computation is offering wonderful new insights into fundamental physics. It will be very interesting to see how their analysis is received by other physicists. Yet—even if nonlocality were now to be regarded as less of an issue—we're still left with the intriguing issue of why quantum particles are endowed with "locally inaccessible" information.

One possible inference might be that this feature was built into the universe to make it possible to quantum compute. Remember that, as we saw in the opening chapter, the superior power of quantum computing over classical computing comes from the allegedly "nonlocal" features afforded by the phenomenon of entanglement. So could it be that in order for us to exist it is necessary for the universe to be able to quantum compute?

Of course, if we believed that the mind was the product of a quantum program, then the answer would be yes. However, as we have seen, only a few people support that idea. In his book *The Fifth Miracle,* Paul Davies has pointed to another possibility that's probably equally speculative: Perhaps the origin of life itself might have depended in some way on the computational advantages afforded by quantum systems. Scientists have long pondered how life could have got started and, in recent years, the closer they've looked, the harder the problem has become.

In the 1950s things looked rather different when Stanley Miller performed the famous experiment in which he passed an electrical discharge through a mixture of gas and liquid in a bell jar. Starting with simple chemicals, he produced within a few days a black tar material that contained amino acids, the building blocks of all living cells. It looked as though by imitating what were thought to have been the early conditions on Earth, he had produced the starting materials for life in virtually no time at all. Suddenly, the emergence of life seemed almost inevitable, and the only question was why we weren't plagued by critters from outer space.

But in the decades that have followed, scientists have increasingly realized that producing amino acids is still light-years from creating life. The Holy Grail in this subject is to find a really simple self-replicating molecular system. Once such a system comes into being, it's thought molecular Darwinism can swing into action and produce more diverse and more sophisticated self-replicating entities. Evolution will have truly begun. But the problem is that a system needs a minimum level of machinery to be self-replicating, and so far all plausible candidates for the first self-replicating entities appear to be far too complex to have emerged from a cocktail of inanimate molecules simply by chance.

As Davies and others have suggested, the problem can be regarded as a computational one. The amount of time a bunch of simple molecules would require to find a way of assembling themselves into a self-replicating system would appear to be far longer than the few hundred million years it actually took on Earth. The computation required is simply too big. But that's assuming the computation was performed classically. Could the origin of life have depended in some way on the much greater resources of quantum computation? Well, it's an interesting suggestion, but it's hard to see how nature could have overcome the problems of decoherence in the hot conditions of the prebiotic soup. But perhaps life didn't start in such conditions. Who knows?

As we have seen, regarding the universe from a computational standpoint may cast new light on questions about origins. But it may also tell us something about our fate too. I'm thinking of probably the most sensational idea advanced about the computational universe, one proposed by Frank Tipler in his book *The Physics of Immortality.* In it, he argued that if sometime in the future the universe stopped expanding and started recontracting toward a Big Crunch, there would be time for an infinite amount of information processing in the last few dying moments as the universe shrank increasingly rapidly. His scenario, dubbed the *omega-point theory,* which enjoys David Deutsch's qualified support and a lively, if eccentric, following on the Internet, envisages the emergence of a cosmic intelligence so powerful that it would be able to resurrect everyone who has ever lived. And rather than pessimistically waiting for the lights to go out and the walls to cave in, the superbeing would, according to Tipler, live forever because its sense of subjective time would be infinite.

If you haven't the patience to sit it out for untold billions of years to see whether Tipler is right, perhaps you might prefer the following amusing parable as told by Ed Fredkin.

One day, Fredkin muses, you hear about an advertising campaign by the Heaven Machine Corporation, and you become curious to find out what it is. Arriving at the corporation's offices, you are ushered into a room and shown some brochures. The salespeople explain that what they have is a huge computer simulation into which they can load an exact copy of your brain state. Unfortunately, if you decide to accept the company's offer, the process of duplication will destroy your original brain, at which point your life on Earth will have ended.

Instead, however, you are promised eternal life within the machine and, not only that, you are told life there is heavenly. Because you look skeptical, they let you talk to a neighbor of yours who recently made a deal with the Heaven Machine Corporation. A computer screen the size of a wall is unveiled. The picture is initially milky but gradually clears, revealing your neighbor. You say, "Hi Joe, how are you doing?" Joe replies, "Just fantastic. It's really heavenly up here. There are all these amazing people: Einstein, Buddha, Confucius. You wouldn't believe the conversations I've had.

I do as much fishing as I want, and you should see the fish! You remember the guy I played tennis with at the club and how he always used to beat me? Well, now I always beat him! By the way, there's something that I meant to tell you before I came here. I borrowed your lawn mower and it's still in the garage. So just go and take it."

Hearing this, you realize that must be Joe. Perhaps you too should think about accepting a deal with the Heaven Machine Corporation.

Is this really a heavenly future or simply a nightmarish fantasy? Or are we already living inside the Heaven Machine—but with the unfortunate twist that things aren't at all heavenly. Such questions may only ever be answered once we understand the true relationship between minds, machines, and the multiverse. Far-fetched though it might seem, the quest for the quantum computer may have given us one of the most important clues we'll ever get.

Appendices

Appendix A: Logarithms

Logarithms reverse the idea of a power, or exponential. Take the formula:

$$x = 2^y$$

In words this is expressed as "x is equal to 2 to the power y." The equation can be turned around and reexpressed in logarithms as:

$$y = \log_2 x$$

This is expressed in words by saying, "y is the log of x, to the base 2." Now if $y = 3$, then x is 2 *cubed,* which is 8. Therefore, $\log_2 8 = 3$. Similarly, $\log_2 16 = 4$, $\log_2 32 = 5$, and so on. Note that these are base-2 logarithms. Logarithms using bases 10 and the natural number e are also widely used.

Logarithms have the very useful property that when numbers are multiplied, their logarithms are added, and when numbers are divided, their logarithms are subtracted from one another. In other words:

$$\log a \cdot b = \log a + \log b \text{ and } \log a/b = \log a - \log b$$

Appendix B: Minimum Energy Required for Erasing a Bit of Information

Resetting the ball in Figure 2.3 is associated with an entropy change given by Boltzmann's formula $S = k \log W$, where the probability, W, of the state is halved because the ball is forced into one of two different (but equally probable) states. According to this argument the entropy, S, of the system must then change by at least $k \log 2$ because $S = k \log W/2 = k \log W - k \log 2$. Entropy is related to energy by multiplying by the temperature, giving $kT \log 2$ as the amount of heat wasted each time one bit of information is reset.

Appendix C: Testing Many Universes Experimentally

In 1985 David Deutsch published a thought experiment that showed how the many-universes interpretation of quantum mechanics could be tested. The experiment, which he had devised many years earlier, uses what we would nowadays call a quantum computation to reveal its answer. His starting point was to assume we have a fully fledged quantum computer loaded with an artificial intelligence program for consciousness.[1] It's a colossal assumption, of course, and the machine would need to run without any environmental disturbance during its operation because otherwise the results of the experiment would be invalid.

The computer reserves, for the purposes of this experiment, one quantum bit—one qubit—of its memory. Let's call this qubit *Q*. It puts *Q* into a superposition of 0 and 1 by first setting it to 0 and then applying a square root of NOT operation (see page 125).

Having applied √NOT to produce the superposition, the conscious AI machine then looks at *Q* in its memory and observes which state it sees. According to the Copenhagen interpretation, this act of observation will collapse the superposition, causing *Q* to settle on the state 0 or the state 1 with a 50-50 chance. However, according to the many-universes interpretation, both states will continue to coexist, although each will be in a different universe. In one universe the machine will see 0, and in the other universe it will see 1. The machine then records a note declaring that it has observed one value for the qubit and only one value. This is possible even in the many-universes theory because the machine *would only be aware of seeing one value in each universe*—it would not be aware of its clone in the other universe seeing a different value.

Now, in order to avoid spoiling the results later, it is crucial that the AI machine erases all record of the actual value it saw from its memory. The reason is that we need to conduct an interference experiment between the two universes. If the computer remembered the value it saw, the interference would be destroyed. This is analogous to the way the interference fringes in Young's two-slit experiment suddenly disappear if any attempt is made to observe which slit a photon passes through. This requirement to erase any record of "which way" information is a general feature of all quantum interference phenomena. It also partly explains why we couldn't use a human observer to play the role of the AI machine—nobody could guarantee to wipe all trace of the result from his or her brain. (Another, more obvious, reason is that the human brain wouldn't be capable of remaining in a coherent superposition for long enough, for much the same reasons as Schrödinger's cat couldn't.)

In Deutsch's experiment, interference can be brought about rather simply by performing another √NOT operation on qubit *Q*, but this time after the AI machine's observation. What does this do? If the many-universes interpretation is correct, the superposition *Q* remains intact during the observation by the AI machine and the second √NOT transforms the superposition to the *definite* state 1. If the Copenhagen interpretation is correct, however, *Q* collapses onto a definite state of either 0 or 1 during the observation, but when the second √NOT operation is applied, the qubit is transformed back into a *superposition* of 0 and 1.

To complete the experiment, either the computer itself or we, the outside observers, now measure the state of *Q*. If many universes is true, we will see 1 *with certainty,* and if the Copenhagen interpretation is true, we will see either 0 or 1 *at random.* To distinguish between these two outcomes we would need to repeat the experiment a number of times. If we always saw 1, then we could be confident that many universes was correct. If, however, we saw a 0 from time to time, we would have to accept the Copenhagen interpretation.

The crux of the experiment, therefore, is to use a simple quantum computation to test whether consciousness collapses the superposition.

Appendix D: Quantum Circuit for Deutsch's Stock Market Prediction

Before looking at the full circuit to perform Deutsch's program, it's useful to consider one of its components, sometimes known as an *f*-controlled NOT gate. The circuit in Figure D.1 is typical. It consists of a quantum circuit **U** that evaluates some mathematical function *f*. It takes as two inputs the values *x* and 0, calculates *f*(*x*), and places this value reversibly on the input with 0.

So the top qubit remains unaffected, while the bottom qubit returns the answer to the question, what is the value of *f* given *x*. If $|x>$ is a superposition, this circuit is, in effect, the quantum version of the hat full of answers described in Chapter 1 (page 36). The answers are held by the bottom

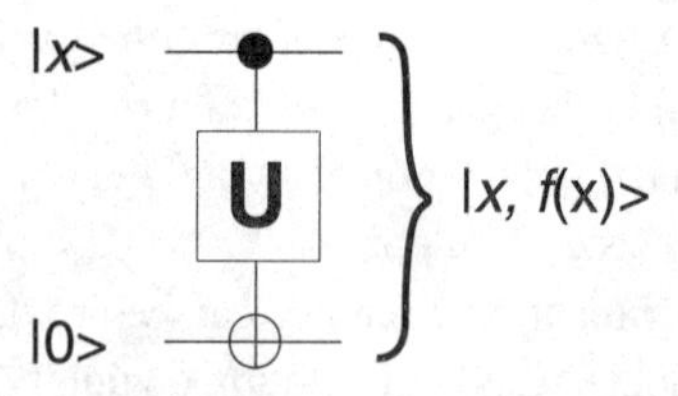

Figure D.1. The *f*-controlled NOT gate

In a typical quantum circuit we will want to evaluate some function using quantum operations (incorporated in the box marked **U**). Because quantum operations are reversible, the output has to be placed on one of the inputs. In this case, the output appears via a quantum-controlled-NOT gate marked by the cross, ⊕, on the lower qubit. The controlled-NOT gate, in effect, copies the output of **U** onto the lower qubit.

qubit, while their associated questions are held on the top qubit. The linkage is via entanglement, and measuring either qubit will destroy the superposition, leaving only one answer and its associated question on the qubits.

So now here, courtesy of Deutsch's colleague Artur Ekert, is the full network that computes Deutsch's 1985 stock market algorithm. Unfortunately, to understand how it works, we need to face up to some strange-looking math, but it really isn't too hard.

The objective is to find out whether *f*(0) is the same as *f*(1)—that is, whether the function *f* is constant for its two input values. Supposedly, *f* is some function that is part of a prediction for the stock market for the following day. It takes slightly less than 24 hours to compute *f* for just one value, so we need to get the answer without having to do two separate runs. Although *f* is likely to be complicated to work out, because its inputs and outputs are limited to just 0s and 1s, it can actually take only one of four forms.

Either *f*(0) and *f*(1) are the same, where

1. $f(0) = f(1) = 0$ *or* 2. $f(0) = f(1) = 1$

Or *f*(0) and *f*(1) are different, where

3. $f(0) = 0; f(1) = 1$ *or* 4. $f(0) = 1; f(1) = 0$

The circuit that computes whether *f* is the same or different is shown in Figure D.2. Reading the circuit from left to right, qubit 1 starts as |0> and passes through an **H** gate, resulting in the superposition |0> + |1>. **U** then computes the values of *f* for both 0 and 1 simultaneously, producing an en-

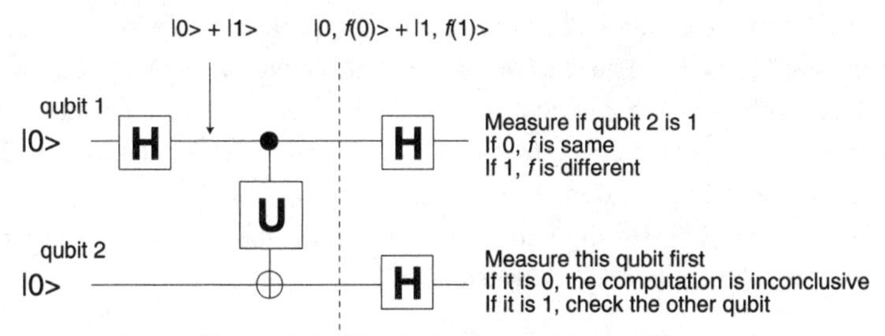

Figure D.2. Circuit for Deutsch's problem

The boxes marked H are Hadamard transforms and U is the main processing element, which calculates $f(x)$, where x is the value on qubit 1 and places the output on qubit 2. U leaves qubit 1 unchanged (though see text).

tangled superposition $|0, f(0)> + |1, f(1)>$, which means in one quantum universe qubit 1 is 0 and qubit 2 is $f(0)$, while in the other universe qubit 1 is 1 and qubit 2 is $f(1)$. The qubits are now each processed by an **H** gate. We measure qubit 2 and if the answer is 0, the computation has failed. If it is 1, we measure qubit 1. If qubit 1 is 0, *f* is the same, and if it is 1, then *f* is different.

How does this circuit work? Consider what happens for the different kinds of function *f* can be. If $f(0)$ and $f(1)$ are the same, as in cases 1 and 2 above, then the output at the dotted line will be, respectively:

1. $\overbrace{(|0> + |1>)}^{\text{qubit1}}\ \overbrace{|0>}^{\text{qubit2}}$
2. $(|0> + |1>)\ |1>$

If they are different, we get:

3. $\overbrace{|0>}^{\text{q1}}\overbrace{|0>}^{\text{q2}} + \overbrace{|1>}^{\text{q1}}\overbrace{|1>}^{\text{q2}}$
4. $|1>|0> + |0>|1>$

If you look carefully at the results in the two types of situations, you'll notice some interesting differences. When *f* is the same (cases 1 and 2), the first qubit is $(|0> + |1>)$ regardless of the content of the second qubit. When *f* is different, on the other hand, the two qubits become entangled, so that the value of the first qubit depends on the value of the second qubit. In case 3, for example, the qubits always give the same value, whereas in case 4 they always have the opposite value.

To detect these subtle differences, though, we need to do some further processing. In fact, we need to perform an interference measurement. This is achieved by passing each qubit through an **H** gate. A crucial feature of these is that the gates erase information about which computational path the circuit actually followed. This works as follows.

After qubit 2 passes through its **H** gate, we get the following quantum states for each of four possible versions of *f*:

$$1.\ \overbrace{(|0>+|1>)}^{\text{qubit 1}}\overbrace{(|0>+|1>)}^{\text{qubit 2}} = \overbrace{|0>|0>}^{\text{qu1 qu2}} + \overbrace{|0>|1>}^{\text{qu1 qu2}} + \overbrace{|1>|0>}^{\text{qu1 qu2}} + \overbrace{|1>|1>}^{\text{qu1 qu2}}$$

$$2.\ (|0>+|1>)(|0>-|1>) = |0>|0> - |0>|1> + |1>|0> - |1>|1>$$

$$3.\ |0>(|0>+|1>) + |1>(|0>-|1>) = |0>|0> + |0>|1> + |1>|0> - |1>|1>$$

$$4.\ |1>(|0>+|1>) + |0>(|0>-|1>) = |0>|0> - |0>|1> + |1>|0> + |1>|1>$$

We now measure qubit 2 and get a definite result of 0 or 1, each with a 50-50 chance. If the answer is 0, we can eliminate the second and fourth terms in each line (which in the many-universes interpretation represent worlds in which the qubit takes the value 1). The quantum state for each version of *f* becomes:

1. $(|0> + |1>)\,|0>$
2. $(|0> + |1>)\,|0>$
3. $(|0> + |1>)\,|0>$
4. $(|0> + |1>)\,|0>$

These states are all indistinguishable, which means the computation fails because it is unable to tell us anything useful about *f*. However, if we get a 1 on the second qubit, then the quantum state becomes:

1. $(|0> + |1>)\,|1>$
2. $-(|0> + |1>)\,|1>$
3. $(|0> - |1>)\,|1>$
4. $(|1> - |0>)\,|1>$

The first qubit is now processed by another **H** gate, which gives:

1. $|0>\,|1>$
2. $-|0>\,|1>$
3. $|1>\,|1>$
4. $-\,|1>\,|1>$

If we now measure the first qubit, we get 0 for cases 1 and 2, and 1 for cases 3 and 4 (note that the negative signs or phases here have no effect on the final measurement). Therefore the value of the first qubit tells us whether *f* is the same or different, just as we first set out to show.

Anyone used to looking at electronic circuits or, for that matter, programming a conventional computer may well be taken aback by some odd features of the quantum network we have just examined.

First, you may wonder why we need two qubits rather than just one. The laws of quantum mechanics are reversible, and so quantum mechanical gates must also be reversible. It is therefore necessary to have enough inputs and outputs to be able to reverse any computations. If we had only one input and one output, and if *f* was a constant function, the computation would not be reversible. This is reminiscent of the fact that in reversible computing we needed extra inputs and outputs on logic elements like the Fredkin gate to ensure reversibility.

Second, a more remarkable feature of the circuit is that we get the final answer on whether *f* is constant or different by reading the qubit that was supposedly left untouched by **U**. This is brought about by the way the first qubit becomes entangled with the second. Making the measurement on the second qubit has a "kickback" effect on the first qubit, though there are no discernible signals involved. The linkage is analogous to that involved in the two arms of the EPR experiment. When we measure one qubit, quantum entanglement ensures that the measurement "influences" the other.

A New and Improved Quantum Network

It is a measure of how strange these circuits are and how unfamiliar their operation is that it took more than ten years for someone to spot a better solution to Deutsch's original quantum algorithm. In 1996 Artur Ekert and colleagues produced a slightly modified version of the previous circuit, which determines whether *f* is constant or different *without fail.* (See Figure D.3.)

The circuit is almost the same as before, apart from two things. First, the input to qubit 2 is set to |0> − |1> instead of |0>, and second, there is no need to send the second qubit through an **H** gate, because we don't even need to look at it. To understand how this circuit works, we need to examine a little more closely the controlled-NOT, or XOR gate used by **U** for placing its answer on qubit 2. Let's remind ourselves of its truth table:

c	t	c'	$t' = c$ XOR t
0	0	0	0
0	1	0	1
1	0	1	1
1	1	1	0

If c, the control qubit, is 0, then we see that t, the target qubit, is left unchanged. So if c is 0, then t remains as $|0> - |1>$. If c is 1, t is flipped, transforming the superposition into $|1> - |0>$. Thus the *phase* of qubit 2 is flipped if qubit 1 is 1. So now, considering the four possible versions of f, we get the following states after the operation **U** (see dotted line in diagram):

1. $(|0> + |1>)(|0> - |1>)$ [if $f(0)$ and $f(1)$ are both 0]
2. $(|0> + |1>)(|1> - |0>)$ [if $f(0)$ and $f(1)$ are both 1]
3. $|0>|(0> - |1>) + |1> |(1> - |0>)$ [if $f(0)= 0, f(1) = 1$]
4. $|0>|(1> - |0>) + |1>|(0> - |1>)$ [if $f(0) = 1, f(1) = 0$]

States 3 and 4 are entangled because the state of qubit 2 depends on qubit 1. States 1 and 2 are, however, not entangled. After putting qubit 1 through an **H** gate, we get (after simplifying the mathematical terms a little):

1. $|0>|(0> - |1>)$
2. $|0>|(0> - |1>)$
3. $|1>|(0> - |1>)$
4. $- |1>|(0> - |1>)$

So now we know that if we observe qubit 1 to be 0, then f is of type 1 or 2 (f is constant), and if we get 1, then f is type 3 or 4 (f is different).

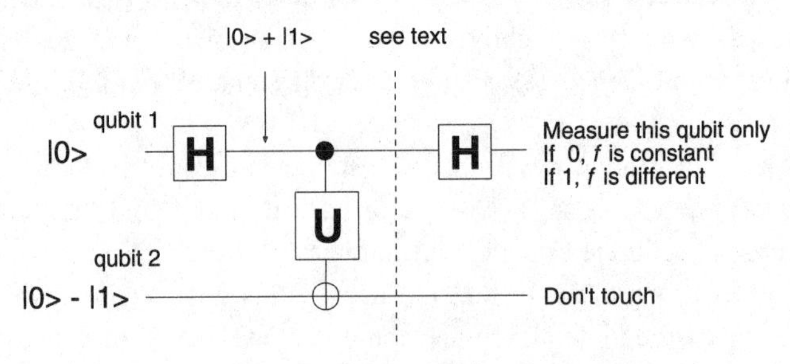

Figure D.3

New and improved circuit for Deutsch's problem.

This particular circuit has an even stranger feature than last time. Although the second qubit is essential for the calculation, once it is computed, it is never referred to again. Although the control input of a classical controlled-NOT gate cannot be affected by the gate, the same is not true for a quantum controlled-NOT gate. Measurements on the target qubit can produce a kickback effect on the control qubit, but, as we've just seen in this case, even if the target qubit is left untouched, the phase of the control qubit can still be reversed. This phase reversal is crucial to the way the circuit works.

Appendix E: The Mathematics Behind RSA

A summary of the RSA encyryption and decryption procedures is this:

(a) Find two very large primes, p and q.
(b) Calculate their product, $n = pq$ (the public modulus).
(c) Choose e, such that $e < n$ and has no factors in common with $(p - 1)(q - 1)$.
(d) Find d, such that $de \equiv 1 \bmod (p - 1)(q - 1)$.
(e) The public exponent is e, and the secret one is d.
(f) The public key is (n, e), and the secret key is (n, d).
(g) Keep p and q secret.
(h) Encrypt m using $c = m^e \bmod pq$, where $m < pq$. (1)
(i) Decrypt using $m = c^d \bmod pq$. (2)

The reason decryption equation (2) $m = c^d \bmod pq$ regenerates the original message depends on a well-known result in number theory proved by the eighteenth-century Swiss mathematician Leonhard Euler. Euler's theorem[1] states that if x and y have no common factors, then:

$$x^{\phi(y)} \equiv 1 \bmod y$$

where $\phi(y)$, known as Euler's function, represents how many numbers less than y are prime to y (that is, share no common factors with y). In the case $y = pq$, where p and q are prime, then $\phi(y) = (p - 1)(q - 1)$ because all numbers from 1 to pq apart from multiples of p and q are prime to pq.

So if we raise the encryption equation (1) $c = m^e \bmod pq$ to the power d and take the result mod pq, we get:

$$
\begin{aligned}
c^d \bmod pq &= (m^e \bmod pq)^d \bmod pq \\
&= m^{de} \bmod pq \\
&= m \cdot m^{de\,-1} \bmod pq \qquad (3)
\end{aligned}
$$

Note that $de - 1$ is divisible by $(p - 1)(q - 1)$, which means it can be written as:

$k(p - 1)(q - 1)$, where k is some fixed integer.

So equation (3) can be rewritten:

$$m \cdot m^{(p-1)(q-1))k} \bmod pq$$

But by Euler's theorem, $m^{(p-1)(q-1)} \equiv 1 \bmod pq$. So (3) reduces to m mod pq, which is the result in (2). QED.

Appendix F: What's the Connection Between Orders and Factoring?

The reason the formula $x^{f/2} - 1$ has a good chance of giving a factor of n can be understood from the following:

By definition, $x^{f+1} \equiv x \bmod n$ because f, the order, tells us when the sequence $x^1, x^2, x^3 \ldots$ repeats. Provided x and n don't share any common factors, we can divide both sides by x to get $x^f \equiv 1 \bmod n$. Subtracting 1 from each side gives $x^f - 1 \equiv 0 \bmod n$. Now, recalling some simple high school algebra, $a^2 - b^2 = (a - b)(a + b)$, we can write:

$$(x^{f/2} - 1)(x^{f/2} + 1) \equiv 0 \bmod n$$

which means that, provided f is even, $x^{f/2} - 1$ and $x^{f/2} + 1$ must be either factors of n or multiples of factors.

Appendix G: Circuit Analysis of Figure 7.1

The network in Figure 7.1 (see page 233), can be understood in the following way. The dotted lines correspond to the state of qubits a, b, and c at times t_1 to t_4, so that as time passes the state of these bits, in effect, travels from left to right.

Now, if input $c = 0$, the first XOR gate has no effect on b, so that at time t_2, b will have been rotated by $45° + 45° = 90°$. If a is also 0, then the second XOR gate has no effect and the state of b is simply rotated back by 90° by the time t_4. The net result is that b is unchanged—giving 0, which is what we expect for 0 AND 0.

If $c = 0$ and $a = 1$, then the effect of the second XOR gate is to invert b. Inversion of 0 and 1 is mathematically the same as subtracting the rotation of the state from 180°. At t_2, b has been rotated by 90°, so subtracting that from 180° gives 90° again, leaving the state unchanged. Therefore, between t_3 and t_4, b is simply rotated back by 90°, returning it to its initial state 0, which is again what we expect for 0 AND 1.

If $c = 1$, then the first XOR gate inverts the first rotation of 45°, subtracting it from 180°, giving a state rotated by 135°. So by t_2, b has rotated another 45°, making 180° in total. If $a = 0$, then the rest of the network simply undoes this rotation, restoring b to its initial state 0. If $a = 1$, then the state is inverted between t_2 and t_3, bringing b back to 0. Between t_3 and t_4, the rotation of the state evolves through the states −45°, 225°, finally settling on 180°, which means b now gives the answer 1, which is what we want for 1 AND 1.

Appendix H: How to Produce a Pure Quantum State in a Cup of Coffee

Imagine the coffee contains molecules (caffeine was suggested, but only as a hypothetical illustration) with four different nuclear spins. There are sixteen possible spin states for each molecule as shown below, and in normal circumstances these would be present in virtually equal numbers. We put a cup of our coffee in a strong magnetic field, whereupon some of the spins align themselves with the magnetic field. The thermal vibrations of the molecules keep most of the spins randomly oriented, but after a while the frequency distribution of each state settles down to a pattern of small deviations from the background average. The size of the deviations from the background depends on how strongly aligned the spins are with the magnetic field and are indicated by the numbers below.

↑↑↑↑	↑↑↑↓	↑↑↓↑	↑↑↓↓	↑↓↑↑	↑↓↑↓	↑↓↓↑	↑↓↓↓
2	1	1	0	1	0	0	−1
↓↑↑↑	↓↑↑↓	↓↑↓↑	↓↑↓↓	↓↓↑↑	↓↓↑↓	↓↓↓↑	↓↓↓↓
1	0	0	−1	0	−1	−1	−2

The deviations are small, so there might be, say, two extra molecules in the state ↑↑↑↑ per million molecules, one extra of ↑↑↑↓, one extra of ↑↑↓↑, and so on. The state ↑↑↑↑ represents the one in which all the spins are aligned with the field, making it the most populated state. Spin states such as ↑↑↓↓ have no net alignment with the magnetic field and are neither boosted nor reduced in number. Such states make no contribution to the NMR signals because their effect is canceled out by other equally represented states (in this case, ↓↓↑↑). Negative deviations associated with states like ↓↓↓↓ signify that they are underrepresented in the general population. Note that these negative deviations *do* make a contribution to the NMR signals, so they cannot be ignored.

To purify the quantum state, we select only those molecules in which spins 1 and 2 point up. This is achieved by making all operations on spins 3 and 4, which become our working qubits, *conditional* on spins 1 and 2 being up. This narrows the population down to the first four states ↑↑↑↑, ↑↑↑↓, ↑↑↓↑, and ↑↑↓↓. The last one, ↑↑↓↓, is neither overrepresented nor underrepresented so we can ignore it. But that still leaves 3 different quantum states in our population, so we haven't yet got a pure state. In Gershenfeld and Chuang's scheme the state is prepared by applying a sequence of radio frequency pulses to swap some of the states. The result they get is the following distribution:

↑↑↑↑	↑↑↑↓	↑↑↓↑	↑↑↓↓	↑↓↑↑	↑↓↑↓	↑↓↓↑	↑↓↓↓
2	0	0	0	−1	1	1	1
↓↑↑↑	↓↑↑↓	↓↑↓↑	↓↑↓↓	↓↓↑↑	↓↓↑↓	↓↓↓↑	↓↓↓↓
1	−1	−1	−1	−2	0	0	0

Now when we restrict ourselves to molecules with spins 1 and 2 pointing up, only the pure state ↑↑↑↑ contributes to the answer, because the other three states, ↑↑↑↓, ↑↑↓↑, and ↑↑↓↓, are no longer present in surplus or deficit, so they do not contribute to the NMR signals. We now have ourselves a two-bit quantum processor. And as MIT's Seth Lloyd once said, "A two-bit microprocessor is a two-bit microprocessor." But it is possible to extend this scheme beyond two bits.

Notes

Chapter 1

1. Arthur C. Clarke, *The Lost Worlds of 2001,* New York: Signet, 1972.
2. Deutsch prefers to call it a conclusion rather than a belief because it is based on arguments to which there are in his opinion no rational counterarguments.
3. The comparison is a loose one, though. A key aspect of the quantum universe is that different branches of the quantum universe have different "weights"—some branches are much more strongly represented than others. Also, it's possible for different branches to overlap with one another, causing interference effects. Neither of these features is reflected in Borges's story.
4. Tim Folger, "The Best Computer in All Possible Worlds," *Discover,* October 1995.
5. David Deutsch, "Quantum Theory: the Church-Turing Principle and the Universal Quantum Computer," *Proceedings of the Royal Society of London,* A400 (1985), pp. 97–117.
6. David Deutsch, *The Fabric of Reality* (New York: Penguin, 1997), p. 217. Reprinted with permission from Penguin © 1997.
7. Gary Stix, "Toward Point One," *Scientific American,* February 1995, p. 90.
8. Bill Gates, Nathan Myhrvold, and Peter N. Rinearson, *The Road Ahead* (New York: Viking Penguin, 1996), p. 33.
9. Workshop in quantum computing at Southampton University, England, September 1997.

Chapter 2

1. Claude Shannon, "A Mathematical Theory of Communication," reprinted in C. E. Shannon and Warren Weaver (eds.), *A Mathematical Theory of Communication* (Urbana, Ill.: University of Illinois Press, 1963).
2. Shannon's formula was actually a little more sophisticated. He considered the information content of a stream of n independent symbols whose probabilities of appearing are $p_1, p_2, p_3 \ldots p_n$. The average information per symbol is given by:

$$H = -[p_1 \log p_1 + p_2 \log p_2 + \ldots + p_n \log p_n] \text{ or } \Sigma\, p_i \log p_i$$

3. The formula $S = k \log W$ was cast by Max Planck, who based it upon Boltzmann's earlier work. See Engelbot Broda, *Ludwig Boltzmann: Man, Physicist, Philosopher,* Larry Gay (trans.) (Woodbridge, Conn.: Ox Bow Press, 1983).
4. Quoted in Kenneth Denbigh, "How Subjective Is Entropy?" reprinted in *Maxwell's Demon,* Harvey Leff and Andrew Rex, p. 113.
5. Harvey Leff and Andrew Rex, *Maxwell's Demon* (Princeton, N.J.: Princeton University Press, 1990), p. 6.
6. Ibid., p. 16. Reprinted with permission of IOP Publishing © 1990.
7. J. Von Neumann, Fourth University of Illinois lecture, in *Theory of Self-Reproducing Automata,* A. W. Burks, ed. (Urbana, Ill.: University of Illinois Press, 1966), p. 66.
8. R. Landauer, "Irreversibility and Heat Generation in the Computing Process," *IBM Journal of Research and Development* 3 (1961): 183–191.
9. C. H. Bennett, "Notes on the history of reversible computation," *IBM Journal of Research & Development* 32, 16–23 (1988); reprinted in *Maxwell's Demon,* Harvey Leff and Andrew Rex, p. 283.
10. C. H. Bennett, "The thermodynamics of computation—a review," *International Journal of Theoretical Physics* 21 (1982): pp. 905–40.
11. Yves Lecerf, *Comptes Rendus* Hebdomadaires des Séances de L'academie des Sciences, 257 (1963), pp. 2597–2600.
12. Paul Davies, *About Time,* p. 209. The "almost" refers to the fact that certain subatomic particles, called kaons, exhibit a limited form of time asymmetry, but they are very much the exception to the rule.
13. One might object that Heisenberg's famous uncertainty principle in quantum mechanics already places a limit on the accuracy of such information. Although this is true up to a point, some quantities in quantum mechanics are still continuous. In the case of spatial position, the uncertainty principle limits its accuracy only when combined with knowledge about momentum. If we are willing to sacrifice information about the speed or momentum of an object, there is *no* limit to the accuracy with which its position can be specified.
14. Robert Wright, *Three Scientists and Their Gods* (New York: Times Books, 1988), p. 47. Reprinted by permission.
15. The NOR gate, which can be made by following an OR gate with a NOT gate, is also universal.
16. Feynman diagrams are pictorial representations of atomic or quantum interactions.
17. In the case of electronic signals this constraint doesn't appear to be important, although Fredkin maintains that it will be when computers work at the level of individual electrons.

18. Ed Fredkin and Tom Toffoli, "Conservative Logic," *International Journal of Theoretical Physics* 21 (1982), pp. 219–253.
19. Robert Wright, *Three Scientists and Their Gods,* p. 55. Reprinted by permission.
20. Tommaso Toffoli, "Physics and Computation," *International Journal of Theoretical Physics* 21 (1982): 165–175.
21. Workshop on Physics and Computation: PhysComp '96, Boston, Mass., November 22–24, 1996.
22. E. Fredkin, "A New Cosmogony," IEEE Proceedings of the Physics of Computation Workshop, October 1992. In his paper, Fredkin argues that the missing workload points toward the idea that the Planck scale is not the appropriate scale on which the universe's CA works. There must be, he claims, some much larger scale.

Chapter 3

1. Richard Feynman, "Simulating Physics with Computers," *International Journey of Physics,* 21 (6–7) (1982), pp. 467–488.
2. Exceptions include superconductivity and superfluidity.
3. See, for example, Abraham Pais, *Inward Bound* (Oxford: Oxford University Press, 1988), p. 248.
4. Walter Moore, *Schrödinger: Life and Thought* (Cambridge: Cambridge University Press, 1992), p. 219.
5. *Inward Bound,* p. 250.
6. The absolute square is represented mathematically by $|\psi|^2$ and is required because ψ is a generally complex number. Squaring a complex number usually gives another complex number. The absolute square, on the other hand, is always a positive real number (see footnote 5 of chapter 4).
7. *Inward Bound,* p. 257.
8. This problem was also independently solved by Carl Eckart at Caltech.
9. There are, in fact, techniques using classical cryptography that will work satisfactorily for playing poker and other games over the phone or Internet.
10. The setting was inspired by Roger Penrose's Quintessential Trinkets dodecahedron device, described in his book *Shadows of the Mind.* His device, though more elegant, is trickier to explain, and the Morphic Resonator is more closely modeled on Feynman's argument. The use of the name Morphic Resonator is not entirely serious.
11. For simplicity of the presentation, 30 degrees is a good choice, but the strongest correlation in defiance of classical rules occurs at 22.5 degrees.
12. David Mermin, "Is the moon there when nobody looks?" *Physics Today* 38 (4) (1985): 38.

13. Benioff's model was a classical computer built out of quantum mechanical components, so it wasn't really what we nowadays would call a quantum computer. On the other hand, Deustch's paper proposing a test of the many-universes interpretation (see Chapter 3, pages 111ff.) used what we would now call a quantum computer and was written and circulated as a preprint in 1978. It was not published, though, until 1984.
14. Richard Feynman, "Quantum Mechanical Computers," *Optics News,* February 1985, pp. 11–20.
15. After publishing his 1985 paper, David Deutsch was convinced that "designer Hamiltonians" would not be needed, because he saw that quantum physics was "computation friendly." He believed that computation could be built out of almost any kind of interaction, a notion he later expressed in a conjecture suggesting that "almost all three-bit quantum gates are universal." Much later, he and Adriano Barenco proved that an even stronger claim was true—that almost all *two*-bit quantum gates were universal. See Chapter 7, page 232.
16. Quantum theory is often described as the most accurate theory ever but in *The Nature of Space and Time* by Stephen Hawking and Roger Penrose (Princeton, N.J.: Princeton University Press, page 61), Penrose points out that general relativity theory has now been tested to one part in 10^{14}, thanks to the decaying orbit of two neutron stars observed in the Hulse-Taylor binary pulsar system PSR1913 + 16. Quantum theory has been proven accurate to about one part in 10^{11}.
17. *The Ghost in the Atom* (Cambridge: Cambridge University Press, 1993), p. 61.
18. See, for example, Chris Isham, *Lectures on Quantum Theory,* (World Scientific Pub. Co., 1995), p. 84.
19. Except in special circumstances, such as the quantum eraser experiments mentioned earlier.
20. *Subtle Is the Lord* (Oxford: Oxford University Press, 1983), p. 5.
21. *Schrödinger: Life and Thought,* p. 308.
22. Except some hidden-variable interpretations, which were disproved by experiments of Alain Aspect et al.
23. Church's version of the thesis involved a mathematical technique known as the lambda calculus rather than Turing machines.
24. David Deutsch, *The Fabric of Reality* (New York: Allen Lane, 1997), p. 132. The reason for dropping Church's name is to emphasize the physical aspect of the principle. Church approached universality from a purely mathematical point of view. Turing, although a mathematician, came closer to giving it a physical perspective by introducing the notion of a Turing machine.
25. Ibid., p. 132.
26. André Berthiaume and Gilles Brassard, "The Quantum Challenge to Structural

Complexity Theory," 7th IEEE Conference on Structure in Complexity Theory, June 1992.

Chapter 4

1. *The Fabric of Reality,* p. 210.
2. André Berthiaume and Gilles Brassard, "Oracle Quantum Computing," *Journal of Modern Optics* 41 (1994): 2521–35. Reprinted with permission from Taylor & Francis Ltd., © 1994. http://www.tandf.co.uk/journals
3. Ibid.
4. The state $|0> + |1>$ is properly written with a factor $1/\sqrt{2}$ in front (i.e., $1/\sqrt{2}$ $(|0> + 1>)$ to ensure that a^2 and b^2 each give the correct probability of ½. For simplicity of presentation I have ignored these *normalization* factors.
5. Complex numbers are numbers that include the imaginary number *i,* which is the square root of -1. They take the general form $a + ib$ where a and b are real numbers. As noted in footnote 5 of Chapter 3, the absolute square (given by $a^2 + b^2$) is always a positive real number, which is much easier to interpret physically.
6. See, for example, Richard Feynman, "Simulating Physics With Computers," *International Journal of Physics* 21, nos. 6/7 (1982): 479–80.
7. July/August 1995.
8. His gate was based on the idea of an irrational power of the Toffoli gate. Two Toffoli gates in a row are equivalent to a Toffoli gate squared. An irrational power extends the idea to (Toffoli gate)$^\alpha$, where α is an irrational number such as $\sqrt{2}$. See David Deutsch, "Quantum Computational Network," *Proceedings of the Royal Society of London,* A425 (1868) (September 1989), pp. 73–90.
9. D. Deutsch and R. Jozsa, "Rapid Solution of Problems by Quantum Computation," *Proceedings of the Royal Society of London,* A439 (1907) (December 8, 1992), pp. 552–58.
10. See, for example, David Harel, *Algorithmics* (Reading, Mass.: Addison-Wesley 1992), p. 182.
11. Ian Stewart, *From Here to Infinity* (Oxford: Oxford University Press, 1996), p. 259.
12. Theodore Baker, John Gill, and Robert Solovay, "Relativizations of the P = ?NP Question," *SIAM Journal of Computing* 4 (1975): 431–42.
13. Note, P and NP for oracle machines are *not* necessarily the same as P and NP for conventional computers. Theorists use the notation P^X and NP^X for the *relativized* classes that apply to a computer with oracle X to distinguish them

from the classes P and NP. Consider a computer with an oracle X that can solve the traveling salesman problem. It would, by NP-completeness, be able to solve all NP problems in polynomial time. So its class P^X would include both P and NP.

Chapter 5

1. Ian Stewart, *From Here to Infinity,* p. 13.
2. Don Zagier, "The first 50 Million Prime Numbers," *Mathematical Intelligencer* 0 (1977): 7–19.
3. The size of n can be expressed mathematically as $\log n$.
4. Gary L. Miller, "Riemann's Hypothesis and Tests for Primality," *Journal of Computer and System Sciences* 13 (1976): 300–17. The algorithm takes of the order of $\log^4 n$ or L^4 steps, where L is the number of digits in n.
5. The hypothesis concerns the behavior of a remarkable mathematical function known as the Riemann zeta function, which can be expressed either as an infinite sum involving the integers:

$$z(s) = 1 + 1/2^s + 1/3^s + 1/4^s + 1/5^s + \ldots$$

or as an infinite product of terms involving prime numbers:

$$z = 1/\,(1 - 1/2^s)\,(1 - 1/3^s)\,(1 - 1/5^s)\,(1 - 1/7^s)\,(1 - 1/11^s)\ldots$$

Mathematicians are interested in the behavior of this function when s takes complex values (that is, values with both real and imaginary components). The above equations are only valid when the real value of s is greater than 1, but the definitions can be modified to work for any value. Of particular interest are the "zeros" of the function—that is, values of s for which $z = 0$. It turns out that apart from "trivial" zeros at negative values of s, an infinite number of zeros appear in a "critical strip" between 0 and 1. The Riemann hypothesis is that all of the zeros appear on the line defined by:

$$\text{real part of } s = \tfrac{1}{2}$$

After the discovery of a proof of Fermat's last theorem in 1994, the Riemann hypothesis remains perhaps the biggest unsolved problem in mathematics. The extended Riemann hypothesis involves a generalization of the zeta function to other parts of number theory.
6. Simson Garfinkel, *PGP: Pretty Good Privacy* (Cambridge, Mass.: O'Reilly & Associates, 1995), p. 74.
7. Private keys can be quite a bit shorter than public ones because in many cases the only known way of attacking private-key algorithms without knowing the

private key is a brute-force search: try every possible combination of key until you get a decrypted message that makes sense. With public-key systems it is possible to take shortcuts, such as the number field sieve. For this reason public-key systems need significantly longer keys.

8. M. Gardner, "Mathematical Games," *Scientific American,* August 1977, p. 120–24.
9. The calculation producing this estimate was later shown to have been too high by a factor of 10.
10. A. K. Lenstra, H. W. Lenstra Jr., M. S. Manasse, and J. M. Pollard, "The Factorization of the Ninth Fermat Number," *Mathematics of Computation* 61 (1993): pp. 319–49.
11. Derek Atkins, Michael Graff, Arjen K. Lenstra, and Paul C. Leyland, "The Magic Words Are Squeamish Ossifrage." *American Scientist* 82 (1) July–August 1994, pp. 312–16.
12. Gary Taubes, "Small Army of Code-Breakers Conquers a 129-Digit Giant," *Science* 264 (1994): 776.
13. The running time grows roughly according to the expression $e^{(L^{1/3}(\log L)^{2/3})}$.
14. The reason two delta pulses transform into a sine wave is that a sine wave consists of just one frequency, so when it is plotted on a frequency spectrum or graph, it appears as a single spike. The Fourier transform produces a second spike in the negative direction because mathematically negative frequencies are equivalent to positive frequencies.
15. See, for example, Thomas Cormen, Charles Leiserson, and Ronald Rivest, *Introduction to Algorithms,* p. 800.
16. Peter Shor, "Polynomial-Time Algorithms for Prime Factorization and Discrete Logarithms on a Quantum Computer," *SIAM Journal of Computing* 26 (1997), pp. 1484–1509.
17. The fraction 1/7, for example, expressed as a decimal is 0.142857142857 recurring. This is equivalent to finding the order of 10 modulo 7—that is, 10, 10^2, 10^3, 10^4. . .—which gives the sequence 326451326451 . . . Both sequences repeat every six digits.
18. A. Barenco and A. Ekert, "Quantum Computation," *Acta Physica Slovaca* 45 (1995), pp. 205–216.
19. Adriano Barenco, Artur Ekert, Kalle-Antti Suominen, and Päivi Törmä, "Approximate Quantum Fourier Transform and Decoherence," *Physical Review A* 54 (1), July 1996, pp. 139–46.

Chapter 6

1. Gilles Brassard, “The Impending Demise of RSA?” RSA Laboratories’ *CryptoBytes* 1, no. 1 (1995). Asher Peres suggested this quote to Gilles Brassard.
2. The paper is often referenced as an abbreviation of Bennett and Brassard, 1984. C. Bennett and G. Brassard, Proceedings of IEEE International Conference on Computers, Systems and Signal Processing, Bangalore, India (IEEE, New York, 1984), p. 175.
3. This remains unsolved, though some experimentalists are coming close to producing such a source.
4. A. Ekert, “Quantum Cryptography Based on Bell’s Theorem,” *Physical Review Letters* 67 (August 5, 1991), pp. 661–63. Reprinted with permission from the American Physical Society, © 1991.
5. A. Ekert, J. Rarity, P. Tapster, G. Palma, “Practical Quantum Cryptography Based on Two-Photon Interferometry,” *Physical Review Letters* 69 (1992), pp. 1293–95.
6. W. K. Wootters and W. H. Zurek, *Nature* 299, 802 (1982).
7. See, for example, *Applied Cryptography* 2nd edition by Bruce Schneier, page 116.
8. Actually, in Wiesner’s scheme Bob only receives a random selection of bits from one or other of the messages.
9. Charles Bennett, Gilles Brassard, Claude Crépeau, and Marie-Helene Skubiszewska, “Practical Quantum Oblivious Transfer,” *Advances in Cryptology,* Crypto ’91 Proceedings, August 1991, pp. 351–66.
10. G. Brassard, C. Crépeau, “25 Years of Quantum Cryptography,” *SIGACT News* 27 (3), (September 1996), pp. 13–24.
11. Crypto ’90.
12. G. Brassard, C. Crépeau, R. Jozsa, and D. Langlois, “A Quantum Bit Commitment Scheme Provably Unbreakable by Both Parties,” *34th Symposium on Foundations of Computer Science,* IEEE, 1993, pp. 42–52.
13. G. Brassard, C. Crépeau, “25 Years of Quantum Cryptography.”
14. Artur Ekert, “Quantum Cryptography Based on Bell’s Theorem,” *Physical Review Letters* 67 (1991), pp. 661–663.
15. D. Deutsch, A. Ekert, R. Jozsa, C. Macchiavello, S. Popescu, A. Sanpera, “Quantum Privacy Amplification and the Security of Quantum Cryptography Over Noisy Channels,” *Physical Review Letters* 77 (1996): p. 2818.
16. C. H. Bennett, G. Brassard, S. Popescu, B. Schumacher, J. A. Smolin, W. K. Wootters, “Purification of Noisy Entanglement and Faithful Teleportation via Noisy Channels,” *Physical Review Letters* 76 (1996): p. 722.
17. Teleportation.
18. G. Brassard, S. Braunstein, R. Cleve, “Teleportation as a Quantum Computation,” *Physica D,* 120 (1–2) (September 1, 1998), pp. 43–47.
19. Qubits a', b', and c' will all need to be measured in the end to perform the ex-

periment, so the question is how soon can they be measured. In this case, a′ and b′ only participate as control bits and it makes no difference whether they are measured before or after the controlled-NOT gates.

20. Lawrence Krauss. *The Physics of Star Trek* (New York: Basic Books, 1995), pp. 65–83.

Chapter 7

1. Gerard Milburn, *Schrödinger's Machines,* p. 179.
2. Rolf Landauer, "Information Is Physical," *Physics Today,* May 1991, p. 28.
3. Proceedings of the Drexel-4 Symposium on Quantum Nonintegrability, Quantum-Classical Correspondence, edited by D. H. Feng and B-L. Hu.
4. As noted in Chapter 4, Deutsch's gate is an irrational power of the Toffoli gate. The quantum version of the Toffoli gate itself is *not* universal.
5. David DiVincenzo, "Two-Bit Gates Are Universal for Quantum Computation," *Physical Review A* (February 1995): 1015–22.
   Adriano Barenco, "A Universal Two-Bit Gate for Quantum Computation," *Proceedings of the Royal Society of London* A 449: (1937) (June 8, 1995), pp. 679–83.
   T. Sleator and H. Weinfurter, "Realizable Universal Quantum Logic Gates," *Physical Review Letters,* 74 (May 15, 1995): 4,087–90.
6. Seth Lloyd, "Almost Any Quantum Logic Gate Is Universal," *Physical Review Letters* 75: (2) (July 10, 1995), pp. 346–349.
7. D. Deutsch, A. Barenco, A. Ekert, "Universality in Quantum Computation," *Proceedings of the Royal Society of London* A 449: (1937), (June 8, 1995), pp. 669–77. Reprinted with permission from the Royal Society of London, © 1995.
8. A. Barenco, C. Bennett, R. Cleve, D. Di Vincenzo, N. Margolus, P. Short, T. Sleator, J. Smolin, H. Weinfurter, "Elementary Gates for Quantum Computation, *Physics Review* A.52 (1995), 3457.
9. The rotation angles are given in terms of spin orientation. If we were using polarized photons, these rotations would be exactly half the ones here—that is, 22.5 degrees. See Chapter 4, p. 129.
10. The phase shift, although it is added to the second qubit, actually affects *both* qubits because the quantum state $|11>$, which is the only state changed by the gate, represents the state of both qubits rather than just the second.
11. Robert Griffiths and Chi-Sheng Niu, "Semiclassical Fourier Transform for Quantum Computation," *Physical Review Letters* 76 (1996), pp. 3228–31.
12. Then at Los Alamos National Laboratory.
13. *Science* 261 (17 September 1993): 1,569–71.

14. W. Teich, K. Obermayer, and G. Mahler, *Physical Review B* 37, no. 14 (1988): pp. 8,111–20.
15. Rolf Landauer, "Dissipation and Noise Immunity in Computation and Communication," *Nature* 335 (October 27, 1988): p. 783. Reprinted with permission from *Nature.* Copyright 1988 Macmillan Magazines Limited.
16. David Deutsch prefers to put the idea the other way around by saying that the environment has to be protected from the quantum state. That's because in the many-universes interpretation, when a quantum state decoheres, the environment becomes entangled with the quantum state, causing the universe to differentiate into multiple realities.
17. Ed Regis, *Nano* (New York: Little Brown, 1996), pp. 155, 158.
18. After the German physicist, Wolfgang Paul, who shared the Nobel Prize for physics with Dehmelt and Norman Ramsey in 1989 for their pioneering work on ion traps.
19. J. L. Cirac and P. Zoller, "Quantum Computations with Cold Trapped Ions," *Physical Review Letters* (May 15, 1995), pp. 4091–94.
20. Q. A. Turchette, C. J. Hood, W. Lange, H. Mabuchi, and H. J. Kimble, "Measurement of Conditional Phase Shifts for Quantum Logic," *Physical Review Letters* (December 1995).
21. Bertram Schwarzschild, "Labs Demonstrate Logic Gates for Quantum Computation," *Physics Today* (March 1996): 21–33.
22. Serge Haroche and Jean-Michel Raimond, "Quantum Computing: Dream or Nightmare?" *Physics Today* (August 1996), pp. 51–52.
23. *Science* 275 (17 January 1997): 307–09 and 351–56.
24. *Proceedings of the National Academy of Sciences* 94 (1997): 1,634–39.
25 See, for example, David DiVincenzo, "Quantum Computation," *Science* 270 (October 13, 1995), pp. 255–61.
26. Warren S. Warren, "The Usefulness of NMR Quantum Computing," *Science* 277 (September 12, 1999), pp. 1688–90.
27. Some of these figures have been adapted from a similar table appearing in David DiVincenzo, ibid.

Chapter 8

1. John Preskill, "Quantum Computing: Pro and Con," *Proceedings of the Royal Society of London* A454 N1969 (January 8, 1998), pp. 469–98.
2. It's better in other ways, though. In classical systems errors accumulate exponentially with time, whereas in quantum systems, errors accumulate additively and do not tend to grow with time. David Deutsch points to this as yet another

computation-friendly feature of physics. The downside, though, is that the decoherence rate typically increases exponentially *with the number of bits involved,* and that presents a similarly daunting barrier to that for classical analog computation.

3. A. Berthiaume, D. Deutsch, and R. Josza, *Proceedings of the Workshop on Physics and Computation, PhysComp '94,* p. 60.
4. S. F. Huelga, C. Machiavello, T. Pellizzari, A. Ekert, "Improvement of Frequency Standards with Quantum Entanglement," *Physical Review Letters* 38 (4) (May 1, 1997), pp. 249–54.
5. P. Shor, *Physical Review A* 52 (1995): 2493–96.
6. George Johnson, "Quantum Theorists Try to Surpass Digital Computing," *New York Times,* February 18, 1997.
7. David Mermin, "What's Wrong With These Elements of Reality?" *Physics Today,* June 1990.
8. R. Feynman, "Simulating Physics with Computers," *International Journal of Physics* 21 (6/7) (1982), pp. 467–488.
9. Gilles Brassard, "Searching a Quantum Phone Book," *Science,* 31 January 1997.
10. David Deutsch, Richard Jozsa, and others have described the shake as being really a *rotation,* through Hilbert space, from a place that merely leans in the direction of the right answer to a place that points directly at the right answer.
11. Charles Bennett, Ethan Bernstein, Gilles Brassard, and Umesh Vazirani, "Strengths and Weaknesses of Quantum Computing," quant-ph/9701001.
12. R. Cleve, A. Ekert, C. Macciavello, and M. Mosca, "Quantum Algorithms Revisited," *Proceedings of the Royal Society of London* A454 (1998), pp. 339–354.

Chapter 9

1. In the Foreword to *Schrödinger's Machines,* Gerard Milburn, pp. viii–ix.
2. Andrew Steane, "Quantum Computing," *Reports on Progress in Physics* 61 (1998), pp. 117–173. Reprinted with permission from IOP Publishing, © 1998.
3. David Deutsch, "Quantum Mechanics Near Closed Timelike Lines," *Physics Review D* 44 10 (1991), 3197–217. See also D. Deutsch, M. Lockwood, "The Quantum Physics of Time-Travel," *Scientific American* (March 1994), pp. 50–56.
4. 1 nanometer = 10^{-9} meters, which is about the width of ten atoms in a crystal lattice.
5. Ed Regis, *Nano,* p. 227.

6. See, for example, *http://alife.santafe.edu.*
7. Quoted in "Mind Children," Steve Alan Edwards, *21 C* 23, p. 48.
8. L. Adleman, "Molecular computation of solutions to combinatorial problems," *Science* 266 (November 1994): 1021–24.
9. Interview with Len Adleman by Diane Krieger, *Networker,* University of Southern California, September/October 1996. Networker@USC.
10. Lila Kari, "DNA Computing: The Arrival of Biological Mathematics," *The Mathematical Intelligencer,* 19(2) (1997), pp. 1–22.
11. D. Beaver, "Factoring: The DNA Solution," in *Advances in Cryptology—Proceedings of the 4th International Conference on the Theory and Applications of Cryptology,* Asiacrypt '94, 1994, pp. 419–23. Springer-Verlag, 1995.
12. Douglas Hofstadter, *Gödel, Escher, Bach: An Eternal Golden Braid* (New York: Basic Books, 1979), pp. 476–477.
13. Roger Penrose, *Shadows of the Mind,* p. 331.
14. Ibid., p. 401.
15. Private correspondence, December 21, 1997.
16. P. C. W. Davies, "Why Is the World So Comprehensible" in *Complexity, Entropy and the Physics of Information,* edited by Wojciech Zurek, p. 66. Reprinted with permission of Perseus Books, © 1990.
17. David Deutsch, *The Fabric of Reality,* p. 135.
18. R. P. Feynman, "Simulating Physics with Computers," *International Journal of Theoretical Physics* 21, nos. 6/7 (1982): 486.
19. Ibid.
20. John Horgan, *The End of Science,* p. 211.
21. Murray Gell-Mann, *The Quark and the Jaguar: Adventures in the Simple and the Complex* (New York: W. H. Freeman & Company, 1995), p. 138. Copyright 1994 by Murray Gell-Mann. Reprinted with permission from W. H. Freeman & Company, New York.
22. Deutsch, *The Fabric of Reality,* p. 48.
23. Gell-Mann, *The Quark and the Jaguar,* p. 149. Copyright 1994 by Murray Gell-Mann. Reprinted with permission from W. H. Freeman & Company, New York.
24. David Lindley, *Where Does the Weirdness Go?* p. 220.
25. E. Fredkin, "A New Cosmogony," *Proceedings of the Physics of Computation Workshop,* October 2–4, 1992. Reprinted with permission from IEEE, © 1992.
26. E. Fredkin, "Digital Mechanics," *Physica D* 45 (1990): 254–70. Reprinted with permission of Elsevier Science.
27. Lindley, *Where Does the Weirdness Go?* p. 110.
28. Penrose, *Shadows of the Mind,* p. 312. Reprinted with permission from Oxford University Press, © 1997.

29. David Deutsch, "Quantum Theory of Probability and Decisions," *Proceedings of the Royal Society of London* A455 (1999): 3129–38.
30. Penrose, *Shadows of the Mind,* p. 312.

Appendix C

1. Note that this doesn't in any way imply that quantum computing is necessary for consciousness. If we assume that classical computation is sufficient to simulate consciousness, the same program could run in a reversible form on a quantum computer.

Appendix E

1. For proof, see virtually any introductory text on number theory. A good example is William LeVeque's *Elementary Theory of Numbers,* page 64.

Bibliography

Michael Brooks. *Quantum Computing and Communications.* New York: Springer-Verlag, 1999.

Scott H. Clearwater and Colin P. Williams. *Explorations in Quantum Computing.* New York: Springer-Verlag, 1997.

Paul Davies. *The Fifth Miracle: The Search for the Origin and Meaning of Life.* New York: Simon & Schuster, 1999.

Paul Davies and Julian Brown, eds. *The Ghost in the Atom: A Discussion of the Mysteries of Quantum Physics.* Cambridge: Cambridge University Press, 1993.

David Deutsch. *The Fabric of Reality.* New York: Penguin, 1997.

Murray Gell-Mann. *The Quark and the Jaguar: Adventures in the Simple and the Complex.* New York: W. H. Freeman & Company, 1995.

David Harel. *Algorithmics: the Spirit of Computing.* Reading, Mass.: Addison-Wesley, 1992.

Anthony Hey and Robin Allen, eds. *Feynman Lectures on Computation.* Reading, Mass.: Addison-Wesley, 1996.

Anthony J. G. Hey. *Feynman and Computation: Exploring the Limits of Computers.* Sydney: Perseus Books, 1998.

Douglas Hofstadter. *Gödel, Escher, Bach: An Eternal Golden Braid.* New York: Basic Books, 1979.

Lawrence Krauss. *The Physics of Star Trek.* New York: Basic Books, 1995.

Harvey Leff and Andrew Rex (eds). *Maxwell's Demon: Entropy, Information, Computing.* Princeton, N.J.: Princeton University Press, 1990.

David Lindley. *Where Does the Weirdness Go?: Why Quantum Mechanics Is Strange, but Not As Strange As You Think.* New York: Basic Books, 1996.

Gerard Milburn. *The Feynman Processor: Quantum Entanglement and the Computing Revolution.* Malibu, Calif.: Perseus Books, 1998.

———. *Schrödinger's Machines: The Quantum Technology Reshaping Everyday Life.* New York: W. H. Freeman & Company, 1997.

Walter Moore. *Schrödinger: Life and Thought.* Cambridge: Cambridge University Press, 1989.

Roger Penrose. *The Emperor's New Mind: Concerning Computers, Minds, and the Laws of Physics.* Oxford: Oxford University Press, 1989.

———. *Shadows of the Mind: A Search for the Missing Science of Consciousness.* Oxford: Oxford University Press, 1996.

Edward Regis. *Nano: The Emerging Science of Nanotechnology.* Boston: Little Brown & Company, 1996.

Bruce Schneier. *Applied Cryptography,* 2nd ed. New York: John Wiley and Sons, 1994.

Simon Singh. *The Code Book: The Evolution of Secrecy from Mary Queen of Scots to Quantum Cryptography.* New York: Doubleday, 1999.

Frank Tipler. *The Physics of Immortality: Modern Cosmology, God and the Resurrection of the Dead.* New York: Anchor Books, 1995.

Robert Wright. *Three Scientists and Their Gods: Looking for Meaning in an Age of Information.* New York: Times Books, 1988.

Index

T